Texts in Applied Mathematics 16

Texts in Applied Mathematics

Eric F. Van de Velde

Concurrent Scientific Computing

With 47 Illustrations

Springer-Verlag
New York Berlin Heidelberg London Paris
Tokyo Hong Kong Barcelona Budapest

Eric F. Van de Velde
Applied Mathematics 217-50
California Institute of Technology
Pasadena, CA 91125
USA

Series Editors

F. John	J.E. Marsden	L. Sirovich
Courant Institute of	Department of	Division of Applied
Mathematical Sciences	Mathematics	Mathematics
New York University	University of California	Brown University
New York, NY 10012	Berkeley, CA 94720	Providence, RI 02912
USA	USA	USA

M. Golubitsky	W. Jäger
Department of Mathematics	Department of Applied Mathematics
University of Houston	Universität Heidelberg
Houston, TX 77204	Im Neuenheimer Feld 294
USA	Heidelberg, FRG 6900

Mathematics Subject Classification (1992): 42A38, 65T20, 68Q10

Library of Congress Cataloging-in-Publication Data
Van de Velde, Eric F.
    Concurrent scientific computing / Eric F. Van de Velde.
      p.   cm. — (Texts in applied mathematics ; 16)
    Includes bibliographical references and index.
    ISBN 0-387-94195-9
      1. Parallel processing (Electronic computers)   I. Title.
    II. Series.
QA76.58.V35   1994
519.4′0285′52 — dc20   93-43289

Printed on acid-free paper.

Production managed by Ellen Seham; manufacturing supervised by Genieve Shaw.
Photocomposed pages prepared from the author's LaTeX file.
Printed and bound by R.R. Donnelley and Sons, Harrisonburg, VA.
Printed in the United States of America.

9 8 7 6 5 4 3 2 1

ISBN 0-387-94195-9 Springer-Verlag New York Berlin Heidelberg
ISBN 3-540-94195-9 Springer-Verlag Berlin Heidelberg New York

Series Preface

Mathematics is playing an ever more important role in the physical and biological sciences, provoking a blurring of boundaries between scientific disciplines and a resurgence of interest in the modern as well as the classical techniques of applied mathematics. This renewal of interest, both in research and teaching, has led to the establishment of the series: *Texts in Applied Mathematics (TAM)*.

The development of new courses is a natural consequence of a high level of excitement on the research frontier as newer techniques, such as numerical and symbolic computer systems, dynamical systems, and chaos, mix with and reinforce the traditional methods of applied mathematics. Thus, the purpose of this textbook series is to meet the current and future needs of these advances and encourage the teaching of new courses.

TAM will publish textbooks suitable for use in advanced undergraduate and beginning graduate courses, and will complement the *Applied Mathematical Sciences (AMS)* series, which will focus on advanced textbooks and research level monographs.

Preface

A successful concurrent numerical simulation requires physics and mathematics to develop and analyze the model, numerical analysis to develop solution methods, and computer science to develop a concurrent implementation. No single course can or should cover all these disciplines. Instead, this course on concurrent scientific computing focuses on a topic that is not covered or is insufficiently covered by other disciplines: the algorithmic structure of numerical methods. Because traditional computer-science courses usually focus on combinatorial computations, they avoid most numerical methods. Traditional numerical-analysis courses, on the other hand, ignore algorithmic issues and almost always hide the step that transforms the mathematical description of a numerical method into an implementation. It is precisely this step that merits special attention when implementing a numerical method on concurrent computers.

Efficient concurrent programs will be derived by stepwise refinement. This technique has its origins in formal theory of programming. Although the formal aspects have been toned down, the program derivations remain more formal and more detailed than is customary in scientific computing. Some details are simply too important to be ignored: many ambitious projects based on high-level programming concepts have failed, because they do not address the low-level requirements of concurrent scientific computing. The detailed program derivations are necessarily limited to a small sample of numerical methods. The collection of algorithms is sufficiently representative, however, and almost every fundamental technique of concurrent scientific computing is discussed somewhere in this book.

Although the emphasis is on algorithms, enough mathematical background material is provided to make the text self-sufficient and to gain an intuitive grasp of the numerical methods involved. This is important, because implementing numerical methods without understanding their limitations is probably the most frequently occurring mistake. All too often, one considers a switch to concurrent computers, while a switch to another numerical method would result in faster and more reliable computations. Because the text is organized by algorithms, the mathematical derivations occasionally refer to concepts that are only introduced in later chapters.

In the computing literature, the amount of terminology exceeds the number of concepts. All too often, ambiguous and arcane terminology is used to confuse and impress. Good terminology should clarify and simplify. With the faint hope of starting a movement, I have eliminated from this text as much duplicate and ambiguous terminology and as many acronyms as possible. One visible victim is the term "parallel," a geometric concept that applies to concurrent computing only in some vague analogy. Because it says what it means, I favor the term "concurrent."

Students of this course should already have a general knowledge of numerical linear algebra, numerical analysis, and sequential programming. At Caltech, this text is the core of a course aimed at advanced undergraduates and beginning graduate students. It should be supplemented with practical materials and laboratory sessions that bridge the gap between textbook and actual implementation. As concurrent operating systems converge, it will become feasible to offer standard "bridges," but for now this must be left to an instructor familiar with local computing resources.

Eric F. Van de Velde
Pasadena

Acknowledgments

This book would not have been possible without the support and encouragement of Dr. Herbert B. Keller. It has been and continues to be a privilege to work with him.

Dr. K. Mani Chandy motivated me to apply stepwise refinement to programs of concurrent scientific computing. This ultimately led to the writing of this book.

Dr. James W. Demmel, Dr. Paul F. Fischer, Dr. Joel N. Franklin, Dr. David L. Harrar II, Dr. Gautam M. Shroff, Dr. Stefan Vandewalle, Dr. Roy D. Williams, and Dr. Gregoire S. Winckelmans suggested improvements in content and in presentation and pointed out errors in the manuscript.

This book is based on the lecture notes of the course "Concurrent Scientific Computing" that I have been teaching at Caltech since 1988. I apologize to those students who silently suffered through material that was not fully developed. I am grateful to those students who protested and complained, because their feedback was crucial for writing this book.

I could always count on my friends and colleagues of Applied Mathematics for a break from the tedium of writing. I thank Dr. Kirk E. Brattkus, Dr. Cheryl Carey, Dr. Donald J. Estep, Dr. David L. Harrar II, Dr. Steve Roy Karmesin, Dr. Herbert B. Keller, Dr. Daniel I. Meiron, Dr. Ellen Randall, and Dr. Philip G. Saffman for the gossip sessions, the political, legal, scientific, and moral debates, and the coffee breaks, the lunches, and the dinners.

JoAnn Boyd organized and thoroughly verified an extensive bibliographic data base, which was used to compile the bibliography of this book.

The electronic versions of the bibliographies of Golub and Van Loan [40] and Ortega, Voigt, and Romine [69] simplified the task of assembling a bibliographic data base. Following their example, the data base is available on the Internet through netlib@ornl.gov.

This work is supported by the Center for Research on Parallel Computation, which was established by the National Science Foundation under Cooperative Agreements Nos. CCR-8809615, and CCR-9120008.

Contents

List of Figures

List of Tables

1
The Basics

We begin our study of concurrent scientific computing with a study of the two most elementary scientific-computing problems: compute the sum of two vectors and compute the inner product of two vectors. These two trivial example problems are sufficient to introduce almost all concepts needed for our journey into concurrent scientific computing.

Reasoning and communicating ideas about programs is always difficult, because even the simplest program transformations can lead to a notational quagmire. Good notation reduces the burden but does not eliminate it. Informal program descriptions help intuition, but lack precision. Programs in a standard programming language are precise specifications, but the real issues are buried by technical detail. Avoiding either extreme, we shall use a modified version of the UNITY notation. Our goal is to specify programs precisely for rigorous algorithmic analysis but to avoid the technical detail of notations geared for consumption by computers.

Notation is important, but it is not the whole story. The fundamentals are valid and must not be ignored, whatever notation is used. In fact, most programs were implemented and tested in either C or FORTRAN on various concurrent computers. Although notational details differ from one language to the next, the concurrent programs still rely on the fundamental concepts introduced in this chapter: data distribution and recursive doubling.

1.1 Notation

Given two vectors $\vec{x}$ and $\vec{y}$ and two scalars α and β, we wish to compute the vector $\vec{z}$ such that

$$\vec{z} = \alpha\vec{x} + \beta\vec{y}.$$

Program Vector-Sum-1 is an algorithmic form of this mathematical equation. It is a program that specifies how to obtain values of the components of $\vec{z}$ from the input data α, β, $\vec{x}$, and $\vec{y}$.

> **program** Vector-Sum-1
> **declare**
> m : integer ;
> x, y, z : array $[0..M-1]$ of real
> **initially**
> $\langle\; ;\; m \;:\; 0 \le m < M \;::\; x[m], y[m] = \tilde{x}_m, \tilde{y}_m \;\rangle$
> **assign**
> $\langle\; \| \; m \;:\; 0 \le m < M \;::\; z[m] := \alpha x[m] + \beta y[m] \;\rangle$
> **end**

The keyword **program** and the program name head the program specification. The keyword **end** signals the end of the program specification. In between, we distinguish three sections of a program: the **declare**, the **initially**, and the **assign** section.

In the **declare** section, the types of all variables are declared by means of a Pascal-like syntax. The variables x, y, and z are arrays of length M of real variables; the first entry in the array has index 0, the last has index $M-1$. Note that M, α, and β have not been declared. Undeclared variables are assumed to be known constant values of known type. In this case, M is an integer, and α and β are reals.

The **assign** section is the body of the program. The body of program Vector-Sum-1 consists of one *quantification*, an expression bracketed by $\langle$ and $\rangle$. It specifies that, for every index m that satisfies $0 \le m < M$, the array entry $z[m]$ is assigned the value $\alpha x[m] + \beta y[m]$. When an assignment is performed, the value of the expression on the right-hand side of the assignment symbol $(:=)$ is put in the location specified on the left-hand side.

The concurrent quantification

$$\langle\; \| \; m \;:\; 0 \le m < M \;::\; z[m] := \alpha x[m] + \beta y[m] \;\rangle$$

is a short-hand notation for the sequence of assignments

$z[0] := \alpha x[0] + \beta y[0]$ $\|$
$z[1] := \alpha x[1] + \beta y[1]$ $\|$
$\cdots$ $\|$
$z[M-1] := \alpha x[M-1] + \beta y[M-1]$

These assignments are separated by the *concurrent separator* ||, which means that all assignments in this sequence may be performed in any order and/or concurrently. The concurrent quantification and the sequence of assignments separated by the concurrent separator are also equivalent to the *simultaneous assignment*

$$z[0], \ldots, z[M-1] := \alpha x[0] + \beta y[0], \ldots, \alpha x[M-1] + \beta y[M-1]$$

All right-hand sides of a sequence of assignments separated by the concurrent separator or, equivalently, all right-hand sides of one simultaneous assignment are computed before any individual assignment is performed. For example, with initial values $a = 1$ and $b = 2$, evaluating either

$$a := b \; || \; b := a$$

or

$$a, b := b, a$$

leads to $a = 2$ and $b = 1$. Either of the above statements switch the values of a and b without using a temporary variable.

The sequential evaluation of a sequence of expressions is specified by the *sequential separator*, denoted by a semicolon. In this case, the right-hand side of an assignment is evaluated after all previous assignments in the sequence are performed. With the same initial values as before, the evaluation of

$$a := b \; ; \; b := a$$

leads to $a = b = 2$.

The **initially** section is not executed, but specifies the initial state of the data space. No assignments occur in the **initially** section, which is stressed by using the equality (=) instead of the assignment symbol (:=). This difference of equality and assignment symbols aside, the **initially** and **assign** sections have identical syntax. In program Vector-Sum-1, the **initially** section specifies how some known values $\tilde{x}_m$ and $\tilde{y}_m$ are, initially, stored in the arrays x and y. In this particular instance, the simplest mapping is specified: $\tilde{x}_m$ to $x[m]$ and $\tilde{y}_m$ to $y[m]$. The expression

$$x[m], y[m] = \tilde{x}_m, \tilde{y}_m$$

is a *simultaneous equality* and is equivalent to the pair

$$x[m] = \tilde{x}_m \; || \; y[m] = \tilde{y}_m$$

The tilde notation differentiates initializing values from variable names. The values $\tilde{x}_m$ and $\tilde{y}_m$ are undeclared and, therefore, constants that are assumed known.

The inner product of two vectors $\vec{x}$ and $\vec{y}$ is a scalar σ, and

$$\sigma = \sum_{m=0}^{M-1} x_m y_m.$$

Program Inner-Product-1 casts this equation into an algorithmic form.

program Inner-Product-1
declare
      m : integer ;
      σ : real ;
      x, y : array $[0..M - 1]$ of real
initially
      $\langle\ ;\ m\ :\ 0 \le m < M\ ::\ x[m], y[m] = \tilde{x}_m, \tilde{y}_m\ \rangle$
assign
      $\sigma := \langle\ +\ m\ :\ 0 \le m < M\ ::\ x[m]y[m]\ \rangle$
end

The **declare** and **initially** sections are identical to those of program Vector-Sum-1. Only the assignment in the **assign** section is new, although its meaning should be obvious: the variable σ is assigned the sum of all expressions $x[m]y[m]$, where $0 \le m < M$. The *addition separator* allows that the expressions bracketed by the quantification be evaluated in any order and/or concurrently. Assignments are prohibited within quantifications separated by the addition separator. However, the following combination of an assignment and a quantification with the addition separator is perfectly valid:

$$x[0] := \langle\ +\ m\ :\ 0 \le m < M\ ::\ x[m]\ \rangle$$

This is equivalent to and more elegant than

$$t := \langle\ +\ m\ :\ 0 \le m < M\ ::\ x[m]\ \rangle\ ;\ x[0] := t$$

1.2 Multicomputers

A multicomputer is a collection of *nodes* connected by a *communication network*. Each node is an independent sequential computer consisting of a *processor* and a *memory*. Figure 1.1 sketches the architecture of a multicomputer. The computational model of multicomputers is based on this prototypical view. All nodes of the multicomputer are simultaneously active, and each node runs one or more *sequential processes*. The multicomputer environment maps sequential processes to nodes and information exchanges between processes to *communication channels* between nodes. All nodes are identical, and the properties of a process do not depend on its location (the node on which the process is running). In particular, processes running

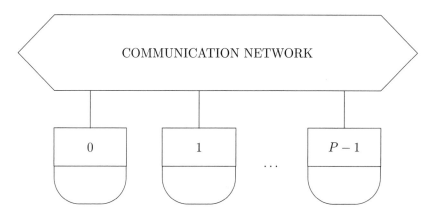

FIGURE 1.1. Prototypical multicomputer.

on the same node do not have any extra properties over processes running on different nodes. In principle, a process may migrate from one node to another during computations, but we shall not use this possibility.

On multicomputers, the only available means of information exchange between processes requires the cooperation between those processes. To share its data, a process must make its data explicitly available by executing a send instruction. This puts the data in a message, which is subsequently made available to another process or to a group of processes. A process can read its messages by executing receive instructions. An important property of the multicomputer computational model is that message order between any two processes is preserved: if process A sends two messages to process B, then it is guaranteed that the first message is received before the second message. However, the order of messages arriving from different source processes is not prescribed: if processes A and B both send a message to process C, then the order in which these two messages are received by process C is unpredictable.

The above multicomputer computational model can be incorporated into our notation as follows. The header

$$\mathcal{P} \parallel p \text{ program } \text{program-name}$$

specifies a multicomputer program. The set $\mathcal{P}$ is an arbitrary finite set, and p is called a *process identifier*. The value of p can be any element of $\mathcal{P}$. In this chapter, the set $\mathcal{P}$ will always be an *index range*

$$\mathcal{P} = 0..P - 1 = \{p : 0 \leq p < P\}.$$

The header then specifies a multicomputer computation with P processes, and each process is associated with one value of $p \in 0..P - 1$. One can now refer to process p. The precise meaning of the header notation will be further explored in Sections 1.3 and 1.4.

Process identifiers are used in the two basic communication primitives **send** and **receive**. The **send** primitive puts a sequence of values in a message and sends the message to a destination process. It takes the form

$$\textbf{send } \{v_0, v_1, \ldots, v_N\} \textbf{ to } q$$

In this expression, q is the process identifier of the destination process. Instead of a destination-process identifier, the keyword **all** may be used to *broadcast* the message to all processes except the sending process itself.

The sequence of values in the message is bracketed by { and }. These curly brackets may be omitted if the sequence contains just one element. The set notation is somewhat inappropriate, because the values in a message have a particular order. We shall compound this notational lapse by using the set notation even more extensively, as in

$$\textbf{send } \{v[n] \ : \ 0 \le n \le N\} \textbf{ to } q$$

to send the sequence of values $v[0], v[1], \ldots, v[N]$ to process q.

The **receive** primitive

$$\textbf{receive } \{v_0, v_1, \ldots, v_N\} \textbf{ from } q$$

receives a sequence of values from process q and assigns them to the sequence of variables specified in between the curly brackets. The keyword sequence **from all** is not allowed.

1.3 Data Distribution

We now explore how to incorporate the computational model of a multicomputer into program Vector-Sum-1. This program is so easy that a multicomputer version can be written down immediately. It is only a question of explaining the notation. However, more important than the multicomputer program itself is establishing a systematic program-derivation procedure that is also applicable to other, more complicated, multicomputer programs.

It is helpful to start with the end result: program Vector-Sum-4 is the multicomputer version of program Vector-Sum-1.

$0..P - 1 \parallel p$ **program** Vector-Sum-4
declare
    $i \ : \$ integer ;
    $x, \ y, \ z \ : \$ array $[\mathcal{I}_p]$ of real
initially
    $\langle \ ; \ i \ : \ i \in \mathcal{I}_p \ :: \ x[i], y[i] = \tilde{x}_{\mu^{-1}(p,i)}, \tilde{y}_{\mu^{-1}(p,i)} \ \rangle$
assign
    $\langle \ ; \ i \ : \ i \in \mathcal{I}_p \ :: \ z[i] := \alpha x[i] + \beta y[i] \ \rangle$
end

The header introduces a multicomputer program with P processes. The declaration of the integer variable i makes available one integer variable named i in each process. The P integer variables i of program Vector-Sum-4 correspond to the one integer variable m of program Vector-Sum-1. Similarly, the declaration of the array variables x, y, and z makes available three arrays in each process. The range of indices over which the arrays are declared varies from process to process. In process p, the array indices are elements of the *index set* $\mathcal{I}_p$. This index set is, in principle, an arbitrary set of integers. Usually, however, $\mathcal{I}_p$ is an index range $0..I_p - 1$. The index sets $\mathcal{I}_p$, the index-range limits I_p, and α and β are not declared and are, therefore, constants that are assumed known.

The **initially** section relates variables of program Vector-Sum-4 to those of program Vector-Sum-1. In program Vector-Sum-1, the array entry $x[m]$ was initialized with value $\tilde{x}_m$. This simple mapping is impossible for program Vector-Sum-4, where the set of M values $\{\tilde{x}_m : 0 \leq m < M\}$ must be distributed over the M available array entries of the P arrays named x (there is one such array in each process). This is called *data distribution*, and it is one fundamental aspect of multicomputer programming.

Technically, the data distribution is accomplished by means of a bijective map of the *global index* m to a pair of indices (p, i), where p is the process identifier and i the *local index*. Given such a map, the value $\tilde{x}_m$ can be stored in array entry $x[i]$ of process p.

Definition 1 *A P-fold distribution of an index set $\mathcal{M}$ is a bijective map $\mu(m) = (p, i)$ such that:*

$$\mu :: \mathcal{M} \to \{(p, i) : 0 \leq p < P \text{ and } i \in \mathcal{I}_p\} : m \to (p, i)$$

and

$$\mu^{-1} :: \{(p, i) : 0 \leq p < P \text{ and } i \in \mathcal{I}_p\} \to \mathcal{M} : (p, i) \to m.$$

In the **initially** section of program Vector-Sum-4, only the local index and the process identifier are available, and the inverse map is used to compute the global index of the initializing values.

The **assign** section of program Vector-Sum-4 is an immediate translation of that of program Vector-Sum-1. The concurrent separator was replaced by a sequential one, because the individual processes of a multicomputer computation must be sequential. In our model, concurrency can only be achieved by running many sequential processes simultaneously.

For future use, we introduce some additional quantities related to a data distribution μ. Let $\mathcal{M} = 0..M - 1$ be the set of global indices. For a given $p \in 0..P - 1$, the subset $\mathcal{M}_p$ is the set of global indices mapped to process p by the P-fold data distribution μ. More formally, we have that

$$\mathcal{M}_p = \bigcup_{i \in \mathcal{I}_p} \{\mu^{-1}(p, i)\}.$$

Because the P subsets $\mathcal{M}_p$ with $0 \le p < P$ satisfy

$$\mathcal{M} = \bigcup_{p=0}^{P-1} \mathcal{M}_p \ \text{ and } \ \mathcal{M}_p \bigcap \mathcal{M}_q = \emptyset \text{ if } p \ne q, \tag{1.1}$$

they define a partition of $\mathcal{M}$. Because μ is a bijective map, the number of elements of $\mathcal{M}_p$ equals the number of elements of $\mathcal{I}_p$, or

$$\forall p \in 0..P-1: \ M_p = \#\mathcal{M}_p = \#\mathcal{I}_p = I_p.$$

We use the standard symbol $\#$ for the cardinality operator of sets.

1.4   Program Development

It is intuitively obvious that programs Vector-Sum-1 and Vector-Sum-4 are equivalent in some sense. That is hardly a solid foundation for a generalizable approach to developing multicomputer programs. In this section, we shall develop a rigorous procedure to transform program Vector-Sum-1 into program Vector-Sum-4. The importance of this section lies not so much with the proofs, which are trivial, but with the principle that there is a rigorous technique to transform a specification program into a multicomputer program. The same technique will be used to derive other, more complicated, multicomputer programs.

As a first step in the program transformation, perform a *P-fold duplication* of program Vector-Sum-1 to obtain program Vector-Sum-2.

> **program** Vector-Sum-2
> **declare**
> $\quad p$: integer ;
> $\quad m$: array $[0..P-1]$ of integer ;
> $\quad x, y, z$: array $[0..P-1]$ of array $[0..M-1]$ of real
> **initially**
> $\quad \langle\ \|\ p\ :\ p \in 0..P-1\ ::$
> $\qquad \langle\ ;\ m[p]\ :\ 0 \le m[p] < M\ ::$
> $\qquad\quad x[p][m[p]], y[p][m[p]] = \tilde{x}_{m[p]}, \tilde{y}_{m[p]}$
> $\qquad \rangle$
> $\quad \rangle$
> **assign**
> $\quad \langle\ \|\ p\ :\ p \in 0..P-1\ ::$
> $\qquad \langle\ \|\ m[p]\ :\ 0 \le m[p] < M\ ::$
> $\qquad\quad z[p][m[p]] := \alpha x[p][m[p]] + \beta y[p][m[p]]$
> $\qquad \rangle$
> $\quad \rangle$
> **end**

In program Vector-Sum-2, *every* variable except p is an array over the index range $0..P - 1$. The **assign** section is a concurrent quantification over p. The expressions bracketed by this quantification are assignments executed by process p, where $p \in 0..P - 1$. To ensure that the data spaces of the P processes are separated, only variables with first index p are allowed within the quantification over p.

It is useful to introduce a short-hand notation for duplicated programs. In fact, we have already used it in Section 1.3. The header "$0..P - 1 \parallel p$ **program**" denotes the P-fold duplication of a program. Program Vector-Sum-2 can, therefore, also be specified as follows.

$0..P - 1 \parallel p$ **program** Vector-Sum-2
declare
      m : integer ;
      x, y, z : array $[0..M - 1]$ of real
initially
      $\langle \; ; \; m \; : \; 0 \leq m < M \; :: \; x[m], y[m] = \tilde{x}_m, \tilde{y}_m \; \rangle$
assign
      $\langle \; \parallel \; m \; : \; 0 \leq m < M \; :: \; z[m] := \alpha x[m] + \beta y[m] \; \rangle$
end

The former version is an *explicitly duplicated program*, and the latter is an *implicitly duplicated program*. The latter is merely a short-hand notation for the former; both versions are equivalent in all other aspects. Occasionally, it will come in handy to make the duplication explicit on one or two variables only. On those variables, we shall use the process identifier as an extra index. For example, array entry $x[p][m]$ is entry number m of array x in process p. Situations where a mixed use of implicit and explicit process identifiers leads to ambiguity will be avoided.

The duplication transformation can be useful only if the meaning of the program is, in some sense, preserved. A terminating program maps the *initial state* of its data space to a *final state*. This map is the essence of a program. Two programs that compute the same map are equivalent. We say that program A is an implementation of or implements program B if the map of program B can be retrieved from the map of program A.

Lemma 1 *Program Vector-Sum-2 implements program Vector-Sum-1.*

In program Vector-Sum-1, let $x_m^{(0)}$ and $y_m^{(0)}$ denote the initial values of the variables $x[m]$ and $y[m]$, respectively, and let $z_m^{(\infty)}$ be the final values of the variables $z[m]$. This can also be denoted as follows:

$$\forall m \in \mathcal{M} : \begin{cases} x_m^{(0)} & = & \text{value}(x[m]) \text{ at } \textbf{initially} \\ y_m^{(0)} & = & \text{value}(y[m]) \text{ at } \textbf{initially} \\ z_m^{(\infty)} & = & \text{value}(z[m]) \text{ at } \textbf{end.} \end{cases}$$

Program Vector-Sum-1 computes the map:

$$\forall m \in 0..M - 1 : \quad z_m^{(\infty)} = \alpha x_m^{(0)} + \beta y_m^{(0)}. \tag{1.2}$$

Note that this is a relation between values and that it is irrelevant which variables contain these values.

Each instance of the duplicated program is trivially equivalent to program Vector-Sum-1. We must show that the map between initial and final state of program Vector-Sum-1 can be retrieved from the map of its duplication. This is easily done using the data-distribution concept.

Any P-fold distribution μ of the index set $\mathcal{M} = 0..M - 1$ defines a partition of $\mathcal{M}$ that consists of P disjoint sets $\mathcal{M}_p$. Using this partition, we retrieve the following values from the variables of program Vector-Sum-2:

$$\forall p \in 0..P - 1, \forall m \in \mathcal{M}_p : \quad \begin{cases} x_m^{(0)} & = & \text{value}(x[p][m]) \text{ at } \textbf{initially} \\ y_m^{(0)} & = & \text{value}(y[p][m]) \text{ at } \textbf{initially} \\ z_m^{(\infty)} & = & \text{value}(z[p][m]) \text{ at } \textbf{end}. \end{cases}$$

It is easily seen that these satisfy Equation 1.2.   □

The *selection step* transforms program Vector-Sum-2 into Vector-Sum-3. Quantifications over the index set $\mathcal{M}$ are replaced by quantifications over $\mathcal{M}_p$. The subsets $\mathcal{M}_p$ are defined by the data distribution μ that is used to extract the values $x_m^{(0)}$, $y_m^{(0)}$, and $z_m^{(\infty)}$ from program variables.

> 0..P − 1 ‖ p **program** Vector-Sum-3
> **declare**
> m : integer ;
> x, y, z : array $[\mathcal{M}_p]$ of real
> **initially**
> $\langle$; m : $m \in \mathcal{M}_p$ :: $x[m], y[m] = \tilde{x}_m, \tilde{y}_m$ $\rangle$
> **assign**
> $\langle$; m : $m \in \mathcal{M}_p$ :: $z[m] := \alpha x[m] + \beta y[m]$ $\rangle$
> **end**

The selection step incorporates the data distribution into the program and distributes the work over the processes. (We also replaced a concurrent by a sequential quantification. This is trivially justified in this instance, because the assignments bracketed by the quantification are mutually independent.)

Lemma 2 *Program Vector-Sum-3 implements program Vector-Sum-1.*

The proof this lemma is almost identical to that of Lemma 1. The only difference is that one must use one particular P-fold distribution μ: the one used in program Vector-Sum-3. In Lemma 1, one was allowed to choose *any* data distribution.   □

Program Vector-Sum-3 satisfies the computational model of multicomputers, but arrays defined over arbitrary index sets, like $\mathcal{M}_p$, are difficult and/or inefficient to implement in practical programming environments. This implementation difficulty forces us to perform one more step in the program development: the conversion of global into local indices. This step transforms program Vector-Sum-3 into program Vector-Sum-4 (see Section 1.3). There is no fundamental difference between arrays defined over $\mathcal{M}_p$ and those defined over $\mathcal{I}_p$. There is an important practical difference, however. The local-index sets $\mathcal{I}_p$ are usually index ranges, say $\mathcal{I}_p = 0..I_p - 1$, and arrays defined over index ranges have efficient representations.

Lemma 3 *Program Vector-Sum-4 implements program Vector-Sum-1.*

We only need to adapt the proof of Lemma 2 to incorporate local indices. Retrieve the following values from the variables of program Vector-Sum-4:

$$\forall m \in \mathcal{M} : (p, i) = \mu(m) \text{ and } \begin{cases} x_m^{(0)} & = & \text{value}(x[p][i]) \text{ at } \textbf{initially} \\ y_m^{(0)} & = & \text{value}(y[p][i]) \text{ at } \textbf{initially} \\ z_m^{(\infty)} & = & \text{value}(z[p][i]) \text{ at } \textbf{end}. \end{cases}$$

These values satisfy Equation 1.2, which implies the lemma.   □

To summarize, the transformation of program Vector-Sum-1 into program Vector-Sum-4 is achieved in three steps:

1. *a* **duplication step** *sets up the process structure,*

2. *a* **selection step** *distributes the work and the data over the processes,*

3. *a* **conversion of global into local indices**.

1.5 Communication

Can the same procedure be used to derive other multicomputer programs? As a next step, consider the computation of the inner product of two vectors. The (P-fold) duplication of program Inner-Product-1 leads to program Inner-Product-2.

```
0..P − 1 ∥ p program Inner-Product-2
declare
      m : integer ;
      σ : real ;
      x, y : array [0..M − 1] of real
initially
      ⟨ ; m : 0 ≤ m < M :: x[m], y[m] = x̃_m, ỹ_m ⟩
assign
      σ := ⟨ + m : 0 ≤ m < M :: x[m]y[m] ⟩
end
```

Lemma 4 *Program Inner-Product-2 implements program Inner-Product-1.*

Program Inner-Product-1 establishes the map

$$\sigma^{(\infty)} = \sum_{m=0}^{M-1} x_m^{(0)} y_m^{(0)}, \tag{1.3}$$

where (0) and (∞) tag the initial and final states of the data space, respectively. Program Inner-Product-2 establishes the same map, provided the values are extracted from the program variables as follows. *Any P-fold distribution μ of the index set $\mathcal{M} = 0..M-1$ defines a partition of $\mathcal{M}$ that consists of P disjoint sets $\mathcal{M}_p$. Using such a partition, choose

$$\forall p \in 0..P-1, \; \forall m \in \mathcal{M}_p : \begin{cases} x_m^{(0)} & = \text{value}(x[p][m]) \text{ at } \textbf{initially} \\ y_m^{(0)} & = \text{value}(y[p][m]) \text{ at } \textbf{initially} \end{cases}$$

and, for *an arbitrary value of $p \in 0..P-1$*, let

$$\sigma^{(\infty)} = \text{value}(\sigma[p]) \text{ at } \textbf{end}.$$

These values satisfy Equation 1.3, which establishes the lemma.   □
Compared with Lemma 1, there is an extra degree of arbitrariness here: not only the data distribution μ but also the process p that defines the value $\sigma^{(\infty)}$ is left unspecified.

Once again, the duplication step is followed by the selection step. Here, we impose a specific data distribution on the arrays x and y and eliminate superfluous arithmetic from program Inner-Product-2. Because the addition separator neither specifies an order on expression evaluation nor implies any order in summing the values of these expressions, the **assign** section of program Inner-Product-2 is equivalent to:

```
σ := ⟨ + q  :  0 ≤ q < P  ::
          ⟨ + m  :  m ∈ M_q  ::  x[m]y[m] ⟩
      ⟩
```

The sets $\mathcal{M}_q$, where $0 \le q < P$, are defined by a P-fold distribution μ of the index set $\mathcal{M} = 0..M-1$. Because addition is associative and commutative, the inner quantifications may be evaluated first to obtain:

```
⟨ ‖  q  :  0 ≤ q < P  ::
      w[q] := ⟨ + m  :  m ∈ M_q  ::  x[m]y[m] ⟩
⟩ ;
σ := ⟨ + q  :  0 ≤ q < P  ::  w[q] ⟩
```

Remember that this is the body of a duplicated program.

When the P-fold duplication and the process identifiers of the variables σ and ω are made explicit, the same program segment reads as follows:

$$\langle \parallel p : 0 \le p < P ::$$
$$\langle \parallel q : 0 \le q < P ::$$
$$\omega[p][q] := \langle + m : m \in \mathcal{M}_q :: x[m]y[m] \rangle$$
$$\rangle ;$$
$$\sigma[p] := \langle + q : 0 \le q < P :: \omega[p][q] \rangle$$
$$\rangle$$

Clearly, the sum over a particular index set $\mathcal{M}_p$ should be computed just once, and not in every process. One possible strategy is to compute the sum over $\mathcal{M}_p$ only in process p, where it is assigned to the variable $\omega[p][p]$. The values of variables $\omega[p][q]$ with $p \ne q$ are subsequently obtained by assignment instead of arithmetic. Here is one possible formulation.

$$\langle \parallel p : 0 \le p < P ::$$
$$\omega[p][p] := \langle + m : m \in \mathcal{M}_p :: x[m]y[m] \rangle$$
$$\rangle ;$$
$$\langle \parallel p : 0 \le p < P ::$$
$$\langle \parallel q : 0 \le q < P \text{ and } q \ne p :: \omega[p][q] := \omega[q][q] \rangle ;$$
$$\sigma[p] := \langle + q : 0 \le q < P :: \omega[p][q] \rangle$$
$$\rangle$$

The concurrent quantification over p is interrupted to guarantee that all values $\omega[p][p]$ have been computed before being used. Such a synchronization point is called a *barrier*, and it guarantees that all processes reach a particular state before proceeding further. The barrier symbol (; ;) specifies a barrier in the body of an implicitly duplicated program. The following is equivalent to the above explicitly duplicated program segment:

$$\omega[p][p] := \langle + m : m \in \mathcal{M}_p :: x[m]y[m] \rangle ;;$$
$$\langle \parallel q : 0 \le q < P \text{ and } q \ne p :: \omega[p][q] := \omega[q][q] \rangle ;$$
$$\sigma := \langle + q : 0 \le q < P :: \omega[p][q] \rangle$$

In this implicitly duplicated program segment, the process identifier of the ω-variables remains explicit.

The assignments $\omega[p][q] := \omega[q][q]$ with $p \ne q$ are not allowed by the computational model of multicomputers. Assignments involving variables of two different processes must be split into two parts. The first part, done in process q, sends the current value of $\omega[q][q]$ to process p. The second part, done in process p, waits until this value arrives and assigns it to $\omega[p][q]$. The two basic communication primitives can be used to accomplish this.

Combining notation of Section 1.2 for **send** and **receive** primitives, the local indices obtained from the data distribution μ as in Section 1.4, and the previously developed code segments, we obtain program Inner-Product-3.

$0..P - 1 \parallel p$ **program** Inner-Product-3
declare
$\qquad i$: integer ;
$\qquad \sigma$: real ;
$\qquad \omega$: array $[0..P - 1]$ of real ;
$\qquad x, y$: array $[\mathcal{I}_p]$ of real
initially
$\qquad \langle\ ;\ i\ :\ i \in \mathcal{I}_p\ ::\ x[i], y[i] = x^{(0)}_{\mu^{-1}(p,i)}, y^{(0)}_{\mu^{-1}(p,i)}\ \rangle$
assign
$\qquad \omega[p] := \langle\ +\ i\ :\ i \in \mathcal{I}_p\ ::\ x[i]y[i]\ \rangle$;
$\qquad$ **send** $\omega[p]$ **to all** ;
$\qquad \langle\ ;\ q\ :\ 0 \le q < P$ **and** $q \ne p\ ::$
$\qquad\qquad$ **receive** $\omega[q]$ **from** q
$\qquad \rangle$;
$\qquad \sigma := \langle\ +\ q\ :\ 0 \le q < P\ ::\ \omega[q]\ \rangle$
end

The process identifier of all variables is implicit. The variable ω is an array of length P in every process. Program Inner-Product-3 no longer contains a barrier; it has become superfluous, because the value of $\omega[p]$ is sent out only after it has been properly initialized and the receive primitive waits until the message arrives.

1.6   Performance Analysis

Most applications have an unbounded appetite for memory and processor time. The efficient use of computing resources is, therefore, always a concern in scientific computing. The efficiency determines the feasible range of problem sizes and the feasible range of applications. Performance considerations are always important when using concurrent computers. In fact, increased performance is usually the principal and sometimes the only motivation for using concurrent computers.

However, the quest for performance has its limitations. Any program is the end result of many design decisions, which were made on the basis of many other criteria besides performance, such as numerical stability, robustness, application range, portability, etc. Whether or not some of these properties should be sacrificed for performance should be judged based on the merits of each individual case. One property can never be sacrificed: correctness. Programs that do not meet their specifications are the most inefficient of all!

1.6.1   Computations and Execution Times

A *computation* is a *program* with particular *input data* executed on a given *computer* at a particular *moment in time*. A computation occurs each time a program is executed on a computer. Therefore, one program is always associated with infinitely many computations. The time elapsed between the start and the termination of a computation is called its *execution time*. On a multicompuer, the execution time of a computation is the time elapsed between the instant when all nodes begin execution and the instant when the last node terminates.

Performance analysis is the study of execution times of computations. Performance analysis is difficult, because execution time is a function on the extremely complicated space of computations. The coordinates of a computation in this space are all the parameters that define a computation: program, input data, computer, and moment in time. A program and its input data define the problem and its solution procedure. In a performance-analysis context, one often refers separately to "the problem" and "the program." The latter then refers to all aspects that define the solution procedure; this is not just the program text, but also the inputs that determine how the problem is solved. Similarly, "the problem" is defined by a combination of program text and input data. Consider, for example, program Vector-Sum-4. A problem is defined by the values α, β, M, and $\tilde{x}_m$ and $\tilde{y}_m$ for $0 \leq m < M$. For each problem, the program text describes infinitely many different solution procedures: one for each value of the number of processes P and for each possible data distribution μ.

The first task of any performance analysis is to simplify the space of computations such that only a few important parameters identify each computation. Typically, the space of computations is simplified to the extent that a computation is defined by just one independent parameter, called the *study parameter*. When analyzing how the execution time depends on this parameter, an attempt is made to keep all other parameters fixed. That is often impossible, and the non-study parameters are divided into two categories. The *fixed parameters* are really kept unchanged from one computation to the next. The *dependent parameters* are some function of the study parameter. How to choose study, fixed, and dependent parameters depends on the question one wants answered. In Sections 1.6.2 and 1.6.3, the study parameter is the number of nodes of the multicomputer. In Section 12.6, a more general variation in computer hardware is allowed.

In all our studies, we shall ignore the moment in time as a parameter. Computations that are identical except for the moment in time when they are executed can have widely varying execution times, particularly on computers shared by many users. This problem is avoided by always obtaining single-user execution times. The statistical noise that remains is usually sufficiently small to be ignored.

1.6.2 Speed-up: Theory

Assume we are given a certain problem to solve and a particular multi-computer consisting of P identical nodes. We wish to analyze how the execution time required to solve this problem depends on the number of nodes p, where $1 \leq p \leq P$. For now, we shall ignore the fact that memory restrictions often prohibit solving the given problem if only a few nodes are available.

In this analysis, the number of nodes is the study parameter, and the multicomputer and the problem are fixed parameters. All other aspects defining the computation are dependent parameters chosen as follows.

Imagine running all possible p-node computations that solve the problem. The minimal execution time observed on p nodes is called the *best p-node time* and is denoted by T_p^*. This execution time was achieved by the *best p-node computation* and the *best p-node program*. The case $p = 1$ is special, and we shall refer to *the best sequential time, computation, and program*. For every value of p, choose the program (in the sense discussed above) that corresponds to the best p-node time.

If the number of nodes p is the sole study parameter, if the problem and the multicomputer are fixed parameters, and if all other parameters are chosen on the basis of optimality, the following theorem holds.

Theorem 1 *On a given multicomputer and for a fixed problem, the best p-node computation is at most p times faster than the best sequential computation, or*

$$T_p^* \geq T_1^*/p.$$

The proof is by contradiction. Using the best p-node computation, we can construct a sequential computation by running all processes of the p-node computation on one node. In the sequential computation, exactly one process is running at every instant. If the running process cannot continue, control is switched to a process that can. (We omit the proof of the fact that there is always at least one process that can continue.) If switching control between processes is free (this requires an idealized node with a free context switch), this new sequential computation has an execution time of at most pT_p^*. If the best p-node computation violates the theorem, there exists a sequential computation with execution time $pT_p^* < T_1^*$. This contradicts the statement that T_1^* is the best sequential time and, hence, proves the theorem.   □

In Figure 1.2, Theorem 1 is represented graphically. Each diamond corresponds to a computation, and its coordinates are the number of nodes p and the execution time t of the computation. As before, problem and multicomputer are fixed parameters. For each number of nodes p, there is an infinity of computations; obviously, only a few of those are represented. For each number of nodes p, there is a best computation, which is not necessar-

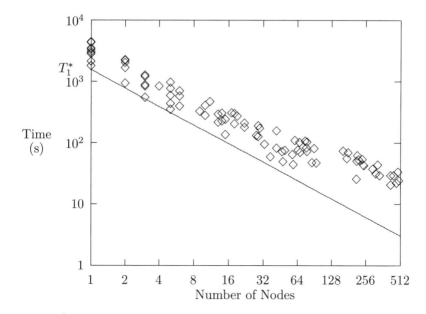

FIGURE 1.2. A logarithmic execution-time plot for a fixed problem solved on a particular multicomputer.

ily unique. The diamond with coordinates $(p, t) = (1, T_1^*)$ corresponds to the best sequential computation. Theorem 1 implies that all computations lie on or above the curve defined by the equation

$$t = T_1^*/p. \tag{1.4}$$

Points on this curve correspond to p-node computations that are exactly p times faster than the best sequential computation. Because of the logarithmic scales used in Figure 1.2, this curve is represented by a straight line with slope -1 through $(p, t) = (1, T_1^*)$. This line is called the *line of perfect speed-up*. In the figure, the slope of the line of perfect speed-up appears different from -1, because of different scaling factors for p and t axes.

Actual computations are compared against the line of perfect speed-up by means of two frequently used measures: the *speed-up* and the *efficiency*.

Definition 2 *The speed-up of a p-node computation with execution time T_p is given by:*

$$S_p = T_1^*/T_p,$$

where T_1^* is the best sequential time obtained for the same problem on the same multicomputer.

Definition 3 *The efficiency of a p-node computation with speed-up S_p is given by:*

$$\eta_p = S_p/p.$$

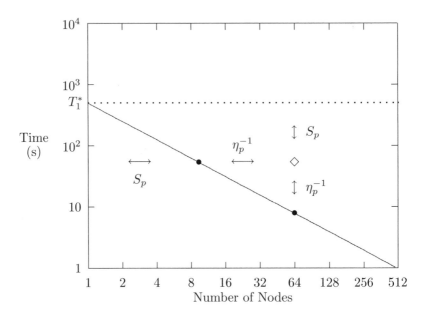

FIGURE 1.3. Speed-up and efficiency of a computation in a logarithmic execution-time plot of a fixed problem solved on a particular multicomputer.

A p-node computation on the line of perfect speed-up has a speed-up p and an efficiency 1 (or 100%). Theorem 1 states that $S_p \leq p$ and $\eta_p \leq 1$ for all actual p-node computations.

In a logarithmic execution-time plot, the speed-up and efficiency of represented computations are easily found graphically. In Figure 1.3, the solid line is the line of perfect speed-up, and the diamond represents a p-node computation with execution time T_p. It follows from:

$$\log T_p - \log \left(\frac{T_1^*}{p} \right) = \log \eta_p^{-1}$$

that the vertical distance of the diamond to the line of perfect speed-up is the logarithm of the inverse of efficiency. This confirms that the more efficient computations are closer to the line of perfect speed-up. Similarly, the logarithm of the speed-up is the vertical distance between the diamond and the horizontal line $t = T_1^*$. Because the slope of the line of perfect speed-up is -1, vertical and horizontal distances to the line of perfect speed-up are equal. Speed-up and efficiency are, therefore, also represented graphically by the horizontal distances indicated. In Figure 1.3, vertical and horizontal distances appear different, and the slope of the line of perfect speed-up appears different from -1, because p- and t-axes have different scaling factors.

1.6.3 Speed-up: Practice

In most practical performance analyses, performance data are obtained for just one sequential and just one multicomputer program. Nevertheless, the family of possible computations is still very large because the text of one multicomputer program corresponds to many possible solution procedures: number of processes, data distributions, process mesh (see Chapter 2), process placement (see Chapter 12), etc., can all be varied. To reduce the number of computations further, additional constraints are often imposed. Such additional constraints include setting the number of processes equal to the number of nodes and specifying particular data distributions, process meshes, and process placements as a function of the number of nodes. The program remains a dependent parameter, but it is no longer chosen on the basis of optimality over the space of all possible programs that solve the given problem. As a result, Theorem 1 is not applicable to the obtained execution times. Nevertheless, its bound remains a useful guide.

To simplify the discussion, assume that enough constraints were imposed such that just one p-node computation is left for each value of $p > 1$. The execution time of this p-node computation is denoted by T_p. There remain two different sequential computations: one based on the original sequential program with execution time T_1^S and one based on the multicomputer program, but run on one node, with execution time T_1. It is typical that $T_1^S < T_1$. Figure 1.4 summarizes the data in a logarithmic execution-time plot.

The time T_1^S is somewhat arbitrary, because it depends on the sequential program that happened to be used. Usually, it is assumed that T_1^S is sufficiently near the unknown best sequential time T_1^* to compute speed-ups and efficiencies with reasonable accuracy. In the logarithmic execution-time plot, the line with slope -1 through $(p, t) = (1, T_1^S)$ is called the *line of linear speed-up*. This line is just as arbitrary as the sequential-execution time T_1^S. A computation that falls below the line of linear speed-up has *superlinear speed-up*. Theorem 1 implies that superlinear speed-up can occur only if the sequential computation is suboptimal. Superlinear speed-up is never the result of a "superoptimal" concurrent computation. The occurrence of superlinear speed-up should be used as a warning sign to be extra suspicious of the sequential-execution time used for the performance analysis. For an elaboration on superlinear speed-up, see Section 1.8.

Often, multicomputer computations with a small number of nodes lie near another line with slope -1: the *line of scalability*. It contains the point with coordinates $(p, t) = (1, T_1)$, which represents a sequential computation with the multicomputer program. The line of scalability does not have any special theoretical properties, and it is neither a lower nor an upper bound for the execution time of multicomputer computations. Usually, however, computations lie above the line of scalability.

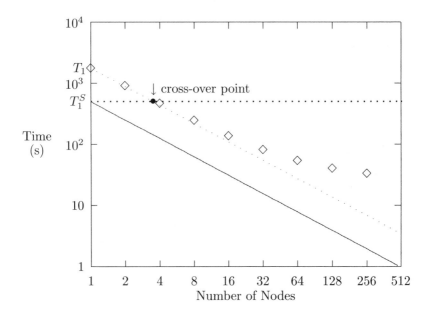

FIGURE 1.4. A realistic logarithmic execution-time plot for a family of multicomputer computations based on the same multicomputer program and one sequential computation based on a sequential program.

The number of nodes at the *cross-over point*, which we denote by p_c, is just sufficient to make up for the difference between T_1 and T_1^S. If p_c is small, the cross-over point is usually near the intersection of the line of scalability and the horizontal line $t = T_1^S$. As a rule, the superposition of overheads when the number of nodes is increased causes all computations to fall above the line of scalability. Moreover, the distance between the computations and the line of scalability usually increases with the number of nodes. The efficiency at the cross-over point is, therefore, a heuristic upper bound for the efficiency of computations that solve the given problem with the given multicomputer program. From $T_1^S = T_{p_c}$, it follows that the *cross-over efficiency* is given by:

$$\eta_{p_c} = \frac{T_1^S}{p_c T_{p_c}} = \frac{1}{p_c}.$$

A program with cross-over point at $p_c = 4$ has a cross-over efficiency of 25% and, therefore, an expected maximum efficiency of 25%.

In scientific computations, there are three important sources of inefficiency: floating-point overhead, communication overhead, and load imbalance. Floating-point overhead typically causes the difference between T_1 and T_1^S; graphically, it is the (vertical or horizontal) distance between the lines of linear speed-up and scalability. In Section 1.6.2, the global efficiency loss was related to the (vertical or horizontal) distance between the computation and the line of perfect speed-up, here approximated by the line of linear speed-up. Subtracting floating-point overhead from global overhead, we see that the combined effect of communication overhead and load imbalance is represented graphically by the (vertical or horizontal) distance between the computation and the line of scalability. In Figure 1.4, computations using a small number of nodes suffer mainly from floating-point overhead: the computations lie on or near the line of scalability. As the number of nodes increases, the distance between the computations and the line of scalability increases, which shows that communication overhead and load imbalance become more significant.

1.6.4 Performance Modeling

To obtain an execution time, one must perform a computation on a computer. In theoretical performance analyses, a model computer is used to estimate execution times. A familiar model for sequential computers is to count the number of floating-point operations. The operation count is assumed to be proportional to the execution time of the computation. Such a model can be used for the nodes of a multicomputer. However, a multicomputer-performance model also requires a communication model and some assumptions regarding the concurrent execution of processes. Performance models are often too inaccurate to predict the performance of actual computers and are not a substitute for computational experiments. However, the execution-time formulas obtained via models are often better to gain a qualitative insight, because the sources of inefficiency are obvious when comparing terms corresponding to different parts of the computation.

Performance is the result of a subtle interplay between program and environment. The more details are known about an environment, the better the performance analysis can be for that particular environment, and the more irrelevant it becomes for other environments. We shall use some basic performance assumptions that are sufficiently accurate for qualitative assessments of the performance of multicomputer programs. Note that the properties of the performance model are NEVER used to prove correctness of programs, only to model their execution times.

In our multicomputer-performance model, every process is run on a dedicated node. Such processes are called *independent processes*. Computations with independent processes do not suffer from interference between processes due to process scheduling. For performance-analysis purposes, it is assumed that all nodes begin execution at the same instant. The execution time of a multicomputer computation is the time elapsed between the instant when all nodes begin execution and the instant when the last node terminates. All nodes of the multicomputer have identical performance characteristics. Every floating-point operation requires a time τ_A, the *arithmetic time*. The time required by integer arithmetic is ignored.

The exchange of two messages of length L between any two processes takes a time $\tau_C(L)$, the *message-exchange time*. The length of a message is measured by the number of floating-point words that fit into the message. In one message-exchange operation, one message is sent by one process and received by a second process, while another message of the same length is sent in the reverse direction. In our simple performance model, it is assumed that these events occur simultaneously and that there is a simple linear relationship between the message-exchange time τ_C and message length L:

$$\tau_C(L) = \tau_S + \beta L. \tag{1.5}$$

The time τ_S is called the *latency* or *start-up time*; it is the time required to exchange a message of length 0. The *band width* is the maximal rate (in floating-point words per second) at which messages can be exchanged, and it is given by β^{-1}. This definition of band width is non-standard, because the unit of message length is a floating-point word and because a message exchange instead of a simple send–receive pair is the fundamental communication unit.

Our performance model is a major simplification of reality and ignores many practical issues. For example, it is implicitly assumed that no process ever has to wait for a message. The model also makes no provisions for overlapping communication and computation within a single process, a feasible feature of many current multicomputers. Some of these practical communication issues will be explored in Section 12.3.

1.6.5 Vector Operations

Programs Vector-Sum-4 and Inner-Product-3 satisfy the computational model for multicomputers. However, are these programs efficient?

Associate the following quantities with the distribution μ used by both programs:

$$M = \#\mathcal{M} = \sum_{p=0}^{P-1} M_p$$

$$M_p = \#\mathcal{M}_p = \#\mathcal{I}_p = I_p$$

$$\bar{I} = \max_{0 \leq p < P} I_p.$$

Based on the assumptions of the multicomputer-performance model, the execution time of program Vector-Sum-4 is easily estimated by:

$$T_P = \max_{0 \leq p < P} 3I_p \tau_A = 3\bar{I}\tau_A. \tag{1.6}$$

The execution time T_P is minimal for distributions with minimal $\bar{I}$.

Definition 4 *A P-fold distribution μ of an index set $\mathcal{M}$ is* load balanced *if and only if:*

$$\bar{I} = \max_{0 \leq p < P} \#\mathcal{M}_p$$

is minimal over the set of all P-fold distributions of $\mathcal{M}$.

If M is an exact multiple of P, a load-balanced distribution satisfies

$$\forall p \in 0..P-1 : \ I_p = M_p = \bar{I} = \frac{M}{P},$$

and if M is not an exact multiple of P, it satisfies

$$\forall p \in 0..P-1 : \ I_p = \bar{I} \ \text{ or } \ I_p = \bar{I} - 1,$$

where $\bar{I} = \lceil \frac{M}{P} \rceil$.

The speed-up and efficiency of program Vector-Sum-4 are, respectively, given by

$$\begin{aligned} S_P &= \frac{3M\tau_A}{3\bar{I}\tau_A} = \frac{M}{\bar{I}} \\ \eta_P &= M/(P\bar{I}). \end{aligned}$$

If μ is load balanced and M is a multiple of P, we have that $S_P = P$ and $\eta_P = 100\%$. If either assumption is violated, the speed-up is less than P and the efficiency less than 100%.

For a given vector length M, the maximal number of non-idle processes that can be used by program Vector-Sum-4 is M. For $P = M$, a load-balanced distribution allocates an array of length one in each process. The execution time is then given by $T_M = 3\tau_A$, which results in a speed-up $S_M = M$ and an efficiency $\eta_M = 100\%$. However, reducing the number of independent processes by just one, say $P = M - 1$, doubles the execution time ($T_{M-1} = 6\tau_A$), halves the speed-up ($S_{M-1} = M/2$), and almost halves the efficiency ($\eta_{M-1} = M/2(M-1) \approx 50\%$).

Figure 1.5 represents Equation 1.6 in a logarithmic execution-time plot for several values of M, assuming that load-balanced data distributions are used. The unit of time used in the plot is τ_A, the arithmetic time. Note that $T_1 = T_1^S = T_1^*$ for this program. For small numbers of nodes, each plot has a slope -1, indicating perfect speed-up. However, the irregularities in the plots point out the inefficiency of computations with a number of nodes comparable to M. Of course, computations with more than M nodes are the least efficient.

In summary, the performance of program Vector-Sum-4 is strongly tied to the distribution μ being load balanced. Even if load balanced, some inefficiency occurs if M is not a multiple of P. The latter loss of efficiency is negligible if $M \gg P$. However, if $P \approx M$ and M is not a multiple of P, then there is a significant loss of efficiency even for load-balanced distributions. Computations with $P \ll M$ are said to be *coarse grained*, and computations with $P \approx M$ are called *fine grained*.

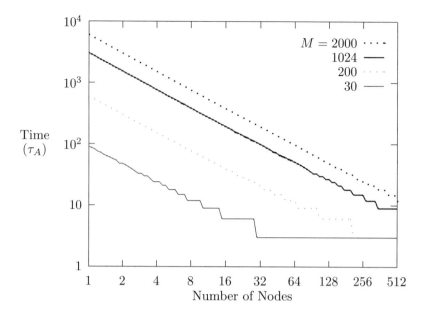

FIGURE 1.5. Logarithmic execution-time plot for program Vector-Sum-4 with load-balanced data distributions.

We now turn to an analysis of program Inner-Product-3. An immediate observation is that the quantification used to receive values for the variables $\omega[p][q]$ with $p \neq q$ should be changed so that messages are received in order of arrival. However, since our performance model assumes no waiting, our conclusions will be independent of such strategies.

For vectors of length M, the execution time of program Inner-Product-3 on P nodes of a multicomputer is given by

$$T_P = (2\bar{I} - 1)\tau_A + (P - 1)\tau_C(1) + (P - 1)\tau_A. \tag{1.7}$$

The first term is the time required for the calculation of the local inner product, the second term is the time required for communication, and the third term is the time for summing the local inner products.

The logarithmic execution-time plot of Figure 1.6 displays Equation 1.7 for two values of M. The plot assumes load-balanced distributions. The unit of time used in the plot is the arithmetic time τ_A. The message-exchange time $\tau_C(1)$ is assumed to be a factor of 50 larger. (For most current multicomputers, the ratio of message-exchange over arithmetic time is larger.) Clearly, if the number of nodes is large, the dimension of the vectors is irrelevant as far as execution time is concerned. The execution time of program Inner-Product-3 is quickly dominated by communication-time considerations. For reference purposes, Figure 1.6 also plots the line of perfect speed-up for each value of M (note that $T_1 = T_1^S = T_1^*$).

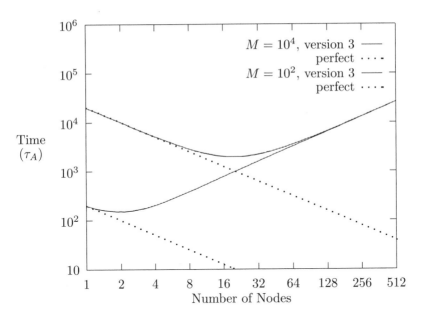

FIGURE 1.6. Logarithmic execution-time plot for program Inner-Product-3 with load-balanced data distribution and $\tau_C(1) = 50\tau_A$.

The finest-grained computation possible with program Inner-Product-3 uses M processes, each assigned one index from the index set $\mathcal{M}$. The execution time for this case is

$$T_M = M\tau_A + (M - 1)\tau_C(1).$$

Comparison with the sequential time $T_1 = (2M - 1)\tau_A$ immediately shows that this program is not suited for fine-grained computations: because it is always the case that $\tau_C(1) \gg \tau_A$, the concurrent computation is actually slower than the sequential computation.

Coarse-grain performance is less disastrous, but it is certainly not a success. Load-balanced distributions minimize only the first term of T_P (Equation 1.7). The remaining two terms are independent of μ. Load balance and coarse granularity ($M \gg P$) justify the assumption that $\bar{I} \approx M/P$ and that

$$T_P \approx (2M/P + P - 2)\tau_A + (P - 1)\tau_C(1).$$

The corresponding speed-up is given by

$$
\begin{aligned}
S_P &\approx \frac{(2M - 1)\tau_A}{(2M/P + P - 2)\tau_A + (P - 1)\tau_C(1)} \\
&\approx P \frac{1}{1 + \frac{(P-1)^2}{2M-1} + \frac{P(P-1)}{2M-1}\frac{\tau_C(1)}{\tau_A}}
\end{aligned}
$$

and is determined by the characteristic $\frac{\tau_C(1)}{\tau_A}$ of the multicomputer, the number of independent processes P, and the problem size M. The formula for the speed-up suggests that the condition

$$P^2 \ll 2M \qquad (1.8)$$

must be satisfied for the computation to be efficient.

1.7   The Recursive-Doubling Algorithm

1.7.1   Specification

Program Inner-Product-3 is inefficient because of the global sum of the local inner products. This part generated the communication and the last quantification of program Inner-Product-3, which remained duplicated in every process. Here, our goal is to implement the global-sum operation more efficiently.

Given P variables $\omega[p]$, where $p \in 0..P-1$, we wish to overwrite each $\omega[p]$ with the sum of all $\omega[p]$-values. The specification program is given by program Recursive-Doubling-1.

$0..P - 1 \parallel p$ **program** Recursive-Doubling-1
declare
      q : integer ;
      ω : real
initially
      $\omega[p] = \tilde{\omega}_p$
assign
      $\omega[p] := \langle\, + q \,:\, 0 \le q < P \,::\, \omega[q] \,\rangle$
end

The specification program is an implicitly duplicated program but with explicit process identifiers on the ω-variables. It is our task to transform program Recursive-Doubling-1 into a multicomputer program.

1.7.2   Implementation

Recursive doubling is pairwise summation of the terms, as shown in Figure 1.7 for the case $P = 8$. Every line in the figure corresponds to a message sent from one process to another.

The communication is best formulated in terms of the binary representation of the process identifiers. For now, assume that $P = 2^D$. Then, the binary representation of every process identifier consists of D bits, numbered 0 to $D - 1$. In step d, process p exchanges values with process q, where q is obtained from p by reversing bit number d of its binary representation. In step 0, for example, the least significant bit is reversed such

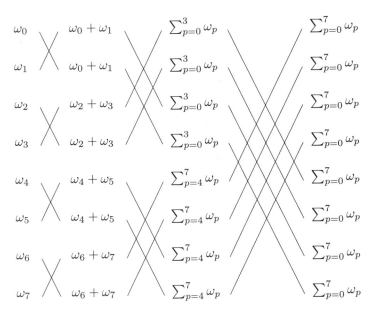

FIGURE 1.7. Concurrent addition of eight values.

that odd and even processes communicate with each other. It is useful to introduce logical operations on integers that act on the bit strings obtained from their binary representation. Using the *exclusive-or* operation on integers, denoted by $\bar{\vee}$, the expression $p\bar{\vee}2^d$ reverses bit number d of p.

Program Recursive-Doubling-2 is a multicomputer implementation of this procedure.

$0..P-1 \parallel p$ **program** Recursive-Doubling-2
declare
  d : integer ;
  ω, h : real
initially
  $\omega = \tilde{\omega}_p$
assign
  $\langle\ ;\ d\ :\ 0 \leq d < D\ ::$
    send ω **to** $p\bar{\vee}2^d$;
    receive h **from** $p\bar{\vee}2^d$;
    $\omega := \omega + h$
  $\rangle$
end

The variable h is assigned a value every time a receive instruction is evaluated. In fact, the send–receive pair implements the assignment with barrier:

$h[p] := \omega[p\bar{\vee}2^d]$;;

However, the above assignment is not permitted by the computational model, because left- and right-hand sides of the assignment refer to variables of different processes, as indicated by the explicit process identifiers.

Lemma 5 *Program Recursive-Doubling-2 implements program Recursive-Doubling-1.*

Every process of program Recursive-Doubling-2 executes the send instruction D times and increments the value of d by one between two successive sends. Let $\omega_p^{(d)}$ be the value of the variable ω when process p executes send instruction number d. This defines $\omega_p^{(d)}$ for all $p \in 0..P - 1$ and for all $d \in 0..D - 1$, where $P = 2^D$. Furthermore, let $\omega_p^{(D)}$ be final-state values.

To prove the lemma, we must show that

$$\forall p \in 0..P - 1 : \omega_p^{(D)} = \sum_{q=0}^{P-1} \omega_q^{(0)},$$

which is the map between initial and final state established by program Recursive-Doubling-1.

Let $\mathcal{D}(p, d)$ be the set of all integers $q \in 0..P - 1$ that differ from p only in bits number 0 through $d - 1$:

$$\mathcal{D}(p, d) = \{q : 0 \leq q < P \text{ and } p \vee (2^d - 1) = q \vee (2^d - 1)\}.$$

The symbol $\vee$ denotes the *logical-or* operation on integers. Hence, the expression $p \vee (2^d - 1)$ is the integer obtained from p by setting its d least significant bits to 1. We claim that the following identities hold:

$$\forall d \in 0..D, \ \forall p \in 0..P - 1 : \ \omega_p^{(d)} = \sum_{q \in \mathcal{D}(p,d)} \omega_q^{(0)}.$$

The identities are true for $d = 0$, because $\mathcal{D}(p, 0) = \{p\}$. Proceeding by induction, we assume that the identities are true for some d. It follows from program Recursive-Doubling-2 that processes p and $r = p \,\bar{\vee}\, 2^d$ exchange the values $\omega_p^{(d)}$ and $\omega_r^{(d)}$ and sum them. In process p, the new value of ω is then given by

$$\omega_p^{(d+1)} = \omega_p^{(d)} + \omega_r^{(d)}. \tag{1.9}$$

If $q \in \mathcal{D}(p, d)$, then bit numbers d of q and p must be equal. Similarly, if $q \in \mathcal{D}(r, d)$, then bit numbers d of q and r must be equal. Since bit numbers d of p and r are different, it follows that $\mathcal{D}(p, d)$ and $\mathcal{D}(r, d)$ are disjoint.

This result, the induction hypothesis, and Equation 1.9 imply that

$$\omega_p^{(d+1)} = \sum_{q \in \mathcal{D}(p,d) \bigcup \mathcal{D}(r,d)} \omega_q^{(0)}.$$

The set $\mathcal{D}(p, d) \bigcup \mathcal{D}(r, d)$ contains all integers $q \in 0..P-1$ that differ from p or r only in bits numbered 0 through $d-1$. Because only bit number d of p and r differ, this is the set $\mathcal{D}(p, d+1)$ of all q that differ from p only in bits numbered 0 through d. This establishes the induction for $d+1$. The lemma follows from the observation that $\mathcal{D}(p, D) = 0..P - 1$.   □

We can now remove the assumption that $P = 2^D$. The general case, $2^D < P < 2^{D+1}$, is reduced to the former by an incomplete recursive-doubling step. In this step, process p with $p \geq 2^D$ communicates its ω-value to process $p - 2^D$, where a pairwise sum is made. Subsequently, the global sum is computed by plain recursive doubling between all 2^D processes. Finally, the computed sum, known only to processes p with $p < 2^D$, is communicated to processes 2^D through $P - 1$. This procedure is precisely formulated in program Recursive-Doubling-3.

```
0..P − 1 ‖ p program Recursive-Doubling-3
declare
        d : integer ;
        ω, h : real
initially
        ω = ω̃_p
assign
        if p ≥ 2^D then send ω to p∨̄2^D ;
        if p < P − 2^D then begin
                receive h from p∨̄2^D ;
                ω := ω + h
        end ;
        if p < 2^D then
                ⟨ ; d : 0 ≤ d < D ::
                        send ω to p∨̄2^d ;
                        receive h from p∨̄2^d ;
                        ω := ω + h
                ⟩
        if p < P − 2^D then send ω to p∨̄2^D ;
        if p ≥ 2^D then receive ω from p∨̄2^D ;
end
```

For brevity, we shall use the basic algorithm with $P = 2^D$ in the future. The general algorithm is almost always applicable, however.

1.7.3 Performance Analysis

Program Recursive-Doubling-2 sums P values distributed over P processes. Under the assumptions of the multicomputer-performance model of Section 1.6.5, its multicomputer-execution time is

$$D(\tau_A + \tau_C(1)).$$

Based on a sequential-execution time of $(P-1)\tau_A$, its speed-up is

$$\frac{P-1}{D}\frac{1}{1+\frac{\tau_C(1)}{\tau_A}}.$$

The speed-up is at most $(P-1)/D$, an upper bound reached only if communication is free ($\tau_C(1)=0$).

In program Inner-Product-4, recursive doubling is used as part of the calculation of the inner product of two vectors.

```
0..P − 1 ‖ p program Inner-Product-4
    declare
        i, d : integer ;
        σ, h : real ;
        x, y : array [I_p] of real
    initially
        ⟨ ; i : i ∈ I_p :: x[i], y[i] = x̃_{μ^{-1}(p,i)}, ỹ_{μ^{-1}(p,i)} ⟩
    assign
        σ := ⟨ + i : i ∈ I_p :: x[i]y[i] ⟩ ;
        ⟨ ; d : 0 ≤ d < D ::
            send σ to p∇̄2^d ;
            receive h from p∇̄2^d ;
            σ := σ + h
        ⟩
    end
```

Its execution time is given by

$$T_P = (2\bar{I}-1)\tau_A + D(\tau_A + \tau_C(1))$$

and, with $T_1^S = (2M-1)\tau_A$, its speed-up by

$$S_P = \frac{(2M-1)\tau_A}{(2\bar{I}-1)\tau_A + D(\tau_A + \tau_C(1))}.$$

In Figure 1.8, programs Inner-Product-3 and Inner-Product-4 are compared for load-balanced data distributions. If the number of nodes is large, the execution time of both programs is dominated by communication. However, the communication effects are postponed significantly when using program Inner-Product-4.

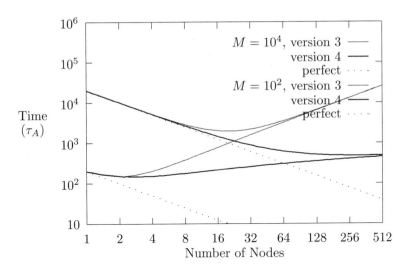

FIGURE 1.8. Logarithmic execution-time plot for programs Inner-Product-3 and Inner-Product-4 with load-balanced data distributions and $\tau_C(1) = 50\tau_A$.

For a coarse-grained computation with a load-balanced distribution, we have that $\bar{I} \approx M/P$, $1 \ll 2M$, $1 \ll 2\bar{I}$, and

$$S_P \approx P/(1 + \frac{D}{2\bar{I}} \frac{\tau_A + \tau_C(1)}{\tau_A}).$$

Hence, a near-maximum speed-up is achieved if $D \ll 2\bar{I}$ or, equivalently, if

$$P \log_2 P \ll 2M.$$

This is a substantially weaker condition than Equation 1.8, which was derived for program Inner-Product-3.

Program Inner-Product-4 also outperforms program Inner-Product-3 for fine-grained computations. With one vector component per process, the execution time is

$$T_M = \tau_A + (\tau_A + \tau_C(1)) \log_2 M.$$

For large M, this leads to a speed-up

$$S_M \approx \frac{2M}{\log_2(2M)} \frac{1}{1 + \frac{\tau_C(1)}{\tau_A}} < \frac{2M}{\log_2(2M)}.$$

The upper bound is obtained by setting $\tau_C(1) = 0$. The efficiency of the computation is bounded by $2/\log_2(2M)$. For $M = 2^{16} = 65,536$, the efficiency is at most 11.7%. If this seems small, remember that 2^{16} nodes are used for only $2^{17} - 1$ floating-point operations!

1.7.4 Numerical Aspects

Summing real numbers is an associative operation; the floating-point sum of floating-point values, however, is not. Consider a computer with six decimal digits for the mantissa and two decimal digits for the exponent. (Of course, actual computers use binary and hexadecimal representations, which is irrelevant to the arguments that follow.) On this computer, the floating-point representation of the real number π is given by 0.314159×10^{01}. Unlike π, the values 1.0 and -1.0 have exact floating-point representations. Let ϵ be a small value with an exact floating-point representation, for example, $\epsilon = 0.123456 \times 10^{-09}$.

Now, compare the floating-point sums $-1.0 + (1.0 + \epsilon)$ and $(-1.0 + 1.0) + \epsilon$. In floating-point arithmetic, we have that

$$1.0 + \epsilon = 1.0,$$

because the change in the ninth significant digit cannot be represented by a mantissa of six digits. It follows that, in floating-point arithmetic,

$$-1.0 + (1.0 + \epsilon) = -1.0 + 1.0 = 0.0$$

and that

$$(-1.0 + 1.0) + \epsilon = 0.0 + \epsilon = \epsilon.$$

This simple example shows that floating-point addition is not associative. The errors introduced by the floating-point representation of real numbers are called *round-off* errors.

The immediate question is whether recursive doubling is numerically stable. When summing a large number of quantities, significant accumulation of round-off errors can occur if small terms are added to large intermediate quantities and/or if large intermediate quantities of opposite sign are summed during the procedure. Both of these problems are avoided with pairwise summation. In fact, pairwise summation is almost always the recommended procedure to compute, sequentially or concurrently, the sum of a large number of quantities. Although artificial problem cases for pairwise summation can be constructed, they rarely occur in practice. For all practical purposes, recursive doubling is assured to be numerically stable.

There is another, somewhat hidden, potential hazard. Suppose the computed sum is used to test for termination of an iteration. Is it possible that, due to round-off errors, the test fails in some processes and succeeds in others? That would surely be catastrophic. Can it be assured that all processes reach **identical** results? The property that saves the day is the commutativity of floating-point addition. Consider processes 0 and 1 in step 0 of the recursive-doubling procedure. The values $\omega_0^{(0)}$ and $\omega_1^{(0)}$ are exchanged and summed in each process. These two floating-point additions differ only in the order of the operands, which does not change round-off errors. It follows that the exchange-and-add operation leads to identical results in both processes. Since the exchange-and-add operation is the only operation used by the recursive-doubling procedure, we conclude that the same result is obtained in every process. Of course, this result is contaminated by round-off errors, but by exactly the same round-off errors in every process. In the above example of summing -1.0, 1.0, and ϵ, either all processes compute the result 0, or all processes compute the result ϵ.

1.8   Notes and References

1.8.1   Communicating Sequential Processes

A formal definition of the computational model implemented by multicomputers can be found in Hoare [52], who called it the model of *communicating sequential processes*. This abstract computational model became a reality with the construction of the Cosmic Cube, which was built in 1982 by Seitz [74]. When implementing an abstract computational model on a real-world computer, many practical considerations come into play. It should be no surprise that there exists a myriad of different operating systems, communication primitives, and programming models, notations, and languages. We shall resist any temptation to compare the real and perceived advantages of available environments. Instead, our focus is on the commonality of all multicomputer systems: the ability to run many sequential processes that can communicate with each other. One cannot deny, however, that concurrent-program development is often frustrating, because targeting a multitude of still-evolving programming environments is inherently bug prone. The forces of the marketplace ensure that these are temporary problems. In the meantime, it is doubly important to focus on the fundamentals.

1.8.2   UNITY

Because our notation is based on UNITY, a good reference for readers of this text is Chandy and Misra [15]. The chapters on "Architectures and Mappings" and "Communicating Processes" should be of particular interest. The original UNITY notation eliminates all temporal structure: a UNITY program is a *set* — not a sequence — of assignments, and the order in which the assignments are evaluated is not important. Such a computational model is perfectly suited to study fundamental synchronization problems. For example, it is possible to study communication primitives in the absence of any temporal structure. We shall assume that these low-level synchronization issues have been addressed in the operating system, and we shall concentrate on issues encountered by application programmers. For this purpose, a programming model with temporal structure is preferable. This is our most radical deviation from the original UNITY.

1.8.3 Stepwise Refinement

Stepwise refinement of programs, while normal practice for combinatorial algorithms, remains the exception in scientific computing. A different attitude towards program development is probably the main source of communication difficulties between computer scientists and users. Highly recommended reading for any programmer are the works of Aho, Hopcroft, and Ullman [1], Dijkstra [23], Knuth [58], and Wirth [86]. The logical complexity of numerical algorithms has significantly increased because of new numerical techniques and because of concurrent computing. The more difficult the programming task, the more important it is to develop general libraries. This shifts the programming effort to developers of libraries. They face an even tougher programming problem, however, because their libraries will be used in mostly unknown environments. This makes it a practical necessity to pursue more formal approaches in program development for scientific computing.

Stepwise refinement forces programmers to make explicit and to question their assumptions. There are tangible benefits. Even the limited formal program development done thus far teaches an important lesson:

concurrency requires global program transformations.

The duplication step involves every single line of the program! Concurrency is NOT the result of shyly changing a few lines of a program. Although the final concurrent program may resemble the original sequential program, the required transformations are of a global nature. That is the fundamental reason why it is so difficult to develop high-level languages and compilers that hide concurrency from the programmer.

1.8.4 Speed-up and Efficiency

Because of Theorem 1, performance results are usually presented by means of speed-up and/or efficiency plots, which display these measures as a function of the number of nodes. Having deviated from tradition by adopting the logarithmic execution-time plot, let us argue the case for execution-time plots and against speed-up and efficiency plots.

The case for execution-time plots is fairly straightforward. The primitive data, the execution times of all computations, can be read directly from the plot, and the derived data, the speed-up and efficiency, are easily obtained graphically from the plot. The logarithmic execution-time plot even contains some clues about the causes of inefficiency for a particular sequence of computations. As discussed, floating-point overhead and the combination of communication overhead and load imbalance are related to specific distances in the plot. In speed-up and efficiency plots, the whole graph can be corrupted by one unreliable estimate for the sequential-execution time T_1^S. In a logarithmic execution-time plot, the time T_1^S has an impact on just one point in the plot. If the sequential computation is too slow for some reason, the resulting speed-up may leave the impression of good performance, independent of whether good performance was actually achieved.

The popularity of speed-up and efficiency plots is partially due to an inflated regard for the importance of Theorem 1. Its simple formulation is deceptive and hides the fact that its result is far removed from the practical world of computing. The number of nodes remains as the only study parameter, because all other computational parameters are eliminated by the clever definition of the best p-node computation as the optimum over the space of all possible p-node computations. This trick simplifies the problem and leads to a nice result in complexity theory. In practical performance analyses, however, the other computational parameters cannot be eliminated. Confusion surrounding Theorem 1 has sparked two controversies that keep surfacing on a regular basis: Amdahl's law and superlinear speed-up.

Amdahl [2] argued that speed-up is limited by a rather pessimistic upper bound. Consider a sequential computation with execution time T_1^S. A fraction of this time, say αT_1^S, consists of "inherently sequential" computations. Now, assume that the remaining fraction, $(1 - \alpha)T_1^S$, can be eliminated entirely by introducing concurrency. Then, we obtain the following lower bound for the p-node execution time:

$$T_p \geq \alpha T_1^S.$$

This leads to an upper bound on the speed-up that is independent of p:

$$S_p \leq \frac{1}{\alpha}.$$

Amdahl's argument shows that it is important to put one's efforts in the right methods, namely those with small sequential fractions. Amdahl's argument also serves to moderate unrealistically high expectations of concurrent computing. However, assigning the term "law" to his observation is an overstatement. Certainly, it does not invalidate the concurrent-computing concept. One must bear in mind that execution times are functions of many computational parameters, among which the number of nodes is just one. The derivation of Amdahl's upper bound relies on the implicit assumption that the sequential fraction α is independent of all other computational parameters. One such parameter is problem size. Gustafson [41] pointed out that α usually decreases as a function of problem size. It follows that the upper bound on the speed-up usually increases as a function of problem size. In other words, concurrent computing is a suitable option for sufficiently large problems.

Superlinear speed-up is another recurring controversy. First, let us remove ambiguity related to terminology. When reporting on the performance of actual programs, speed-up is usually computed with respect to a nonoptimal but "reasonable" sequential-execution time T_1^S; see Section 1.6.3. In this case, superlinear speed-up is a rare occurrence, but it is theoretically and practically possible. There is no controversy here. In more theoretical papers, the term superlinear speed-up is used for computations below the line of perfect speed-up. We shall refer to such computations as having *superperfect speed-up*. The controversy really centers around superperfect speed-up.

Theorem 1 excludes any possibility of superperfect speed-up and, in spite of its elementary proof, some energy is spent to construct and refute counterexamples of the theorem. For an amusing exchange, see [28, 29, 30, 56, 71]. Superperfect speed-up is impossible unless the assumptions of the theorem are violated. Fortunately — or unfortunately — for chasers of superperfect speed-up, Theorem 1 makes many assumptions to violate. For example, it is assumed that every node of the multicomputer has access to an unlimited amount of memory. With node-memory restrictions, the theorem no longer holds. If every node has a certain amount of memory, the total amount of available memory is proportional to the number of nodes. Once the amount of memory and the number of nodes exceed a certain lower bound, it is sometimes possible to switch to another method, perhaps one that trades memory for speed. With node-memory limitations, superperfect speed-up is a rare event, but it is possible, and it is NOT a counterexample to Theorem 1! However, this example does point out a weakness of Theorem 1. By assuming that node memory is unlimited, it does not capture the most important reason for concurrency: increasing the ratio of the number of processors over the total memory size.

Superlinear speed-up is possible with or without memory restrictions. In most cases, however, claims of superlinear speed-up are based on slow sequential computations. There are several reasons why slow sequential computations often occur. Obtaining a good sequential time is frustrating: one must optimize a code purely for the sake of obtaining one execution time. It does not help matters that optimizations performed on the sequential code reduce or eliminate the successes obtained by optimizing the concurrent program! On some concurrent computers, sequential computations are naturally more inefficient because of higher memory-access times; see Section 12.4.

1.8.5   Other Performance Measures

Frequently, memory and/or time restrictions are an impediment to obtaining the sequential-execution times T_1^S and T_1. Estimates for both can be obtained by extrapolation of sequential-execution times for problems smaller than but similar to the problem used for larger values of p. On multicomputers with many nodes, however, problem sizes are so large that even such extrapolations are unreliable. Eventually, we must face the real issue: Why should we even attempt to obtain sequential-execution times? What is their relevance to multicomputer computations with many nodes?

Declaring sequential-execution times irrelevant is, of course, tantamount to declaring speed-up and efficiency irrelevant. Although it is possible to construct performance measures that are independent of any sequential-execution time, they can be misleading too. Consider, for example, *scalability*. A program that results in a sequence of computations that lie on a straight line with negative slope in a logarithmic execution-time plot are called scalable: by increasing the number of nodes, one increases the performance. The slope of this line could be used as a quantification of scalability. The problem with this measure is that inefficient programs can be made scalable by artificially decreasing the node performance. By increasing the arithmetic time τ_A, the ratio $\tau_C(1)/\tau_A$ goes down, and the execution times remain near the line of scalability. If the execution-time plot includes an honest sequential-execution time T_1^S, trickery based on $\tau_C(1)/\tau_A$ is impossible, because the decreased node performance would show up as a distance to the line of linear speed-up. In fact, a low $\tau_C(1)/\tau_A$ ratio often occurs unintentionally and without trickery. New compilers often have the effect of reducing the effective node performance, because they produce insufficiently optimized object code. Not surprisingly, programs written in new languages and compiled by new compilers often outperform established languages in terms of scalability.

Because we understand performance of sequential computers better, it is usually easier to judge whether the sequential-execution time is reasonable for the given problem on the given hardware. The sequential-execution time sets a time scale. That is its fundamental relevance. The sequential-execution time is a performance standard for a node participating in solving the given problem.

Gustafson [41] defines a scaled speed-up, which is obtained by scaling up the problem size with the number of nodes. However, scaling up the problem size can be done in several ways. In his analysis of scaled speed-up, Worley [87] points out that it only has a meaningful interpretation if the execution times are constrained. This is important, because it is useless to obtain an impressive speed-up for a computation with an infeasible concurrent-execution time.

All performance measures are flawed in some way. These flaws do not make them useless. Each measure captures one aspect, but misses many others. One must be aware of this when designing any performance test. Which measures are useful depends on the question one wants answered. A combination of several measures is usually required to get a sufficiently accurate picture of the performance of a program. This is another argument in favor of execution-time plots: given the primitive data, one can always compute whatever measure is appropriate.

Hockney and Jesshope [53, 54] developed an extensive theory on performance analysis and characterization of concurrent computers and programs.

Exercises

Exercise 1 *Implement a general barrier for a multicomputer. How expensive is it?*

Exercise 2 *Given an $N \times N$ matrix in every process, say matrix A_p in process p, write a multicomputer program to compute the product*

$$\prod_{p=0}^{P-1} A_p.$$

Estimate the execution time and display it in a logarithmic execution-time plot. Can you guarantee that all processes reach identical results, in spite of the presence of round-off errors?

Exercise 3 *Consider a sequence of M values ω_m. All have approximately the same magnitude in absolute value. The first $M/2$ values, however, are positive; the remaining are negative. Will program Recursive-Doubling-2 result in significant round-off errors? If so, is there an easy remedy?*

Exercise 4 *Given a sequence of linearly independent vectors $\vec{x}_k \in \mathbb{R}^N$, where $k \in 0..K-1$, find an orthonormal set of vectors $\vec{v}_k \in \mathbb{R}^N$, such that*

$$\text{span}(\vec{x}_0, \ldots, \vec{x}_{K-1}) = \text{span}(\vec{v}_0, \ldots, \vec{v}_{K-1}).$$

The new set of vectors can be computed using modified Gram–Schmidt orthogonalization, which is explained in Section 5.1.2. Develop the concurrent program. Implement and run it on your favorite concurrent computer.

2
Vector and Matrix Operations

The topic of Chapters 2 through 6 is concurrent linear algebra. This expanded coverage is motivated by two factors. First, linear algebra is a frequently used component of scientific computing. Computational problems recast in terms of linear algebra can often be considered solved, because standard libraries can tackle the remaining issues. Second, linear algebra is a rich source for numerical algorithms that are realistic in algorithmic complexity, but can be understood without an extensive mathematical background.

In Sections 2.1, 2.2, and 2.3, row- and column-oriented representations of matrices will be examined. After introducing the vector-shift operation in Section 2.4, a less traditional approach will be taken in Section 2.5, where diagonals will be considered the main substructures of a matrix.

2.1 Matrix–Vector Product

2.1.1 Specification

Our goal is to implement the assignment

$$\vec{y} := A\vec{x},$$

where A is an $M \times N$ matrix, $\vec{x}$ an N-dimensional vector, and $\vec{y}$ an M-dimensional vector. The algorithmic specification of this operation is given by program Matrix-Vector-1.

program Matrix-Vector-1
declare

    m, n : integer ;
    x : array $[0..N - 1]$ of real ;
    y : array $[0..M - 1]$ of real ;
    a : array $[0..M - 1 \times 0..N - 1]$ of real

initially

    $\langle\ ;\ n\ :\ n \in 0..N - 1\ ::\ x[n] = \tilde{x}_n\ \rangle$;
    $\langle\ ;\ (m, n)\ :\ (m, n) \in 0..M - 1 \times 0..N - 1\ ::\ a[m, n] = \tilde{a}_{m,n}\ \rangle$

assign

    $\langle\ \|\ m\ :\ 0 \leq m < M\ ::$
        $y[m]\ :=\ \langle\ + n\ :\ 0 \leq n < N\ ::\ a[m, n]x[n]\ \rangle$
    $\rangle$

end

There is concurrency in both quantifications of the **assign** section. The outer quantification expresses that all components of $\vec{y}$ may be computed concurrently. The inner quantification is the computation of the inner product of the vector $\vec{x}$ with row m of the matrix A.

2.1.2 Implementation

Once again, duplication is the first step of the transformation of the specification program. As introduced in Section 1.2, the header

    $\mathcal{P} \| p$ **program** program-name

defines a multicomputer program in which processes are identified by means of a value $p \in \mathcal{P}$. Thus far, we encountered only the special case $\mathcal{P} = 0..P - 1$. In general, the set $\mathcal{P}$ can have other elements besides integers; any finite set will do. The set $\mathcal{P}$ is usually chosen to impose some structure on the set of processes. We refer to this structure as *process connectivity*.

By choosing $\mathcal{P} = 0..P - 1 \times 0..Q - 1$, elements of $\mathcal{P}$ and values of the process identifiers are coordinate pairs (p, q), and the process connectivity is that of a two-dimensional mesh of processes. This seems an appropriate process connectivity for matrices represented by two-dimensional arrays. Within this process-mesh structure, it is natural to refer to subsets of processes corresponding to the rows and columns of the mesh. *Process row p* is the set of processes

$$\{(p, 0),\ (p, 1),\ \ldots,\ (p, Q - 1)\},$$

while *process column q* is the set of processes

$$\{(0, q),\ (1, q),\ \ldots,\ (P - 1, q)\}.$$

Obviously, there are P process rows and Q process columns.

The duplicated matrix–vector program is introduced by the header

$0..P - 1 \times 0..Q - 1 \parallel (p, q)$ **program** Matrix-Vector-2

but is otherwise identical to program Matrix-Vector-1. The selection step partitions the set $\mathcal{K} = 0..M - 1 \times 0..N - 1$ into $P \times Q$ subsets $\mathcal{K}_{p,q}$ and imposes the resulting data distribution. In principle, any partition $\mathcal{K}_{p,q}$ will do. However, such a level of generality would make it impossible to access efficiently the basic substructures of the matrix: its rows and columns. Although arbitrary data distributions might be interesting from an abstract point of view, reality constrains our choice of data distributions to those obtained by combining arbitrary one-dimensional data distributions.

The *row distribution* μ maps the global index $m \in \mathcal{M} = 0..M - 1$ into a process-row identifier p and a local index i. The row distribution partitions the index set $\mathcal{M}$ into P subsets $\mathcal{M}_p$, where $0 \leq p < P$, each corresponding to a local-index set $\mathcal{I}_p$. Similarly, the *column distribution* ν maps the global index $n \in \mathcal{N} = 0..N - 1$ into a process-column identifier q and a local index j. The column distribution partitions the index set $\mathcal{N}$ into Q subsets $\mathcal{N}_q$, where $0 \leq q < Q$, each corresponding to a local-index set $\mathcal{J}_q$. The two one-dimensional distributions μ and ν are combined into one two-dimensional distribution $\mu \times \nu$, which partitions the set $\mathcal{M} \times \mathcal{N}$ into $P \times Q$ subsets $\mathcal{M}_p \times \mathcal{N}_q$, each corresponding to a local-index set $\mathcal{I}_p \times \mathcal{J}_q$.

In program Matrix-Vector-3, a partial selection has occurred, and only components of $\vec{y}$ for which $m \in \mathcal{M}_p$ are computed in process (p, q). All processes of a process row compute the same components of $\vec{y}$. In the program, the process identifier of the array variables is explicit. The elementary transformation of the inner quantification and the replacement of the concurrent by a sequential separator in the quantification over m are done in preparation for the next step.

$0..P - 1 \times 0..Q - 1 \parallel (p, q)$ **program** Matrix-Vector-3
declare
    m, n, s : integer ;
    x : array $[0..N - 1]$ of real ;
    y : array $[\mathcal{M}_p]$ of real ;
    a : array $[\mathcal{M}_p \times 0..N - 1]$ of real
initially
    $\langle \, ; \, n \, : \, n \in 0..N - 1 \, :: \, x[n] = \tilde{x}_n \, \rangle$;
    $\langle \, ; \, (m, n) \, : \, (m, n) \in \mathcal{M}_p \times 0..N - 1 \, :: \, a[m, n] = \tilde{a}_{m,n} \, \rangle$
assign
    $\langle \, ; \, m \, : \, m \in \mathcal{M}_p \, ::$
        $y[p, q][m] := \langle \, + \, s \, : \, 0 \leq s < Q \, ::$
            $\langle \, + \, n \, : \, n \in \mathcal{N}_s \, :: \, a[p, s][m, n] x[p, s][n] \, \rangle$
        $\rangle$
    $\rangle$
end

The quantification over s can be developed like the inner-product operation in Section 1.5 and replaced by the recursive-doubling procedure of Section 1.7. This is done in program Matrix-Vector-4. The only complication is that the recursive-doubling procedure occurs inside the sequential quantification over m. Note that it would have been incorrect to place the recursive-doubling procedure inside a concurrent quantification. (Why?)

$0..P - 1 \times 0..Q - 1 \parallel (p, q)$ **program** Matrix-Vector-4
declare
      m, n, d : integer ;
      x : array $[\mathcal{N}_q]$ of real ;
      y, t : array $[\mathcal{M}_p]$ of real ;
      a : array $[\mathcal{M}_p \times \mathcal{N}_q]$ of real
initially
      $\langle \; ; \; n \; : \; n \in \mathcal{N}_q \; :: \; x[n] = \tilde{x}_n \; \rangle$;
      $\langle \; ; \; (m, n) \; : \; (m, n) \in \mathcal{M}_p \times \mathcal{N}_q \; :: \; a[m, n] = \tilde{a}_{m,n} \; \rangle$
assign
      $\langle \; ; \; m \; : \; m \in \mathcal{M}_p \; ::$
         $y[m] := \langle \; + \; n \; : \; n \in \mathcal{N}_q \; :: \; a[m, n]x[n] \; \rangle$;
         $\langle \; ; \; d \; : \; 0 \leq d < \log_2 Q \; ::$
            send $y[m]$ **to** $(p, q \bar{\vee} 2^d)$;
            receive $t[m]$ **from** $(p, q \bar{\vee} 2^d)$;
            $y[m] := y[m] + t[m]$
         $\rangle$
      $\rangle$
end

In mapping the vectors $\vec{x}$ and $\vec{y}$ to the distributed arrays x and y, respectively, both vectors are distributed quite differently. The vector $\vec{x}$ is distributed according to the column distribution ν: its components are distributed over the process columns, but duplicated in each process row. The vector $\vec{y}$, on the other hand, is distributed according to the row distribution μ: its components are distributed over the process rows, but duplicated in each process column. Figure 2.1 shows this graphically for the case of a 2×3 process mesh. The row distribution μ splits $\vec{y}$ into two subvectors $\vec{y}_0$ and $\vec{y}_1$. The subvector $\vec{y}_0$, for example, contains all components of $\vec{y}$ mapped to processes $(0, q)$, where $0 \leq q < Q$. The components of $\vec{y}_0$ are not necessarily a contiguous block of components of $\vec{y}$. Similarly, the vector $\vec{x}$ is split into three subvectors, $\vec{x}_0$, $\vec{x}_1$, and $\vec{x}_2$, by the column distribution ν. The combined distribution $\mu \times \nu$ splits the matrix A into the submatrices $A_{p,q}$, where $A_{p,q}$ is stored in process (p, q).

Replace the global by local indices, and bring the quantification over m inside the recursive-doubling procedure to replace $\#\mathcal{M}_p = I_p$ message exchanges of one value by one message exchange of I_p values. The resulting multicomputer program is given by program Matrix-Vector-5.

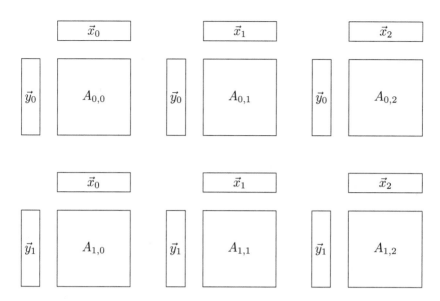

FIGURE 2.1. Data distribution for matrix–vector multiplication on a 2×3 process mesh.

$0..P - 1 \times 0..Q - 1 \parallel (p,q)$ **program** Matrix-Vector-5
declare
$\quad i,\ j,\ d\ :\ $ integer ;
$\quad x\ :\ $ array $[\mathcal{J}_q]$ of real ;
$\quad y,\ t\ :\ $ array $[\mathcal{I}_p]$ of real ;
$\quad a\ :\ $ array $[\mathcal{I}_p \times \mathcal{J}_q]$ of real
initially
$\quad \langle\ ;\ j\ :\ j \in \mathcal{J}_q\ ::\ x[j] = \tilde{x}_{\nu^{-1}(q,j)}\ \rangle\ ;$
$\quad \langle\ ;\ (i,j)\ :\ (i,j) \in \mathcal{I}_p \times \mathcal{J}_q\ ::\ a[i,j] = \tilde{a}_{\mu^{-1}(p,i),\nu^{-1}(q,j)}\ \rangle$
assign
$\quad \langle\ ;\ i\ :\ i \in \mathcal{I}_p\ ::$
$\quad\quad y[i]\ :=\ \langle\ +\ j\ :\ j \in \mathcal{J}_q\ ::\ a[i,j]x[j]\ \rangle$
$\quad \rangle\ ;$
$\quad \langle\ ;\ d\ :\ 0 \le d < \log_2 Q\ ::$
$\quad\quad$ **send** $\{y[i] : i \in \mathcal{I}_p\}$ **to** $(p, q \bar{\vee} 2^d)\ ;$
$\quad\quad$ **receive** $\{t[i] : i \in \mathcal{I}_p\}$ **from** $(p, q \bar{\vee} 2^d)\ ;$
$\quad\quad \langle\ ;\ i\ :\ i \in \mathcal{I}_p\ ::\ y[i] := y[i] + t[i]\ \rangle$
$\quad \rangle$
end

2.1.3   Performance Analysis

Memory use is of some concern because of the duplication of $\vec{x}$ and $\vec{y}$ and the temporary array t. When used in a realistic context, the temporary array t is a frequently reused message buffer, whose memory overhead can be amortized over many operations. The memory allocated for t may, therefore, be ignored. Local-memory use, which is the amount of memory allocated per process, is approximately

$$\frac{MN}{PQ} + \frac{M}{P} + \frac{N}{Q}$$

floating-point words, provided μ and ν are load-balanced distributions. Global-memory use, which is the amount of memory allocated by all processes combined, is then approximately $MN + QM + PN$ floating-point words.

For coarse-grained computations, when $P \ll M$ and $Q \ll N$, the increase in memory use over the sequential program is reasonable. For fine-grained computations, however, program Matrix-Vector-5 may well be unsuitable. The case $P = M$ and $Q = N$ requires an amount of memory equal to three times the size of the matrix. Unless memory is considered a cheap commodity, the program needs to be further developed. As in sequential programming, memory can be traded for speed.

According to the multicomputer-performance model of Section 1.6.4, the exchange of L messages of length 1 requires a time

$$\tau_C(1)L = (\tau_S + \beta)L,$$

while the exchange of one message of length L requires

$$\tau_C(L) = \tau_S + \beta L.$$

On multicomputers with large latency τ_S, one should limit the number of individual sends, as was done by the transformation of program Matrix-Vector-4 into Matrix-Vector-5. If the latency is small, however, combining messages may not be a good idea because of another effect that tends to favor shorter messages and is not included in our performance model. Long messages require longer uninterrupted attention from the communication network, which results in longer waiting times for other messages. The effective message-exchange time may, therefore, be larger for programs in which average messages are longer. This effect is called *network contention*; see Section 12.3. Until then, we shall ignore the possibility of network contention.

The execution time of program Matrix-Vector-5 for multiplying an $M \times N$ matrix and an N-dimensional vector with $P \times Q$ independent processes is estimated by

$$T_{PQ} = \bar{I}(2\bar{J} - 1)\tau_A + ((\tau_S + \beta\bar{I}) + \tau_A\bar{I})\log_2 Q, \qquad (2.1)$$

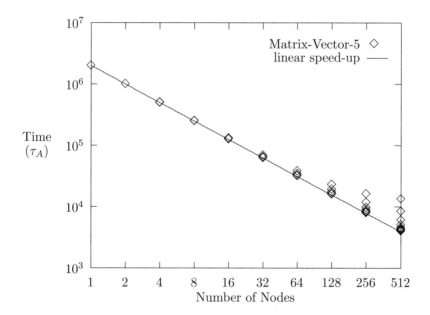

FIGURE 2.2. The performance of program Matrix-Vector-5 for the case $M = N = 1000$ with a load-balanced data distribution. The latency $\tau_S = 50\tau_A$, and the inverse band width $\beta = 0.01\tau_A$ per floating-point word.

where $\bar{I} = \max_{0 \le p < P} \#\mathcal{M}_p$ and $\bar{J} = \max_{0 \le q < Q} \#\mathcal{N}_q$. For load-balanced coarse-grained computations, this simplifies to:

$$T_{PQ} \approx 2\frac{MN}{PQ}\tau_A + \frac{M}{P}(\tau_A + \beta)\log_2 Q + \tau_S \log_2 Q.$$

In Figure 2.2, Equation 2.1 is plotted in a logarithmic execution-time plot for $M = N = 1000$. For each value of the number of nodes, the execution times are displayed for all possible $P \times Q$ process meshes. The line through the best execution times for each number of nodes has slope -1, which indicates near-linear speed-up (note that $T_1 = T_1^S$). For computations with large numbers of nodes, however, the execution times vary considerably from one process mesh to the next.

Using the sequential-execution time

$$T_1^S = T_1 = M(2N - 1)\tau_A \approx 2MN\tau_A$$

and assuming a coarse-grained computation, the speed-up

$$S_{PQ} \approx \frac{PQ}{1 + \frac{Q\log_2 Q}{2N}\left(1 + \frac{\beta}{\tau_A} + \frac{P}{M}\frac{\tau_S}{\tau_A}\right)} \tag{2.2}$$

is obtained. This is nearly linear speed-up provided

$$Q\log_2 Q \ll N \quad \text{and} \quad PQ\log_2 Q \ll MN.$$

The precise limits of efficient computation are determined by the computer-dependent characteristics (β/τ_A) and (τ_S/τ_A). Typically, the ratio (β/τ_A) can be neglected, and it is sufficient to consider only the ratio (τ_S/τ_A).

Equation 2.2 and the program text imply that $Q = 1$ is optimal, because it avoids all communication and the complexity of recursive doubling. However, there are other considerations. With $Q = 1$, the maximal number of processes that can be used is M, and serious load imbalance will occur for computations with a number of nodes that is smaller than but comparable to M. Using a load-balanced row distribution μ for the case $M = PL + R$, there are R processes that compute $L + 1$ components of $\vec{y}$, and $P - R$ processes that compute L components. If $L \leq 5$ or, equivalently, $M \leq 5P$, the load imbalance as the result of computing an extra component is greater than or equal to 20%! By restricting the program to $Q = 1$, the fine-grain limits are reached earlier. Among the computations displayed in Figure 2.2, the best execution times with 512 nodes are obtained with $P \approx Q$. With $M = 1000$, $P = 512$, and $Q = 1$, there is a 50% load imbalance due to fine granularity: most processes compute two components of the vector $\vec{y}$, while 12 processes compute three components. The case $P = 1$ and $Q = 512$ performs the worst. Not only does this case suffer load imbalance, it also involves a maximum of communication.

One could try to obtain the best process mesh for each computer. However, the matrix–vector product is always part of a larger context. This context may constrain your choice of process mesh to "non-optimal" cases. Consider, for example, a program that alternates between the two assignments:

$$\begin{aligned} \vec{y} &:= A\vec{x} \\ \vec{u}^T &:= \vec{v}^T A \end{aligned}$$

Because the vector–matrix product (the second assignment) reverses the roles of P and Q, the optimal performance of the global program is achieved for a certain ratio P/Q, which depends on the relative frequency of matrix–vector and vector–matrix products. However, if each assignment was optimized individually, the first would require a purely row-distributed $(Q = 1)$ and the second a purely column-distributed matrix $(P = 1)$. It makes sense, therefore, to keep one's options open and provide the general case.

Clearly, appropriate judgment is required before ruling out particular algorithms or data distributions on the basis of performance. The above example shows that a large program consisting of individually optimized operations may not be optimal or even near optimal. Over-developing operations like the matrix–vector product with the intent of increasing performance by using every trick possible may actually lead to worse global performance. This is a consequence of our observation in Section 1.8 that concurrency requires global program transformations.

2.1.4 Sparse Matrices

A matrix is called *sparse* if the vast majority of its elements vanishes. Sparse matrices arise naturally when discretizing continuous operators; see Chapter 8, for example. Typically, the number of nonzero elements of a sparse $M \times M$ matrix is $O(M)$. The set of locations of the nonzero elements is called the *fill* of the sparse matrix. Sparse-matrix representations do not store zero elements to save memory and avoid arithmetic with zero elements to save floating-point operations.

We distinguish two broad categories of sparse-matrix representations. *Structured sparse-matrix representations* exploit some regular structure of the fill. Of course, this limits the applicability of the representation to matrices that possess the specified fill structure. If applicable, structured representations are often the most efficient in memory use and floating-point arithmetic. *Unstructured sparse-matrix representations* do not impose any requirements on the fill. However, the bookkeeping costs of unstructured representations can be substantial. For each nonzero entry, extra memory is required to store pointers and/or indices. Moreover, execution times of computations with unstructured representations are dominated not by the floating-point arithmetic, but by the time required to access the nonzero elements.

Many structured representations are difficult to use in multicomputer programs, because the data distribution often destroys the fill structure of the matrix. To use structured representations successfully, the data distribution and the fill structure must be compatible. To achieve this and load balance requires introducing strategies that are unique to each application. In Section 2.5, a structured representation will be introduced for sparse matrices with fill along a few diagonals.

Let us now examine unstructured representations in multicomputer programs. As for full matrices, an element $\alpha_{m,n}$ of the sparse $M \times N$ matrix A is associated with process (p, q) if $(m, n) \in \mathcal{M}_p \times \mathcal{N}_q$. In general, we expect that the local matrices be sparse with fill in random locations. Each of these local matrices can be represented by an unstructured sparse-matrix representation. Define the index sets

$$\mathcal{K}_{p,q} = \{\, (i,j) \; : \; (i,j) \in \mathcal{I}_p \times \mathcal{J}_q \text{ and } \alpha_{\mu^{-1}(p,i),\nu^{-1}(q,j)} \neq 0\},$$

where $0 \leq p < P$ and $0 \leq q < Q$. For each index of $\mathcal{K}_{p,q}$, a matrix element must be stored. In our notation, arrays may be declared over arbitrary index sets. This avoids lots of messy technical detail, which is required when implementing a sparse-matrix package in traditional computer languages. Program Sparse-Matrix-Vector-1 is merely a top-level description, which is nevertheless useful to pinpoint some important performance problems of sparse-matrix representations in a multicomputer context.

$0..P-1 \times 0..Q-1 \parallel (p,q)$ **program** Sparse-Matrix-Vector-1
declare
    $i,\ j,\ d\ :$ integer ;
    $x\ :$ array $[\mathcal{J}_q]$ of real ;
    $y,\ t\ :$ array $[\mathcal{I}_p]$ of real ;
    $a\ :$ array $[\mathcal{K}_{p,q}]$ of real
initially
    $\langle\ ;\ i\ :\ i \in \mathcal{I}_p\ ::\ y[i] = 0\ \rangle\ ;$
    $\langle\ ;\ j\ :\ j \in \mathcal{J}_q\ ::\ x[j] = \tilde{x}_j\ \rangle\ ;$
    $\langle\ ;\ (i,j)\ :\ (i,j) \in \mathcal{K}_{p,q}\ ::\ a[i,j] = \tilde{a}_{\mu^{-1}(p,i),\nu^{-1}(q,j)}\ \rangle$
assign
    $\langle\ ;\ (i,j)\ :\ (i,j) \in \mathcal{K}_{p,q}\ ::$
        $y[i] := y[i] + a[i,j]x[j]$
    $\rangle\ ;$
    $\langle\ ;\ d\ :\ 0 \le d < \log_2 Q\ ::$
        send $\{y[i] : i \in \mathcal{I}_p\}$ **to** $(p, q\bar{\triangledown}2^d)$;
        receive $\{t[i] : i \in \mathcal{I}_p\}$ **from** $(p, q\bar{\triangledown}2^d)$;
        $\langle\ \parallel\ i\ :\ i \in \mathcal{I}_p\ ::\ y[i] := y[i] + t[i]\ \rangle$
    $\rangle$
end

The execution time of program Sparse-Matrix-Vector-1 consists of two parts: the local matrix–vector product and the recursive-doubling procedure. Only the local matrix–vector product differs from the full-matrix case. Let τ_A^* be a unit of time that reflects not only the arithmetic time required for each nonzero matrix element, but also the time required to access a matrix element in the sparse-matrix representation. It is typical that $\tau_A^* \gg \tau_A$. The execution time, under otherwise standard assumptions, is then given by:

$$T_{PQ} = 2 \max_{(p,q)} \#\mathcal{K}_{p,q}\tau_A^* + \frac{M}{P}(\tau_A + \beta)\log_2 Q + \tau_S \log_2 Q.$$

In contrast with other programs examined thus far, the load balance of the computation depends on the data. It is not sufficient that μ and ν are load-balanced distributions to guarantee that the nonzero elements of the matrix are evenly distributed over the processes. Although unlikely, all nonzero elements could be mapped to the same process by a load-balanced data distribution.

Many unstructured sparse-matrix representations are more inefficient for larger than for smaller matrices, because longer lists have to be sorted or searched through. For a given problem of a fixed size, the access time τ_A^* can be significantly larger for the sequential computation than for the multi-computer computation, which uses the smaller distributed sparse matrices. The dependence of τ_A^* on the matrix size may even lead to superlinear speed-up. As is usual, the superlinear speed-up is due to the inefficiency of the sequential computation: represent the large global matrix as a collection of smaller matrices in the sequential computation and the superlinear speed-up will disappear.

In a sparse-matrix context, the duplication of the vectors $\vec{x}$ and $\vec{y}$ may be a problem. Local-memory use is

$$\max_{(p,q)} \#\mathcal{K}_{p,q} + \frac{N}{Q} + \frac{M}{P}$$

floating-point words. If $M = N$, if the nonzero matrix elements are distributed about evenly, and if the total number of nonzero matrix elements is $O(M)$, local-memory use is about

$$O\left(\frac{M}{PQ}\right) + \frac{M}{Q} + \frac{M}{P}$$

floating-point words. The memory required to represent $\vec{x}$ and $\vec{y}$ easily exceeds that required to represent the sparse matrix.

2.2   Vectors and Matrices

In program Matrix-Vector-5, we encountered two types of vectors. *Column-distributed vectors* are duplicated in every process row and distributed over the processes of each process row according to the column distribution. *Row-distributed vectors* are duplicated in every process column and distributed over the processes of each process column according to the row distribution.

Whether a vector is row or column distributed depends on the operations performed with it. In the assignment

$$\vec{y} \ := \ A\vec{x}$$

the vector $\vec{x}$ is column distributed and the vector $\vec{y}$ is row distributed. Conflicts arise, however. In the program segment

$$\begin{aligned} \vec{y} \ &:= \ A\vec{x} \ ; \\ \vec{z} \ &:= \ A\vec{y} \end{aligned}$$

the distribution for $\vec{y}$ required by the first assignment is incompatible with that required by the second assignment. An obvious solution is to replace the second assignment by

$$\begin{aligned} \vec{v} \ &:= \ \vec{y} \ ; \\ \vec{z} \ &:= \ A\vec{v} \end{aligned}$$

where $\vec{v}$ is a column-distributed vector. The assignment $\vec{v} \ := \ \vec{y}$ performs a redistribution of the data over the processes; it is implemented in Section 2.2.1.

Vector operations like vector sum and inner product must be adapted when done on a two-dimensional process mesh. The adaptation of the vector-sum program is trivial. The inner-product program will be adapted in Section 2.2.2.

2.2.1 Vector Conversion

Our goal is to develop a multicomputer program that converts a row-distributed vector of dimension M into a column-distributed vector of dimension N. Although it is required that $M = N$, we keep two separate symbols, one associated with the row distribution (M) and one with the column distribution (N). Our starting point is program Convert-Vector-1.

> $0..P-1 \times 0..Q-1 \parallel (p,q)$ **program** Convert-Vector-1
> **declare**
> > m, n : integer ;
> > v : array $[\mathcal{N}_q]$ of real ;
> > y : array $[0..M-1]$ of real
>
> **initially**
> > $\langle\ ;\ m\ :\ m \in 0..M-1\ ::\ y[m] = \tilde{y}_m\ \rangle$
>
> **assign**
> > $\langle\ \parallel\ n\ :\ n \in \mathcal{N}_q\ ::\ v[p,q][n] := y[p,q][n]\ \rangle$
>
> **end**

Array v is already column distributed, while array y is still fully duplicated. The program is based on global indices, and the process identifiers of the array variables, v and y, are explicit.

After imposing a row distribution, $y[m]$ is available in process (p,q) if and only if $m \in \mathcal{M}_p$. As a result, the assignment $v[p,q][n] := y[p,q][n]$ can be performed in process (p,q) if and only if $n \in \mathcal{M}_p \cap \mathcal{N}_q$. It is, therefore, tempting to replace the concurrent quantification over n by

$$\langle\ \parallel\ n\ :\ n \in \mathcal{N}_q \cap \mathcal{M}_p\ ::\ v[p,q][n] := y[p,q][n]\ \rangle$$

However, what about the other entries? Consider the set of entries

$$\{v[p,q][n] : 0 \le p < P\}$$

for a fixed value of $q \in 0..Q-1$ and a fixed value of $n \in \mathcal{N}_q$. All entries of this set must, eventually, contain identical values. The result of the above quantification is that exactly one entry in this set received a values from y. If all entries were initially zero, it suffices to sum all entries of the set and to assign this sum to each entry.

This argument is further illustrated in Figure 2.3. The above quantification fills in a particular entry of v if the corresponding entry of y is available. After this step, each process has filled in a certain segment of v. Consider now one particular process column. Each process has filled in a different

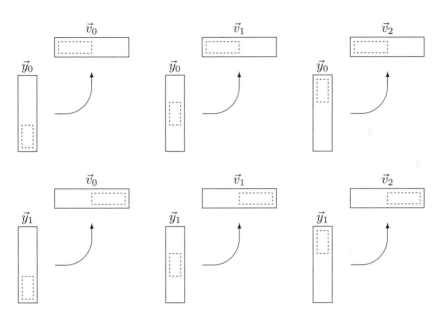

FIGURE 2.3. Conversion of a row-distributed vector $\vec{y}$ into a column-distributed vector $\vec{v}$ on a 2×3 process mesh.

segment of v. If all entries of v were initially zero, the partially assigned arrays can be summed across process columns using recursive doubling.

After converting to local indices, program Convert-Vector-2 is obtained.

$0..P - 1 \times 0..Q - 1 \parallel (p, q)$ **program** Convert-Vector-2
declare
       i, j, d, s : integer ;
       v, t : array $[\mathcal{J}_q]$ of real ;
       y : array $[\mathcal{I}_p]$ of real
initially
       $\langle \ ; j \ : \ j \in \mathcal{J}_q \ :: \ v[j] = 0 \ \rangle$;
       $\langle \ ; i \ : \ i \in \mathcal{I}_p \ :: \ y[i] = \tilde{y}_{\mu^{-1}(p,i)} \rangle$
assign
       $\langle \ ; i \ : \ i \in \mathcal{I}_p \ ::$
           $(s, j) := \nu(\mu^{-1}(p, i))$;
           if $q = s$ **then** $v[j] := y[i]$
       $\rangle$;
       $\langle \ ; d \ : \ 0 \leq d < D \ ::$
           send $\{v[j] : j \in \mathcal{J}_q\}$ **to** $(p \bar{\vee} 2^d, q)$;
           receive $\{t[j] : j \in \mathcal{J}_q\}$ **from** $(p \bar{\vee} 2^d, q)$;
           $\langle \ ; j \ : \ j \in \mathcal{J}_q \ :: \ v[j] := v[j] + t[j] \rangle$
       $\rangle$
end

The first quantification of the **assign** section performs the local assignments for values of the global index $n \in \mathcal{M}_p \cap \mathcal{N}_q$. The global index n corresponding to local index i in process row p is computed with the inverse of the row-distribution map μ. Subsequently, the process column s and local index j corresponding to global index n are computed with the column-distribution map ν. In process column $q = s$, the local assignment is performed.

Each process column subsequently performs a recursive-doubling procedure among its processes. This is expressed by the quantification over d, where it is assumed that $P = 2^D$. As seen in Section 1.7, the requirement that P be a power of 2 is an easily removed restriction.

2.2.2 Inner Product

Programs for vector operations on two-dimensional process meshes are derived from those on one-dimensional meshes by simple Q-fold duplication. Program Inner-Product-5, which computes the inner product of two row-distributed vectors, is a Q-fold duplication of program Inner-Product-4. The program for column-distributed vectors is obtained analogously. The inner product of a row- and a column-distributed vector is left as an exercise.

$0..P-1 \times 0..Q-1 \parallel (p,q)$ **program** Inner-Product-5
declare
$\qquad i,\ d\ :\ $ integer ;
$\qquad \sigma,\ h\ :\ $ real ;
$\qquad x,\ y\ :\ $ array $[\mathcal{I}_p]$ of real
initially
$\qquad \langle\ ;\ i\ :\ i \in \mathcal{I}_p\ ::\ x[i], y[i] = \tilde{x}_{\mu^{-1}(p,i)}, \tilde{y}_{\mu^{-1}(p,i)}\ \rangle$
assign
$\qquad \sigma := \langle\ + i\ :\ i \in \mathcal{I}_p\ ::\ x[i]y[i]\ \rangle$;
$\qquad \langle\ ;\ d\ :\ 0 \le d < D\ ::$
$\qquad\qquad$ **send** σ **to** $(p \bar{\vee} 2^d, q)$;
$\qquad\qquad$ **receive** h **from** $(p \bar{\vee} 2^d, q)$;
$\qquad\qquad \sigma := \sigma + h$
$\qquad \rangle$
end

This program assumes that $P = 2^D$. Program Inner-Product-5 uses $P \times Q$ processes but achieves only an effective P-fold concurrency. The concurrency lost by duplication can be regained, however.

To this end, introduce a Q-fold partition of the local index set $\mathcal{I}_p$, which consists of the sets $\mathcal{K}_{p,q}$, where $0 \le q < Q$. Typically, $\mathcal{I}_p$ is an index range $0..I_p-1$, and the subsets $\mathcal{K}_{p,q}$ divide this range about evenly. A good choice is

$$\mathcal{K}_{p,q} = \{q + kQ : 0 \le k \text{ and } q + kQ < I_p\}.$$

In process (p, q), the partial sum

$$\sum_{i \in \mathcal{K}_{p,q}} x[i]y[i]$$

is computed, and the inner product of the vectors $\vec{x}$ and $\vec{y}$, which is the sum of PQ partial sums, is computed by recursive doubling over all $P \times Q$ processes. Program Inner-Product-6 results.

$0..P - 1 \times 0..Q - 1 \parallel (p, q)$ **program** Inner-Product-6
declare
    i, d, t, u : integer ;
    σ, h : real ;
    x, y : array $[0..I_p - 1]$ of real
initially
    $\langle\ ;\ i\ :\ i \in \mathcal{I}_p\ ::\ x[i], y[i] = \tilde{x}_{\mu^{-1}(p,i)}, \tilde{y}_{\mu^{-1}(p,i)}\ \rangle$
assign
    $\sigma := \langle\ +\ i\ :\ i \in \mathcal{K}_{p,q}\ ::\ x[i]y[i]\ \rangle$;
    $t := pQ + q$;
    $\langle\ ;\ d\ :\ 0 \le d < D\ ::$
        $u := t \bar{\vee} 2^d$;
        send σ **to** $(\lfloor u/Q \rfloor, u \bmod Q)$;
        receive h **from** $(\lfloor u/Q \rfloor, u \bmod Q)$;
        $\sigma := \sigma + h$
    $\rangle$
end

There is a minor notational problem. For recursive doubling, it is convenient to use process identifiers that consist of a single index t, where $0 \le t < PQ$. However, the rest of the program identifies processes by a coordinate pair $(p, q) \in 0..P - 1 \times 0..Q - 1$. To switch between the two kinds of identifiers, we identify the single index t and the coordinate pair (p, q) if $t = Qp + q$ or, equivalently, if $(p, q) = (\lfloor t/Q \rfloor, t \bmod Q)$. (Any other one-to-one map may be used.) In program Inner-Product-6, it is assumed that $PQ = 2^D$.

2.3 Practical Data Distributions

The only restriction on data distributions encountered, thus far, is that data distributions for index sets of the form $\mathcal{M} \times \mathcal{N}$ must be of the form $\mu \times \nu$ for computations on two-dimensional process meshes. The one-dimensional data distributions μ and ν remain arbitrary. Eventually, however, the data distributions must be specified. Earlier observations indicate that load-balanced data distributions are important for performance considerations.

    The *scatter distribution* allocates consecutive vector components to consecutive processes. To distribute the index set $\mathcal{M} = 0..M - 1$ over P processes by the scatter distribution, we let

$$\mu(m) = (p, i), \text{ where } \begin{cases} p = m \bmod P \\ i = \lfloor \frac{m}{P} \rfloor . \end{cases} \tag{2.3}$$

It easily follows that

$$\begin{aligned} \mathcal{M}_p &= \{m : 0 \leq m < M \text{ and } m \bmod P = p\} \\ I_p &= \left\lfloor \frac{M + P - p - 1}{P} \right\rfloor \\ \mathcal{I}_p &= \{i : 0 \leq i < I_p\} \\ \mu^{-1}(p, i) &= iP + p. \end{aligned}$$

The linear distribution allocates consecutive vector components to consecutive local-array entries. The formulas are easy if M is an exact multiple of P, say $M = PL$.

$$\mu(m) = (p, i), \text{ where } \begin{cases} p = \lfloor \frac{m}{L} \rfloor \\ i = m - pL. \end{cases} \tag{2.4}$$

This is called the *perfectly load-balanced linear data distribution*, and it satisfies

$$\begin{aligned} \mathcal{M}_p &= \{m : pL \leq m < (p + 1)L\} \\ I_p &= L \\ \mathcal{I}_p &= \{i : 0 \leq i < L\} \\ \mu^{-1}(p, i) &= pL + i. \end{aligned}$$

The formulas for the general *load-balanced linear distribution* are more complicated. With $M = PL + R$ and $0 \leq R < P$, we have that

$$\begin{aligned} L &= \left\lfloor \frac{M}{P} \right\rfloor \\ R &= M \bmod P \\ \mu(m) &= (p, i), \text{ where } \begin{cases} p = \max(\lfloor \frac{m}{L+1} \rfloor, \lfloor \frac{m-R}{L} \rfloor) \\ i = m - pL - \min(p, R). \end{cases} \tag{2.5} \end{aligned}$$

This maps $L+1$ components to processes 0 through $R-1$ and L components to the remaining processes. Other data-distribution quantities are given by:

$$\begin{aligned} \mathcal{M}_p &= \{m : pL + \min(p, R) \leq m < (p + 1)L + \min(p + 1, R)\} \\ I_p &= \left\lfloor \frac{M + P - p - 1}{P} \right\rfloor \\ \mathcal{I}_p &= \{i : 0 \leq i < I_p\} \\ \mu^{-1}(p, i) &= pL + \min(p, R) + i. \end{aligned}$$

Process	Local	Linear Global	Scatter
0	0	0	0
	1	1	4
	2	2	8
1	0	3	1
	1	4	5
	2	5	9
2	0	6	2
	1	7	6
	2	8	10
3	0	9	3
	1	10	7

TABLE 2.1. A comparison of linear and scatter distributions for $M = 11$ and $P = 4$.

The linear and the scatter distributions are compared in Table 2.1, where the global index m is displayed next to the appropriate local index i. Both data distributions are load balanced. The scatter distribution has the advantage of easier formulas. In programs where μ and/or μ^{-1} are evaluated frequently, this may even have an impact on performance. The linear distribution is more appropriate for programs in which vector components frequently refer to neighboring components.

The *random distribution* maps vector components to array entries by means of a random-number generator. If the amount of work for each index is unpredictable, this is useful to balance the load statistically. The main disadvantage of the random distribution is that a table of length M must be stored in every process. This table is used to compute the maps μ and μ^{-1}, which relate global indices to local indices and process identifiers. Add to this that the random distribution almost never outperforms the scatter distribution, and the conclusion is obvious: the random distribution should be used only in very special circumstances.

2.4 Vector Shift

2.4.1 Definition and Specification

In Section 2.5, the following vector operation will be used extensively: given a vector, shift its components by a certain number of positions. For example, shifting the following vector of dimension 100:

$$[0, 1, 2, 3, 4, 5, \ldots, 95, 96, 97, 98, 99]^T$$

by 10 positions results in the vector:

$$[10, 11, 12, \ldots, 99, 0, 1, 2, \ldots, 9]^T.$$

Mathematically, a shift by one position of an M-dimensional vector is represented by the $M \times M$ matrix

$$S = \begin{bmatrix} 0 & 1 & 0 & \ldots & 0 \\ 0 & 0 & 1 & & 0 \\ & & & \ddots & \\ 0 & 0 & 0 & & 1 \\ 1 & 0 & 0 & \ldots & 0 \end{bmatrix} = [\vec{e}_{M-1} \vec{e}_0 \vec{e}_1 \vec{e}_2 \ldots \vec{e}_{M-2}],$$

where $\vec{e}_m$ is the m-th unit vector (component number m of $\vec{e}_m$ is equal to one, and all other components of $\vec{e}_m$ vanish). The vector $S\vec{x}$ is $\vec{x}$ shifted by one position, and $S^r\vec{x}$ is $\vec{x}$ shifted by r positions.

The vector-shift operator S satisfies the following properties:

$$S^{-1} = S^T \tag{2.6}$$

$$S^M = I \tag{2.7}$$

$$S^m = S^k \Leftrightarrow m = k \bmod M \tag{2.8}$$

$$S^k\vec{x} = [\xi_{(m+k)\bmod M}]_{m=0}^{M-1}. \tag{2.9}$$

Equation 2.6 gives a representation for a shift in the opposite direction. Equation 2.7 implies Equation 2.8. Equation 2.9 demonstrates the effect of a vector shift on the location of the vector components.

We wish to implement the assignment

$$\vec{x} := S^r\vec{x}$$

for arbitrary positive or negative integers r. It is always possible to reduce r to any range of M values by modulo-M arithmetic; see Equations 2.8 and 2.9. In particular, it is possible to reduce r such that $-\frac{M}{2} < r \le \frac{M}{2}$. In this range, only positive shifts need be discussed, since negative shifts can be handled similarly.

The vector-shift operation is specified by program Vector-Shift-1.

```
program Vector-Shift-1
declare
      m : integer ;
      x : array [0..M − 1] of real
initially
      ⟨ ; m : 0 ≤ m < M :: x[m] = x̃_m ⟩
assign
      ⟨ ‖ m : 0 ≤ m < M :: x[m] := x[(m + r) mod M] ⟩
end
```

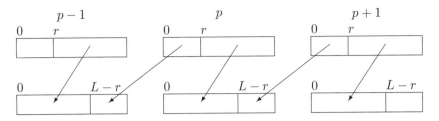

FIGURE 2.4. A shift over r positions with $0 \leq r < L$.

This concurrent quantification shows that one cannot always immediately replace the concurrent by the sequential separator, as was the case in quantifications encountered thus far. With the concurrent separator, all right-hand-side values are evaluated before any assignment is performed. The program would be wrong if one replaced the concurrent by the sequential separator: for $r = 1$, the sequential quantification

$$\langle \, ; \, m \, : \, 0 \leq m < M \, :: \, x[m] := x[(m+1) \bmod M] \, \rangle$$

is equivalent to the sequence of assignments

$$x[0] := x[1]; \; x[1] := x[2]; \; \ldots \; x[M-2] := x[M-1]; \; x[M-1] := x[0]$$

and array entry $x[M-1]$ ends up with the new instead of the old value of $x[0]$.

2.4.2 Implementation

For the purpose of developing a multicomputer implementation, we shall consider only the case of the perfectly load-balanced linear data distribution. Let $M = PL$ and apply Equation 2.4.

First, consider the case $0 \leq r < L$. Figure 2.4 shows what needs to be done. Process p sends its first r values to process $p-1$ and receives r values from process $p+1$. The $L-r$ values that remain in process p must be assigned to the first $L-r$ entries of the local array. Due to the circular nature of the shift, process identifiers must be interpreted modulo P. The following program segment, which is part of a P-fold duplicated program, implements a shift over r positions provided that $0 \leq r < L$.

> **send** $\{x[i] : 0 \leq i < r\}$ **to** $(p-1) \bmod P$;
> $\langle \, ; \, i \, : \, 0 \leq i < L-r \, :: \, x[i] := x[i+r] \, \rangle$;
> **receive** $\{x[i] : L-r \leq i < L\}$ **from** $(p+1) \bmod P$

Second, consider the case where r is an exact multiple of L, or $r = sL$. Here, whole vector segments are shifted among processes, and the following program segment, also part of a duplicated program, is easily obtained:

> **send** $\{x[i] : 0 \le i < L\}$ **to** $(p - s) \bmod P$;
> **receive** $\{x[i] : 0 \le i < L\}$ **from** $(p + s) \bmod P$

Considered by itself, a segment shift could be achieved by a mere change of process identifiers or a renumbering of processes. However, a vector is always shifted with respect to other vectors, and explicit segment shifts remain a necessity.

The general case reduces to a combination of the two previous special cases. Let $r = sL + f$, where $s = \lfloor \frac{r}{L} \rfloor$ and $f = r \bmod L$. The shift operation then reduces to one *fractional shift* over f positions and one *segment shift* over s processes.

The performance analysis of the vector-shift operation can be brief. According to our standard multicomputer-performance model, the cost of a fractional shift over f positions is given by $\tau_C(f)$, and the cost of a segment shift is given by $\tau_C(L)$. Hence, the total cost of a shift over $r = sL + f$ positions is $\tau_C(f) + \tau_C(L) = 2\tau_S + \beta(L + r)$. However, the model on which this estimate is based does not reflect all real costs. In practice, the assignments occurring in the fractional shift can be a serious overhead. Such local copying can be avoided if vectors subject to shifting are stored in floating buffers. Then, a fractional shift is reduced to mere pointer arithmetic. When the vector-shift operation is frequently used, this communication operation must be fully optimized. This usually requires further computer-dependent program transformations; see Exercise 57 of Chapter 12.

2.5   The Diagonal Form of Matrices

2.5.1   *Definition*

For any vector $\vec{x} = [\xi_m] \in \mathbb{R}^M$, define the diagonal matrix

$$
D(\vec{x}) = \begin{bmatrix}
\xi_0 & 0 & 0 & \cdots & 0 \\
0 & \xi_1 & 0 & \cdots & 0 \\
0 & 0 & \xi_2 & & 0 \\
& & & \ddots & \\
0 & 0 & 0 & & \xi_{M-1}
\end{bmatrix} .
$$

The shift operator S applied to this diagonal matrix results in the matrix:

$$
SD(\vec{x}) = \begin{bmatrix}
0 & \xi_1 & 0 & \cdots & 0 \\
0 & 0 & \xi_2 & & 0 \\
& & & \ddots & \\
0 & 0 & 0 & & \xi_{M-1} \\
\xi_0 & 0 & 0 & \cdots & 0
\end{bmatrix} .
$$

When applying S again, we obtain:

$$S^2 D(\vec{x}) = \begin{bmatrix} 0 & 0 & \xi_2 & 0 & \cdots & & 0 \\ 0 & 0 & 0 & \xi_3 & & & 0 \\ & & & & \ddots & & \\ 0 & 0 & 0 & 0 & & & \xi_{M-1} \\ \xi_0 & 0 & 0 & 0 & \cdots & & 0 \\ 0 & \xi_1 & 0 & 0 & \cdots & & 0 \end{bmatrix}.$$

This easily generalizes to other powers of S. With each shift, the diagonal elements are shifted to new elements that form a diagonal-like structure. The shifted diagonal matrices $S^0 D(\vec{x}), S^1 D(\vec{x}), \ldots S^{M-1} D(\vec{x})$ cover all elements of an $M \times M$ matrix. This observation leads to the following definition.

Definition 5 *The* diagonal form *of a matrix* $A = [\alpha_{m,n}] \in \mathbb{R}^{M \times M}$ *is the unique decomposition*

$$A = \sum_{m=0}^{M-1} S^m D(\vec{a}_m). \qquad (2.10)$$

The vector $\vec{a}_m$ *is diagonal number* m *of* A.

With component subscripts interpreted modulo M, it is easily verified that

$$\vec{a}_m = [\alpha_{-m,0}, \ \alpha_{-m+1,1}, \ \ldots, \ \alpha_{-m+k,k}, \ \cdots \alpha_{-m+M-1,M-1}]^T, \qquad (2.11)$$

and that

$$S^m D(\vec{a}_m) = \begin{bmatrix} 0 & \cdots & 0 & \alpha_{0,m} & & & \\ \vdots & & \vdots & & \ddots & & \\ 0 & \cdots & 0 & & & \alpha_{-m-1,M-1} \\ \alpha_{-m,0} & & & 0 & \cdots & 0 \\ & \ddots & & \vdots & & \vdots \\ & & \alpha_{-1,m-1} & 0 & \cdots & 0 \end{bmatrix}.$$

Definition 5 deviates from standard terminology: a diagonal as defined here combines into one structure an upper and a lower diagonal as defined by common usage. In fact, diagonal number m combines the m-th upper and the $(M - m)$-th lower diagonal.

Before proceeding, it is useful to introduce one new vector operation and a few properties. The *component-by-component vector product* of two vectors $\vec{x} = [\xi_m]$ and $\vec{y} = [\eta_m]$ is a vector $\vec{z} = [\zeta_m]$ such that $\zeta_m = \xi_m \eta_m$. In vector notation, the symbol $\otimes$ is used to denote this operation, as in:

$$\vec{z} = \vec{x} \otimes \vec{y}.$$

The following identities are easily established.

$$
\begin{aligned}
\vec{x} \otimes \vec{y} &= D(\vec{x})\vec{y} = D(\vec{y})\vec{x} = \vec{y} \otimes \vec{x} & (2.12) \\
D(\vec{x})D(\vec{y}) &= D(\vec{x} \otimes \vec{y}) & (2.13) \\
SD(\vec{x}) &= D(S\vec{x})S & (2.14) \\
D(\vec{x})S &= SD(S^{-1}\vec{x}) & (2.15) \\
S(\vec{x} \otimes \vec{y}) &= S\vec{x} \otimes S\vec{y}. & (2.16)
\end{aligned}
$$

In Sections 2.5.2 through 2.5.5, the diagonal form is used to develop algorithms for most elementary matrix operations. The basic tools required are the vector sum, the component-by-component vector multiplication, and the vector shift. The multicomputer implementation of the first two is trivial, and that of the vector-shift operation was, at least partially, covered in Section 2.4.

2.5.2 Matrix–Vector Product

From Equations 2.10 and 2.12, it follows that

$$
A\vec{x} = \sum_{m=0}^{M-1} S^m D(\vec{a}_m)\vec{x} = \sum_{m=0}^{M-1} S^m (\vec{a}_m \otimes \vec{x}),
$$

which immediately implies the following implementation of the assignment $\vec{y} := A\vec{x}$:

```
y⃗ := 0⃗ ;
⟨ ; m : 0 ≤ m < M ::
      z⃗ := a⃗_m ⊗ x⃗ ;
      z⃗ := S^m z⃗ ;
      y⃗ := y⃗ + z⃗
⟩
```

Program Matrix-Vector-6 avoids the temporary vector $\vec{z}$ by accumulating the sum directly in $\vec{y}$ and shifting $\vec{y}$ in the opposite direction.

```
program Matrix-Vector-6
declare
      m : integer ;
      x⃗, y⃗ : vector
assign
      y⃗ := 0⃗ ;
      ⟨ ; m : 0 ≤ m < M ::
            y⃗ := y⃗ + a⃗_m ⊗ x⃗ ;
            y⃗ := S^{-1} y⃗
      ⟩
end
```

Upon termination of this program, the vector $\vec{y}$ has been shifted M times and is back in its original position. Program Matrix-Vector-6 is only the starting point of the development of a multicomputer program for computing matrix–vector products based on the diagonal form. The remaining development is straightforward, provided a vector shift on distributed vectors is available.

This matrix–vector algorithm is particularly successful for sparse matrices with diagonal fill. Suppose only diagonals $n_0, n_1, \ldots, n_{R-1}$ are nonzero. The above program is then easily adapted to avoid superfluous arithmetic with zeroes:

$$\vec{y} := \vec{0} \; ;$$
$$\langle \; ; r \; : \; 0 \leq r < R \; :: $$
$$\qquad \vec{y} := \vec{y} + \vec{a}_{n_r} \otimes \vec{x} \; ;$$
$$\qquad \vec{y} := S^{-(n_{r+1}-n_r)}\vec{y}$$
$$\rangle$$

Choose $n_R = M$ to shift the vector $\vec{y}$ back to its original position.

2.5.3 Matrix Transpose

Equations 2.6, 2.10, and 2.15 imply that:

$$
\begin{aligned}
A^T &= \left(\sum_{m=0}^{M-1} S^m D(\vec{a}_m) \right)^T \\
&= \sum_{m=0}^{M-1} D(\vec{a}_m) S^{-m} \\
&= \sum_{m=0}^{M-1} S^{-m} D(S^m \vec{a}_m) \\
&= \sum_{m=0}^{M-1} S^m D(S^{-m} \vec{a}_{-m})
\end{aligned}
$$

The last transition is achieved by reversing the summation and by using that $-m = (M - m) \bmod M$. It follows that the m-th diagonal of A^T is the $(-m)$-th diagonal of A shifted over $-m$ positions. The assignment

$$A := A^T$$

is an in-place matrix transpose; it is implemented by the simple quantification:

$$\langle \; \| \; m \; : \; 0 \leq m < M \; :: \; \vec{a}_m := S^{-m}\vec{a}_{M-m} \; \rangle$$

2.5.4   Vector–Matrix Product

The implementation of the assignment

$$\vec{y}^T := \vec{x}^T A$$

easily follows from Section 2.5.3, because it is equivalent to

$$\vec{y} := A^T \vec{x}.$$

Combining Section 2.5.3 and Equations 2.13, 2.15, and 2.16, it follows that:

$$
\begin{aligned}
A^T \vec{x} &= \sum_{m=0}^{M-1} S^m D(S^{-m}\vec{a}_{-m})\vec{x} \\
&= \sum_{m=0}^{M-1} S^m((S^{-m}\vec{a}_{-m}) \otimes \vec{x}) \\
&= \sum_{m=0}^{M-1} \vec{a}_{-m} \otimes (S^m \vec{x}) \\
&= \sum_{m=0}^{M-1} \vec{a}_m \otimes (S^{-m} \vec{x}).
\end{aligned}
$$

As in Section 2.5.3, the last transition is achieved by reversing the summation and using that $-m = (M - m) \bmod M$. The following program is easily obtained.

```
program Vector-Matrix-1
declare
      m : integer ;
      x⃗, y⃗ : vector
assign
      y⃗ := 0⃗ ;
      ⟨ ; m : 0 ≤ m < M ::
            y⃗ := y⃗ + a⃗ₘ ⊗ x⃗ ;
            x⃗ := S⁻¹x⃗
      ⟩
end
```

As for the matrix–vector product, the vector–matrix product based on the diagonal form is easily adapted for sparse matrices with diagonal fill.

2.5.5   Matrix–Matrix Product

Finally, we consider the matrix assignment

$$C := AB,$$

where A, B, and C are $M \times M$ matrices in diagonal form. Modulo arithmetic on the indices and Equations 2.13 through 2.16 imply that

$$
\begin{aligned}
AB &= \left(\sum_{m=0}^{M-1} S^m D(\vec{a}_m) \right) \left(\sum_{n=0}^{M-1} S^n D(\vec{b}_n) \right) \\
&= \sum_{m=0}^{M-1} \sum_{n=0}^{M-1} S^m D(\vec{a}_m) S^n D(\vec{b}_n) \\
&= \sum_{m=0}^{M-1} \sum_{n=0}^{M-1} S^{m+n} D(S^{-n}\vec{a}_m) D(\vec{b}_n) \\
&= \sum_{k=0}^{M-1} S^k \sum_{\ell=0}^{M-1} D(S^{-\ell}\vec{a}_{k-\ell}) D(\vec{b}_\ell) \\
&= \sum_{k=0}^{M-1} S^k D\left(\sum_{\ell=0}^{M-1} (S^{-\ell}\vec{a}_{k-\ell}) \otimes \vec{b}_\ell \right).
\end{aligned}
$$

The m-th diagonal of C is thus given by

$$
\vec{c}_m = \sum_{\ell=0}^{M-1} (S^{-\ell}\vec{a}_{m-\ell}) \otimes \vec{b}_\ell.
$$

The following program implements the matrix assignment.

```
⟨ ; m : 0 ≤ m < M :: c⃗ₘ := 0⃗ ⟩ ;
⟨ ; ℓ : 0 ≤ ℓ < M ::
    ⟨ ; m : 0 ≤ m < M :: c⃗ₘ := c⃗ₘ + a⃗ₘ₋ℓ ⊗ b⃗ℓ ⟩ ;
    ⟨ ; m : 0 ≤ m < M :: a⃗ₘ := S⁻¹a⃗ₘ ⟩
⟩
```

All diagonals of A are back in place after the operation, because all have shifted over exactly M positions.

2.5.6 Diagonal Form and Column Distribution

The diagonal form of a matrix distributed over P processes is equivalent to a column-distributed matrix. From Equation 2.11, it follows that the k-th component of $\vec{a}_m$ is $\alpha_{-m+k,k}$. Because $\vec{a}_m$ is distributed over P processes by a perfectly load-balanced linear data distribution, process p contains all elements of A in the set:

$$
\{ \alpha_{k-m,k} : 0 \le m < M \text{ and } pL \le k < (p+1)L \}.
$$

Modulo-M arithmetic implies that this set is identical to:

$$
\{ \alpha_{\ell,k} : 0 \le \ell < M \text{ and } pL \le k < (p+1)L \}.
$$

In other words, process p contains columns pL through $(p+1)L-1$. The only difference between a column-distributed matrix and the diagonal form is in the local-index sets. A local permutation converts one distribution into the other; no communication is required.

When applicable, the diagonal form is easy to understand, to program, and to use. It is particularly efficient for sparse matrices with fill along a few diagonals. However, the diagonal form is a one-dimensional distribution. As a result, the diagonal form runs into granularity problems when applied to full matrices: fine-granularity effects become important as soon as the number of processes is comparable to the matrix dimension M.

2.6   Notes and References

Several leading supercomputers were designed around the concept of vector processing, and vector operations have long been a cornerstone of scientific computing. Transforming a program such that it is efficient on pipelined processors is called *vectorization* (see Section 12.1). On most computer systems, assembly-language libraries for vector operations are available. The most widely available standard is the Basic Linear Algebra Subprograms package (BLAS). Originally, BLAS contained vector arithmetic only. Since its inception, it has been extended with matrix–vector and matrix–matrix operations. Efforts are under way to define a concurrent BLAS standard. The evolving BLAS standard is presented in references [24, 25, 50, 60].

The development of efficient and general sparse-matrix representations is a major challenge, even for sequential computing. The destruction of the fill structure by the data distribution makes the issue even harder for concurrent computing. In Section 2.1.4, many difficult technical details were avoided by introducing arrays defined over arbitrary index sets. For an introduction to real-world sparse-matrix technology, see George and Liu [38] and Duff, Erisman, and Reid [27].

The diagonal form was used for vectorization purposes by Madsen, Rodrigue, and Karush [61]. McBryan and Van de Velde [62, 63] introduced it to multicomputer programming.

The matrix–matrix product for row- and column-oriented distributions of matrices is a challenge. Exercise 10 gives a hint on how to proceed. In general, the data distributions of matrices participating in a concurrent matrix–matrix product are severely limited. For matrices with linearly distributed rows and columns, a concurrent matrix-multiplication algorithm was developed and studied by Fox, Otto, and Hey [31].

Exercises

Exercise 5 *Develop a multicomputer program on a $P \times Q$ process mesh that implements the assignment:*

$$\vec{y}^T := \vec{x}^T A.$$

Exercise 6 *Develop a multicomputer program to compute the inner product of a row- and a column-distributed vector on a $P \times Q$ process mesh.*

Exercise 7 *Based on the diagonal form, implement the assignments:*

$$
\begin{aligned}
C &:= AB^T \\
C &:= A^T B \\
C &:= A^T B^T,
\end{aligned}
$$

where A, B, and C are square matrices.

Exercise 8 *Develop a multicomputer program to convert a full matrix in diagonal form into a column-distributed matrix.*

Exercise 9 *Develop a multicomputer program to transpose a row-distributed matrix into a column-distributed matrix via the diagonal form.*

Exercise 10 *Investigate a row- and column-oriented multicomputer implementation of the matrix assignment:*

$$C := AB$$

with A an $M \times N$, B an $N \times K$, and C an $M \times K$ matrix. We need three one-dimensional distributions: μ, ν, and κ. The two-dimensional distributions of A, B, and C are given by $\mu \times \kappa$, $\kappa \times \nu$, and $\mu \times \nu$, respectively.

To choose each of the one-dimensional distributions independently, the specification program is duplicated over a three-dimensional process mesh with dimensions $P \times Q \times R$. In the selection step, μ, ν, and κ are used to distribute data over the respective process-mesh dimensions.

Perform the duplication over $P \times Q \times R$ process mesh of a specification program. Subsequently, perform the selection step. Examine performance and memory use of the resulting multicomputer program. How much duplication remains? Is the duplication acceptable?

To reduce memory overhead, reduce the multicomputer program over a $P \times Q \times R$ process mesh to one over a $P \times Q$ process mesh. Introduce compatibility requirements for the one-dimensional distributions. How much communication is required?

3
Iterative Methods

The power method finds the dominant eigenvalues of a matrix. It will be covered in Section 3.1, not only because it is a useful method, but also because it provides a realistic computational context for the elementary operations of Chapters 1 and 2 without requiring an extensive mathematical background.

The remainder of this chapter summarizes theory that will be useful in later chapters. Skip this part on a first reading, and return to the appropriate sections when necessary. In Section 3.2, standard theory of relaxation methods is summarized. The conjugate-gradient method of Section 3.3 is the best-known Krylov-space method, and we shall use it to solve systems of linear equations with symmetric positive-definite coefficient matrices.

Section 3.4 covers a recently developed Krylov-space method to solve any linear system, not just symmetric positive-definite ones. The generalization of Krylov-space technology is an area of active research, which is motivated in part by the observation that many Krylov-space methods are easily and efficiently implemented on concurrent computers. This justifies covering at least one generalized Krylov-space method, in spite of it requiring mathematical detail that is somewhat removed from the rest of the text.

3.1 The Power Method

3.1.1 Numerics

Let $A \in \mathbb{R}^{M \times M}$ be a matrix with M linearly independent eigenvectors $\vec{v}_m$, where $0 \leq m < M$. Let λ_m be the eigenvalue associated with $\vec{v}_m$. Without loss of generality, we assume that the eigenvalues are numbered such that

$$|\lambda_0| \geq |\lambda_1| \geq \ldots \geq |\lambda_{M-1}|. \tag{3.1}$$

Because A is a real matrix, its complex eigenvalues must occur in complex-conjugate pairs.

Our goal is to compute the largest eigenvalue(s) and corresponding eigenvector(s) of real matrices. We shall consider two cases. The first case is that of a matrix with a *dominant real eigenvalue* λ_0, which has multiplicity one and satisfies

$$|\lambda_0| > |\lambda_1|. \tag{3.2}$$

The second case is that of a matrix with a *dominant complex-conjugate pair of eigenvalues* λ_0 and λ_1, where $\lambda_1 = \bar{\lambda}_0$ and

$$|\lambda_0| = |\lambda_1| > |\lambda_2|. \tag{3.3}$$

Starting from a random real vector $\vec{x}_0$, construct a sequence of real vectors using the recursion

$$\vec{x}_{k+1} = A\vec{x}_k \tag{3.4}$$

for $k \geq 0$. Because the eigenvectors $\vec{v}_m$ are linearly independent, they span the whole space $\mathbb{R}^M$. It follows that

$$\forall k \geq 0 : \quad \vec{x}_k = \sum_{m=0}^{M-1} \lambda_m^k \xi_m \vec{v}_m \tag{3.5}$$

for some (unknown) coefficients ξ_m.

For a matrix A with a dominant real eigenvalue, Equations 3.1, 3.2, and 3.5 imply that

$$\vec{x}_k = \lambda_0^k (\xi_0 \vec{v}_0 + \vec{r}_k)$$

with

$$\lim_{k \to \infty} \vec{r}_k = \vec{0}.$$

If $\xi_0 \neq 0$ and k is sufficiently large, the vectors $\vec{x}_k$ are almost parallel to the eigenvector $\vec{v}_0$, and

$$\vec{x}_{k+1} \approx \lambda_0 \vec{x}_k. \tag{3.6}$$

This approximate vector equation consists of M approximate scalar equations. One method of obtaining an estimate $\hat{\lambda}_0$ for λ_0 is to minimize

$$\Psi(\lambda) = \| \vec{x}_{k+1} - \lambda \vec{x}_k \|^2,$$

where $\parallel \vec{v} \parallel^2 = (\vec{v}, \vec{v})$ and $(\vec{v}, \vec{w}) = \vec{v}^T \vec{w}$. With this choice of inner product and associated norm of real vectors, the function $\Psi(\lambda)$ is a quadratic function of λ. Its minimum is given by

$$\hat{\lambda}_0 = \frac{(\vec{x}_{k+1}, \vec{x}_k)}{(\vec{x}_k, \vec{x}_k)}.$$

Error estimates for the computed eigenvalue $\hat{\lambda}_0$ can be based on the norm of the residual vector:

$$\Psi(\hat{\lambda}_0) = \parallel \vec{x}_{k+1} - \hat{\lambda}_0 \vec{x}_k \parallel^2 = \parallel \vec{x}_{k+1} \parallel^2 - \hat{\lambda}_0(\vec{x}_{k+1}, \vec{x}_k).$$

The above derivation depends on the assumption that $\xi_0 \neq 0$. This is almost always the case, because $\vec{x}_0$ was chosen at random. However, the method will work even if by extreme coincidence $\xi_0 = 0$, because round-off errors accumulated in the iteration will introduce a component along $\vec{v}_0$.

Next, consider the case of a matrix with a dominant complex-conjugate pair of eigenvalues λ_0 and λ_1. The eigenvectors $\vec{v}_0$ and $\vec{v}_1$ are complex conjugate, and for any real vector $\vec{x}_0$, we have that $\xi_1 = \bar{\xi}_0$. Equations 3.1, 3.3, 3.4, and 3.5 imply that

$$\vec{x}_k = |\lambda_0|^k \left(\xi_0 \left(\frac{\lambda_0}{|\lambda_0|} \right)^k \vec{v}_0 + \bar{\xi}_0 \left(\frac{\lambda_1}{|\lambda_0|} \right)^k \vec{v}_1 + \vec{r}_k \right) \qquad (3.7)$$

with

$$\lim_{k \to \infty} \vec{r}_k = \vec{0}.$$

For large k, the iterates $\vec{x}_k$ approximately lie in the two-dimensional subspace spanned by the eigenvectors $\vec{v}_0$ and $\vec{v}_1$. We expect, therefore, that three successive iterates will be almost linearly dependent.

The coefficients α_0 and β_0 of the quadratic

$$p(\lambda) = (\lambda - \lambda_0)(\lambda - \lambda_1) = \lambda^2 + \alpha_0 \lambda + \beta_0$$

are real, because its roots are the complex-conjugate pair λ_0 and λ_1. From

$$\lambda_0^{k+2} + \alpha_0 \lambda_0^{k+1} + \beta_0 \lambda_0^k = 0$$
$$\lambda_1^{k+2} + \alpha_0 \lambda_1^{k+1} + \beta_0 \lambda_1^k = 0,$$

it follows that

$$(\lambda_0^{k+2} \xi_0 \vec{v}_0 + \lambda_1^{k+2} \xi_1 \vec{v}_1) + \alpha_0 (\lambda_0^{k+1} \xi_0 \vec{v}_0 + \lambda_1^{k+1} \xi_1 \vec{v}_1) + \beta_0 (\lambda_0^k \xi_0 \vec{v}_0 + \lambda_1^k \xi_1 \vec{v}_1) = \vec{0}.$$

This and Equation 3.7 imply that, for sufficiently large k, the vectors $\vec{x}_k$, $\vec{x}_{k+1}$, and $\vec{x}_{k+2}$ satisfy the approximate vector equation

$$\vec{x}_{k+2} + \alpha_0 \vec{x}_{k+1} + \beta_0 \vec{x}_k \approx \vec{0},$$

which corresponds to M approximate scalar equations. As in the real case, estimates $\hat{\alpha}_0$ and $\hat{\beta}_0$ for α_0 and β_0, respectively, minimize the norm of the residual vector. The minimization of

$$\Phi(\alpha, \beta) = \| \vec{x}_{k+2} + \alpha\vec{x}_{k+1} + \beta\vec{x}_k \|^2$$

leads to the 2×2 system of equations:

$$\left[\begin{array}{cc} (\vec{x}_{k+1}, \vec{x}_{k+1}) & (\vec{x}_{k+1}, \vec{x}_k) \\ (\vec{x}_{k+1}, \vec{x}_k) & (\vec{x}_k, \vec{x}_k) \end{array} \right] \left[\begin{array}{c} \hat{\alpha}_0 \\ \hat{\beta}_0 \end{array} \right] = \left[\begin{array}{c} -(\vec{x}_{k+2}, \vec{x}_{k+1}) \\ -(\vec{x}_{k+2}, \vec{x}_k) \end{array} \right].$$

Again, the norm of the residual vector of the least-squares fit, $\Phi(\hat{\alpha}_0, \hat{\beta}_0)$, can be used in error estimates. It is an indication that the procedure has not yet converged if the quadratic $p(\lambda)$ based on the estimates $\hat{\alpha}_0$ and $\hat{\beta}_0$ has real roots.

3.1.2 Program

To obtain a practical procedure, the vectors must be scaled to avoid overflow or underflow, and reasonable stopping criteria must be introduced. In program Power, the vectors $\vec{x}$, $\vec{y}$, and $\vec{z}$ represent the successive iterates $\vec{x}_k$, $\vec{x}_{k+1}$, and $\vec{x}_{k+2}$, respectively. The vector $\vec{x}$ is always scaled to unit length. The stopping criteria are based on the least-squares residual, which is computed in the variable ϵ. Each iteration step uses all six inner products possible between the three vectors $\vec{x}$, $\vec{y}$, and $\vec{z}$. However, only the three inner products involving $\vec{z}$ need to be computed, the others can be carried over from the previous iteration step. The results of the computation, the largest eigenvalue and the corresponding eigenvector, are represented by the complex variables λ and $\vec{v}$, respectively. In the complex case, only one half of the complex-conjugate pair is computed. In the program, there are three undeclared known values: $i = \sqrt{-1}$, the maximum number of iteration steps K, and the error tolerance τ.

```
program Power
declare
        k : integer ;
        α, β, ε, π_xx, π_xy, π_xz, π_yy, π_yz, π_zz : real ;
        x⃗, y⃗, z⃗ : real vector ;
        λ : complex ;
        v⃗ : complex vector
assign
        π_xx := x⃗ᵀx⃗ ;
        x⃗ := x⃗/√π_xx ;
        y⃗ := Ax⃗ ;
        π_xy := x⃗ᵀy⃗ ;
        π_yy := y⃗ᵀy⃗ ;
```

$$\langle \; ; \; k \; : \; 0 \leq k < K \; ::$$

$$\quad \text{if } \pi_{yy} - \pi_{xy}^2 < \tau^2 \pi_{yy} \text{ then begin}$$
$$\quad\quad \lambda \; := \; \pi_{xy} \; ;$$
$$\quad\quad \vec{v} \; := \; \vec{y}/\sqrt{\pi_{yy}} \; ;$$
$$\quad\quad \textbf{return}$$
$$\quad \textbf{end} \; ;$$
$$\quad \vec{z} \; := \; A\vec{y} \; ;$$
$$\quad \pi_{xz} \; := \; \vec{z}^T \vec{x} \; ;$$
$$\quad \pi_{yz} \; := \; \vec{z}^T \vec{y} \; ;$$
$$\quad \pi_{zz} \; := \; \vec{z}^T \vec{z} \; ;$$
$$\quad \alpha \; := \; -(\pi_{yz} - \pi_{xy}\pi_{xz})/(\pi_{yy} - \pi_{xy}^2) \; ;$$
$$\quad \beta \; := \; (\pi_{xy}\pi_{yz} - \pi_{yy}\pi_{xz})/(\pi_{yy} - \pi_{xy}^2) \; ;$$
$$\quad \epsilon \; := \; \pi_{zz} + \alpha^2\pi_{yy} + \beta^2 + 2(\alpha\pi_{yz} + \beta\pi_{xz} + \alpha\beta\pi_{xy}) \; ;$$
$$\quad \text{if } \epsilon < \tau^2\pi_{zz} \text{ and } 4\beta - \alpha^2 > 0 \text{ then begin}$$
$$\quad\quad \lambda := -(\alpha - i\sqrt{4\beta - \alpha^2})/2 \; ;$$
$$\quad\quad \vec{v} := (\bar{\lambda}\vec{y} - \vec{z})/\sqrt{\beta\pi_{yy} + \alpha\pi_{yz} + \pi_{zz}} \; ;$$
$$\quad\quad \textbf{return}$$
$$\quad \textbf{end} \; ;$$
$$\quad \vec{x} := \vec{y}/\sqrt{\pi_{yy}} \; ;$$
$$\quad \vec{y} := \vec{z}/\sqrt{\pi_{yy}} \; ;$$
$$\quad \pi_{xy} := \pi_{yz}/\pi_{yy} \; ;$$
$$\quad \pi_{yy} := \pi_{zz}/\pi_{yy}$$
$$\rangle$$

end

All iteration steps except the last one evaluate one matrix–vector product, three inner products, two scalar-times-vector operations, and some purely scalar arithmetic. The multicomputer implementation of the vector operations was already studied in Chapters 1 and 2, provided vectors are represented by appropriately-distributed arrays. The main multicomputer-implementation decision that remains is the representation of the matrix A. We consider three possibilities: row- and/or column-oriented representations, the diagonal form, and implicit representations.

The row- and column-oriented representation distributes the matrix over a $P \times Q$ process mesh by means of a two-dimensional data distribution $\mu \times \nu$. If using row-distributed vectors for $\vec{x}$, $\vec{y}$, and $\vec{z}$, one column-distributed vector needs to be introduced to implement the assignment $\vec{z} := A\vec{y}$; see Section 2.2.1. The alternative is to use row-distributed vectors $\vec{x}$ and $\vec{z}$ and a column-distributed vector $\vec{y}$. In this case, the assignments $\vec{x} := \vec{y}/\sqrt{\pi_{yy}}$ and $\vec{y} := \vec{z}/\sqrt{\pi_{yy}}$ require vector conversions, and the computation of π_{yz} requires an inner-product procedure for differently distributed vectors; see Exercise 6 of Chapter 2.

If program Power is just part of a larger program, it is typical that the computational parameters P, Q, μ, and ν are imposed by the calling program. However, if the power method is the dominating part of the computation, the computational parameters may be chosen for optimal performance of this component. Under the assumption that at most R nodes are available and that at most one process per node is allowed, one should always use load-balanced data distributions. In terms of performance, there is no reason to prefer one load-balanced distribution over another. For coarse-grained computations with load-balanced distributions, there remains only the question of finding the best process mesh or, equivalently, the best P and Q.

One criterion to consider is global-memory use. The representation of A requires an amount of memory that is independent of P and Q. Therefore, we do not take it into account. In one possible implementation, there are three row-distributed vectors ($\vec{x}$, $\vec{y}$, and $\vec{z}$), which are are duplicated Q times, and there is one column-distributed auxiliary vector, which is is duplicated P times. It follows that the vectors require

$$3QM + PM$$

floating-point words. With $PQ = R$, this is minimized for $P = \sqrt{3R}$ and $Q = \sqrt{3R}/3$. In the alternative implementation, $\vec{x}$ and $\vec{z}$ are row-distributed vectors, $\vec{y}$ is a column-distributed vector, and there is no auxiliary vector. This version requires

$$2QM + PM$$

floating-point words, which is minimized for $P = \sqrt{2R}$ and $Q = \sqrt{2R}/2$.

How about execution time? We could combine the performance results of Sections 2.1.3, 2.2.1, and 2.2.2. Although this numerical method is algorithmically straightforward and although the execution-time estimate is based on the many simplifying assumptions of our multicomputer-performance model, quite a complicated formula results. It is clear, however, that the ratio τ_S/τ_A plays an important role. If it is large, one should choose $Q = 1$ to eliminate all communication in the matrix–vector product. There is a limit to increasing P, however, because load imbalance occurs due to fine-grain effects when P approaches M. This may be avoided by increasing Q and keeping $PQ = R$. With two-dimensional distributions, fine-grain effects are postponed until R becomes comparable to M^2.

An alternative representation of A is the diagonal form of Section 2.5. The diagonal form is particularly useful if the matrix A has only a few nonzero diagonals. Then, the matrix–vector product is efficient, while all vector operations are fully concurrent. Because the diagonal form is a one-dimensional matrix distribution, granularity problems are encountered as soon as $R \approx M$. In the sparse case, however, M is likely to be large and fine-grain effects absent.

Often, the matrix A is not known explicitly, and only the procedure to compute the product of the matrix A and an arbitrary vector in $\mathbb{R}^M$ is given. Reconstructing the matrix by computing the matrix–vector product with all M unit vectors is not only inefficient, but also potentially numerically unstable, because the individual matrix elements might be large. It is preferable to simply replace the assignments involving the matrix–vector product by the matrix–vector procedure. In a concurrent environment, this procedure imposes its data distribution on the program implementing the power method. This occurs, for example, when A is the discretization of a three-dimensional partial-differential operator. Then, the vectors are grid functions defined on a three-dimensional grid, and it is convenient to use a three-dimensional process mesh inside the matrix–vector-product procedure. It is then necessary to develop the power method for the process connectivity and data distribution encountered in the matrix–vector product.

This shows one of the challenges of writing concurrent software, particularly software for a multi-purpose library. Say we wish to include the power method in a general (concurrent) library. The data distribution and process connectivity are determined either by the program that calls the power-method program, by the matrix–vector-product program called by the power method, or by global efficiency considerations that involve every component of the whole program. This does not fit the customary modular approach to programming, because the context in which a program is used is as important as the program itself to determine which data distributions and process connectivities are acceptable and/or most efficient.

3.2 Relaxation Methods

Given a nonsingular matrix $A \in \mathbb{R}^{M \times M}$ and a vector $\vec{b} \in \mathbb{R}^M$, we wish to solve the system of equations

$$A\vec{x} = \vec{b}. \tag{3.8}$$

For any coefficient matrix, there exist many *splittings* of the form

$$A = G - H. \tag{3.9}$$

The exact solution of Equation 3.8, which is denoted by $\vec{x}^*$, satisfies

$$G\vec{x}^* = \vec{b} + H\vec{x}^* \tag{3.10}$$

and is, therefore, a fixed point of the iteration

$$G\vec{x}_{k+1} = \vec{b} + H\vec{x}_k. \tag{3.11}$$

This iteration makes sense only if G is invertible.

A *relaxation method* is any iterative solution procedure for Equation 3.8 that can be written in the form of Equation 3.11. Every matrix A can be split in many ways. When choosing a particular splitting and corresponding relaxation method, three questions immediately arise. Under what conditions does the iteration converge? How many iteration steps are required? How much work is required per iteration step? This last question depends on the availability of fast solvers for systems with coefficient matrix G and is, therefore, best analyzed in the context of specific applications; see Chapters 6, 8, and 9.

We now examine the convergence of the relaxation iteration. Subtract Equation 3.11 from Equation 3.10, and obtain that successive error vectors $\vec{v}_k = \vec{x}^* - \vec{x}_k$ satisfy

$$\vec{v}_{k+1} = G^{-1}H\vec{v}_k = (G^{-1}H)^{k+1}\vec{v}_0. \tag{3.12}$$

The matrix $G^{-1}H$ is called the *iteration matrix*. Its *spectral radius*, which is its largest eigenvalue in absolute value, is denoted by $\rho(G^{-1}H)$. Comparing Equations 3.12 and 3.4 and using Equation 3.6, we expect that $\rho(G^{-1}H)$ approximates the factor by which the norm of the error vector $\vec{v}_k$ is multiplied to obtain the norm of $\vec{v}_{k+1}$. (To make this rigorous, one must also take into account that $G^{-1}H$ might not have a full set of eigenvectors or might not have a dominant eigenvalue.) The spectral radius $\rho(G^{-1}H)$ is also called the *convergence factor* of the relaxation method based on the splitting of Equation 3.9.

It can be proven that the relaxation iteration converges for all initial guesses if the convergence factor is less than one. More formally,

$$\textbf{if } \rho(G^{-1}H) < 1 \textbf{ then } \forall \vec{v}_0 \in \mathbb{R}^M : \lim_{k \to \infty} (G^{-1}H)^k \vec{v}_0 = \vec{0}.$$

Whereas $\rho(G^{-1}H) < 1$ is sufficient for global convergence, fast convergence is achieved only if $\rho(G^{-1}H) \ll 1$. Because

$$\| \vec{v}_k \| \approx \rho(G^{-1}H)^k \| \vec{v}_0 \|,$$

the number of iteration steps to reduce the norm of the error from $O(1)$ to $O(\epsilon)$ is given by

$$K \geq \frac{\log \epsilon}{\log \rho(G^{-1}H)}.$$

In a *Jacobi relaxation*, row m of Equation 3.8 and the vector $\vec{x}_k$ are used to compute component m of $\vec{x}_{k+1}$. With $A = [\alpha_{m,n}]_{m=0,n=0}^{M-1,M-1}$, $\vec{b} = [\beta_m]_{m=0}^{M-1}$, $\vec{x}_k = [\xi_m^k]_{m=0}^{M-1}$, and $\vec{x}_{k+1} = [\xi_m^{k+1}]_{m=0}^{M-1}$, the Jacobi relaxation computes component m of $\vec{x}_{k+1}$ as follows:

$$\xi_m^{k+1} = \frac{1}{\alpha_{m,m}} \left(\beta_m - \sum_{n=0}^{m-1} \alpha_{m,n}\xi_n^k - \sum_{n=m+1}^{M-1} \alpha_{m,n}\xi_n^k \right). \tag{3.13}$$

Let L be the strictly lower-triangular, U the strictly upper-triangular, and D the diagonal part of A. Then, we have that

$$A = L + D + U \tag{3.14}$$

and

$$\vec{x}_{k+1} = D^{-1}\left(\vec{b} - (L+U)\vec{x}_k\right).$$ (3.15)

The Jacobi method is, therefore, defined by the splitting:

$$\begin{aligned} A &= G_J - H_J \\ G_J &= D \\ H_J &= -(L+U). \end{aligned}$$ (3.16)

It can be proven that the Jacobi relaxation converges if the coefficient matrix A is strictly diagonally dominant.

Also in a *Gauss-Seidel relaxation*, row m of Equation 3.8 is used to update component m of the vector of unknowns. Unlike the Jacobi relaxation, however, the Gauss-Seidel relaxation uses the last-computed value for each of the other unknowns, not the value of the previous iteration step. A Gauss-Seidel relaxation depends, therefore, on the order in which the unknowns and equations are traversed. Sweeping through the equations and unknowns in the order of their subscripts, Gauss-Seidel relaxation for component m is defined by:

$$\xi_m^{k+1} = \frac{1}{\alpha_{m,m}}\left(\beta_m - \sum_{n=0}^{m-1}\alpha_{m,n}\xi_n^{k+1} - \sum_{n=m+1}^{M-1}\alpha_{m,n}\xi_n^{k}\right).$$ (3.17)

With L, D, and U as in Equation 3.14, this corresponds to:

$$\vec{x}_{k+1} = (D+L)^{-1}\left(\vec{b} - U\vec{x}_k\right),$$ (3.18)

and the Gauss-Seidel relaxation is defined by the splitting:

$$\begin{aligned} A &= G_{GS} - H_{GS} \\ G_{GS} &= D+L \\ H_{GS} &= -U. \end{aligned}$$ (3.19)

It can be proven that the Gauss-Seidel relaxation converges for systems with symmetric positive-definite coefficient matrices.

Successive Over-Relaxation (SOR) is a modification of the basic Gauss-Seidel relaxation. It attempts to accelerate the convergence by taking a larger step in the direction suggested by an elementary Gauss-Seidel step. One SOR step for component m is defined by:

$$\xi_m^{k+1} = \frac{\omega}{\alpha_{m,m}}\left(\beta_m - \sum_{n=0}^{m-1}\alpha_{m,n}\xi_n^{k+1} - \sum_{n=m+1}^{M-1}\alpha_{m,n}\xi_n^{k}\right) + (1-\omega)\xi_m^{k}.$$ (3.20)

This corresponds to the matrix formulation

$$(D+\omega L)\vec{x}_{k+1} = \omega\vec{b} + ((1-\omega)D - \omega U)\vec{x}_k,$$ (3.21)

which is based on the SOR splitting of the matrix A given by

$$
\begin{aligned}
A &= G_{SOR} - H_{SOR} \\
G_{SOR} &= (D + \omega L)/\omega \\
H_{SOR} &= ((1 - \omega)D - \omega U)/\omega.
\end{aligned} \tag{3.22}
$$

An important aspect of SOR theory is the determination of good and/or optimal values for the over-relaxation parameter ω. A necessary condition for convergence is that $0 < \omega < 2$.

The *Symmetric Successive Over-Relaxation* method (SSOR) alternates forward and backward SOR steps. (In a backward step, the components of the iteration vector and the equations of the system are traversed in reverse order.) In matrix formulation, one SSOR step is defined by

$$
\begin{aligned}
(D + \omega L)\vec{x}_{k+\frac{1}{2}} &= \omega \vec{b} + ((1 - \omega)D - \omega U)\vec{x}_k \\
(D + \omega U)\vec{x}_{k+1} &= \omega \vec{b} + ((1 - \omega)D - \omega L)\vec{x}_{k+\frac{1}{2}}
\end{aligned}
$$

and corresponds to the splitting:

$$
\begin{aligned}
A &= G_{SSOR} - H_{SSOR} \\
G_{SSOR} &= \frac{1}{\omega(2-\omega)}(D + \omega L)D^{-1}(D + \omega U) \\
H_{SSOR} &= \frac{1}{\omega(2-\omega)}((1 - \omega)D - \omega L)D^{-1}((1 - \omega)D - \omega U).
\end{aligned} \tag{3.23}
$$

If A is a symmetric positive-definite matrix and if $0 < \omega < 2$, it can be proven that G_{SSOR} and $G_{SSOR}^{-1}H_{SSOR}$ are symmetric positive-definite matrices.

The computation or estimation of the spectral radius of $G^{-1}H$ usually requires detailed knowledge of the specific problem. Here, the professed generality of relaxation methods ends. Relaxation methods are most successful for sparse coefficient matrices of particular structure. The development of programs for and further study of relaxation methods are, therefore, postponed until we encounter problems of a suitable structure.

3.3   The Conjugate-Gradient Method

3.3.1   Subspace Minimization

Let $\mathcal{F}(\vec{x})$ be a quadratic functional

$$
\mathcal{F}(\vec{x}) = \frac{1}{2}\vec{x}^T A\vec{x} - \vec{x}^T \vec{b} \tag{3.24}
$$

with $A \in \mathbb{R}^{M \times M}$ a symmetric positive-definite matrix. The minimum $\vec{x}^*$ of $\mathcal{F}(\vec{x})$ over $\mathbb{R}^M$ is unique and occurs where the gradient of $\mathcal{F}(\vec{x})$ vanishes:

$$
\nabla \mathcal{F}(\vec{x}^*) = A\vec{x}^* - \vec{b} = -\vec{r}^* = \vec{0}.
$$

This quadratic minimization problem is thus equivalent to solving the system of linear equations $A\vec{x} = \vec{b}$.

To reduce the problem size, we minimize $\mathcal{F}(\vec{x})$ over small subsets of $\mathbb{R}^M$, say sets of the form $\vec{x}_0 + \mathcal{S}_n$ with

$$\mathcal{S}_n = \text{span}(\vec{p}_0, \vec{p}_1, \ldots, \vec{p}_{n-1}) \tag{3.25}$$

and the vectors $\vec{p}_k$ arbitrary. This is called *subspace minimization*. The minimum of $\mathcal{F}(\vec{x})$ over $\vec{x}_0 + \mathcal{S}_n$ is a vector of the form

$$\vec{x}_n = \vec{x}_0 + \sum_{k=0}^{n-1} \xi_k \vec{p}_k, \tag{3.26}$$

and the corresponding residual vector $\vec{r}_n = \vec{b} - A\vec{x}_n$ is given by:

$$\vec{r}_n = \vec{r}_0 - \sum_{k=0}^{n-1} \xi_k A\vec{p}_k. \tag{3.27}$$

Our task is to compute the unknown coefficients ξ_k.

Lemma 6 *The vector $\vec{x}_n$ minimizes $\mathcal{F}(\vec{x})$ over $\vec{x}_0 + \mathcal{S}_n$ if and only if*

$$\forall k \in 0..n-1 : \quad \vec{p}_k^T \vec{r}_n = 0. \tag{3.28}$$

The vector $\vec{x}_n$ minimizes $\mathcal{F}(\vec{x})$ over $\vec{x}_0 + \mathcal{S}_n$ if and only if for all $\epsilon > 0$ and for all nonzero $\vec{v} \in \mathcal{S}_n$,

$$\mathcal{F}(\vec{x}_n) < \mathcal{F}(\vec{x}_n + \epsilon\vec{v}) = \mathcal{F}(\vec{x}_n) + \epsilon\vec{v}^T(A\vec{x}_n - \vec{b}) + \frac{1}{2}\epsilon^2\vec{v}^T A\vec{v}.$$

Because A is a positive-definite matrix, the term in ϵ^2 is strictly positive for all $\vec{v} \neq \vec{0}$. If the linear term in ϵ does not vanish, we can always choose the direction of $\vec{v}$ such that the linear term is negative. It is then possible to find some $\epsilon > 0$ for which the sum of the terms in ϵ is less than zero. For these values of ϵ, the inequality would not be satisfied.

Hence, the vector $\vec{x}_n$ is the minimum if and only if

$$\forall \vec{v} \in \mathcal{S}_n : \quad \vec{v}^T(\vec{b} - A\vec{x}_n) = \vec{v}^T\vec{r}_n = 0. \tag{3.29}$$

Equations 3.28 and 3.29 are equivalent, because the vectors $\vec{p}_k$ span the space $\mathcal{S}_n$. □

Equations 3.27 and 3.28 imply that the unknowns ξ_j satisfy the $n \times n$ linear system of equations

$$\forall k \in 0..n-1 : \quad \sum_{j=0}^{n-1} (\vec{p}_k^T A\vec{p}_j)\xi_j = \vec{p}_k^T \vec{r}_0. \tag{3.30}$$

To compute the vector $\vec{x}_n$, this system is solved, and the computed values of ξ_j are substituted in Equation 3.26. Of course, we must avoid singular and near-singular systems.

3.3.2   The Steepest-Descent Method

The linear system defined by Equation 3.30 is trivially solved if $n = 1$. This case is called a *line minimization*. The vector $\vec{x}_1$ is given by:

$$\vec{x}_1 = \vec{x}_0 + \frac{\vec{p}_0^T \vec{r}_0}{\vec{p}_0^T A \vec{p}_0} \vec{p}_0.$$

By converting this formula into an iteration, we reduce the minimization of $\mathcal{F}(\vec{x})$ over $\mathbb{R}^M$ to a sequence of line minimizations. Given a set of search directions $\vec{p}_k$ for $k \geq 0$ and a random vector $\vec{x}_0$, we obtain the recursion:

$$\forall k \geq 0 : \ \vec{x}_{k+1} = \vec{x}_k + \frac{\vec{p}_k^T \vec{r}_k}{\vec{p}_k^T A \vec{p}_k} \vec{p}_k,$$

which translates into a *successive-line-minimization* method defined by:

$$
\begin{aligned}
&\vec{r}_0 := \vec{b} - A\vec{x}_0 \ ; \\
&\langle \ ; \ k \ : \ 0 \leq k < K \ :: \\
&\qquad \alpha_k := \vec{p}_k^T \vec{r}_k / \vec{p}_k^T A \vec{p}_k \ ; \\
&\qquad \vec{x}_{k+1} := \vec{x}_k + \alpha_k \vec{p}_k \ ; \\
&\qquad \vec{r}_{k+1} := \vec{r}_k - \alpha_k A \vec{p}_k \\
&\rangle
\end{aligned}
$$

The *steepest-descent method* is derived from successive line minimization by choosing the search direction $\vec{p}_k = \vec{r}_k = -\nabla\mathcal{F}(\vec{x}_k)$. At $\vec{x} = \vec{x}_k$, this is the direction in which $\mathcal{F}(\vec{x})$ decreases the most, hence the name of the method. One steepest-descent-iteration step strictly reduces the functional value as long as $\vec{x}_k$ has not reached the exact minimum $\vec{x}^*$ of $\mathcal{F}(\vec{x})$, because

$$\mathcal{F}(\vec{x}_{k+1}) = \mathcal{F}(\vec{x}_k) - \frac{1}{2}\alpha_k^2 \vec{r}_k^T A \vec{r}_k < \mathcal{F}(\vec{x}_k)$$

if $\vec{p}_k = \vec{r}_k \neq \vec{0}$. This implies that the steepest-descent method is globally convergent: for any initial guess $\vec{x}_0$, the only fixed point of the steepest-descent iteration is the exact minimum $\vec{x}^*$ and the sequence of vectors $\vec{x}_k$ converges to this fixed point. This fact is offset, however, by computational experience: the steepest-descent method suffers from a decreasing convergence rate after a few iteration steps, because the search directions become nearly linearly dependent.

A simple induction argument shows that the search directions of the steepest-descent method satisfy:

$$\vec{p}_k \in \mathcal{K}_k(\vec{r}_0, A) = \text{span}(\vec{r}_0, A\vec{r}_0, \ldots, A^k \vec{r}_0). \tag{3.31}$$

The space $\mathcal{K}_k(\vec{r}_0, A)$ is called a *degree-k Krylov space of A at $\vec{r}_0$*.

3.3.3 Krylov-Subspace Minimization

Although of the same form as Equation 3.26, solutions computed by successive line minimization or steepest descent are not subspace minima. We now develop a subspace-minimization method for the quadratic functional $\mathcal{F}(\vec{x})$ using a subspace $\mathcal{S}_n$ and search directions $\vec{p}_k$ inspired by the steepest-descent method. Equation 3.31 leads us to choose the search directions $\vec{p}_k$ such that

$$\vec{p}_k = p_k(A)\vec{r}_0, \tag{3.32}$$

with $p_k(t)$ a polynomial of strict degree k. Let $\vec{x}_n$ be the minimum of $\mathcal{F}(\vec{x})$ over the set $\vec{x}_0 + \mathcal{S}_n$ with $\mathcal{S}_n$ defined by Equation 3.25. Our choice of search directions $\vec{p}_k$ implies that

$$\forall n \geq 0 : \ \mathcal{S}_n = \text{span}(\vec{p}_0, \vec{p}_1, \dots, \vec{p}_{n-1}) = \mathcal{K}_{n-1}(\vec{r}_0, A). \tag{3.33}$$

Equations 3.26, 3.27, and 3.33 imply that

$$\vec{x}_n \ \in \ \vec{x}_0 + \mathcal{K}_{n-1}(\vec{r}_0, A) \tag{3.34}$$
$$\vec{r}_n \ \in \ \mathcal{K}_n(\vec{r}_0, A). \tag{3.35}$$

Subspace minimization with search directions given by Equation 3.32 transforms the linear system of Equation 3.30 into:

$$\forall k \in 0..n-1 : \ \sum_{j=0}^{n-1} ((p_k(A)\vec{r}_0)^T A p_j(A)\vec{r}_0)\xi_j = \vec{p}_k^T \vec{r}_0. \tag{3.36}$$

Our goal is to find search directions $\vec{p}_k$ or, equivalently, polynomials $p_k(t)$ such that this system is nonsingular and easily solved.

Let $\mathcal{P}_n$ be the $(n+1)$-dimensional vector space of all polynomials of degree less than or equal to n. Provided A is a symmetric positive-definite matrix and $\vec{r}_0$ satisfies some technical restrictions, the expression

$$(p(t), q(t)) = (p(A)\vec{r}_0)^T A q(A)\vec{r}_0 \tag{3.37}$$

defines an inner product on $\mathcal{P}_n$. Moreover, the coefficient matrix of the system defined by Equation 3.36 is given by

$$G = [\gamma_{k,j}] = [(p_k(t), p_j(t))].$$

It follows that the matrix G is a diagonal matrix if the polynomials $p_k(t)$ are an orthogonal basis of $\mathcal{P}_n$ with respect to the inner product defined by Equation 3.37. (Two polynomials are orthogonal if their inner product vanishes.)

Let the set $\{p_0(t), p_1(t), \ldots, p_n(t)\}$ be an orthogonal basis of $\mathcal{P}_n$, and let $p_k(t)$ be a polynomial of strict degree k. A classical result from the theory of orthogonal polynomials states that the polynomials $p_k(t)$ are connected by a three-term recursion of the form:

$$\forall k \in 0..n-1: \ p_{k+1}(t) = \lambda_k(tp_k(t) - \mu_k p_k(t) - \nu_k p_{k-1}(t)). \tag{3.38}$$

This recursion is initialized by setting $p_{-1}(t)$ equal to zero and $p_0(t)$ equal to a nonzero constant. The coefficients μ_k and ν_k are given by:

$$\mu_k = (tp_k(t), p_k(t))/(p_k(t), p_k(t)) \tag{3.39}$$
$$\nu_k = (tp_k(t), p_{k-1}(t))/(p_{k-1}(t), p_{k-1}(t)), \tag{3.40}$$

and the coefficients λ_k are scaling factors.

The three-term recursion that connects successive orthogonal polynomials $p_k(t)$ implies a three-term recursion for the search directions $\vec{p}_k$. Substituting Equation 3.32 into Equations 3.38, 3.39, and 3.40, we find that

$$\vec{p}_{k+1} = \lambda_k(A\vec{p}_k - \mu_k \vec{p}_k - \nu_k \vec{p}_{k-1}) \tag{3.41}$$
$$\mu_k = (A\vec{p}_k)^T A\vec{p}_k / \vec{p}_k^T A\vec{p}_k \tag{3.42}$$
$$\nu_k = (A\vec{p}_k)^T A\vec{p}_{k-1} / \vec{p}_{k-1}^T A\vec{p}_{k-1}. \tag{3.43}$$

As before, λ_k is a scaling factor. The orthogonality of the polynomials $p_k(t)$ with respect to the inner product defined by Equation 3.37 implies that

$$\vec{p}_k^T A\vec{p}_j = 0 \ \text{ if } \ k \neq j. \tag{3.44}$$

The vectors $\vec{p}_k$ are said to be *A-conjugate* or *conjugate with respect to A*.

The A-conjugate search directions generated by the three-term recursion of Equation 3.41 lead to a diagonal coefficient matrix in the subspace minimization. The linear system defined by Equation 3.30 or, given our specific search directions, by Equation 3.36, therefore, reduces to

$$\forall k \in 0..n-1: \ \xi_k = \frac{\vec{p}_k^T \vec{r}_0}{\vec{p}_k^T A\vec{p}_k}. \tag{3.45}$$

Note that ξ_k is independent of n. This implies that $\vec{x}_{n+1}$ differs from $\vec{x}_n$ only by a component along $\vec{p}_n$ and can be obtained as a simple update of $\vec{x}_n$. More precisely, we have that:

$$\vec{x}_{n+1} = \vec{x}_n + \xi_n \vec{p}_n. \tag{3.46}$$

3.3.4   The Program

In Equations 3.41 through 3.46, we have all the ingredients to develop a program. The following iteration is obtained:

$\vec{r}_0 := \vec{b} - A\vec{x}_0$;
$\vec{p}_0 := \vec{r}_0$;
$\vec{p}_{-1} := \vec{0}$;
$\langle$; k : $0 \le k < K$ ::
$\quad \vec{w}_k := A\vec{p}_k$;
$\quad \xi_k := \vec{p}_k^T \vec{r}_0 / \vec{w}_k^T \vec{p}_k$;
$\quad \vec{x}_{k+1} := \vec{x}_k + \xi_k \vec{p}_k$;
$\quad \vec{r}_{k+1} := \vec{r}_k - \xi_k \vec{w}_k$;
$\quad \mu_k := \vec{w}_k^T \vec{w}_k / \vec{w}_k^T \vec{p}_k$;
$\quad \nu_k := \vec{w}_k^T \vec{w}_{k-1} / \vec{w}_{k-1}^T \vec{p}_{k-1}$;
$\quad \vec{p}_{k+1} := \vec{w}_k - \mu_k \vec{p}_k - \nu_k \vec{p}_{k-1}$;
$\quad$ scale $\vec{p}_{k+1}$ if needed
$\rangle$

Although the k-th iteration step finds the minimum of $\mathcal{F}(\vec{x})$ over $\vec{x}_0 + \mathcal{K}_k(\vec{r}_0, A)$, it does not require solving a $(k+1) \times (k+1)$ system of equations: the A-conjugate directions reduce the multidimensional subspace minimization to a sequence of line minimizations.

The following lemmas reduce the operation count of the above iteration.

Lemma 7 $\forall k \in 0..n-1: \vec{r}_k^T \vec{r}_n = 0$.

With $\mathcal{S}_n = \mathcal{K}_{n-1}(\vec{r}_0, A)$, Equation 3.29 implies that:

$$\forall \vec{v} \in \mathcal{K}_{n-1}(\vec{r}_0, A) : \vec{v}^T \vec{r}_n = 0. \tag{3.47}$$

From Equation 3.35, we have that

$$\forall k \in 0..n-1: \vec{r}_k \in \mathcal{K}_k(\vec{r}_0, A) \subset \mathcal{K}_{n-1}(\vec{r}_0, A).$$

This and Equation 3.47 imply the lemma. $\square$

Lemma 8 $\forall k \ge 0: \vec{p}_k^T \vec{r}_0 = \vec{p}_k^T \vec{r}_k$.

This follows immediately from Equations 3.27 and 3.44. $\square$

Lemma 9 $\forall k \ge 0: \vec{p}_{k+1} = \vec{r}_{k+1} + \beta_k \vec{p}_k$, with $\beta_k = -\dfrac{\vec{r}_{k+1}^T A \vec{p}_k}{\vec{p}_k^T A \vec{p}_k}$.

Substitute $\xi_k A\vec{p}_k = \vec{r}_k - \vec{r}_{k+1}$ into Equation 3.41 to obtain that

$$\vec{p}_{k+1} = \lambda_k ((\vec{r}_k - \vec{r}_{k+1})/\xi_k - \mu_k \vec{p}_k - \nu_k \vec{p}_{k-1}).$$

Setting the scaling factor $\lambda_k = -\xi_k$, Equations 3.33 and 3.35 imply that

$$\vec{p}_{k+1} = \vec{r}_{k+1} + \sum_{j=0}^{k} \beta_j \vec{p}_j.$$

Applying Equations 3.44 and 3.47, we find that $\beta_j = 0$ for $j \in 0..k-1$ and that β_k is given by the value in the statement of the lemma. $\square$

Lemma 10 $\forall k \geq 0 : \quad \vec{p}_k^T \vec{r}_k = \vec{r}_k^T \vec{r}_k.$

The case $k = 0$ is trivial, because $\vec{p}_0 = \vec{r}_0$. For $k > 0$, Lemma 9 implies that

$$\vec{p}_k^T \vec{r}_k = \vec{r}_k^T \vec{r}_k + \beta_{k-1} \vec{p}_{k-1}^T \vec{r}_k.$$

The second term of the right-hand side vanishes, however, because of Equation 3.47. This implies the lemma.   □

Lemma 11 $\forall k \geq 0 : \quad \beta_k = \vec{r}_{k+1}^T \vec{r}_{k+1} / \vec{r}_k^T \vec{r}_k.$

The result follows by substituting $\xi_k A \vec{p}_k = \vec{r}_k - \vec{r}_{k+1}$ into the numerator of the formula for β_k given in Lemma 9.   □

Program Conjugate-Gradient incorporates Lemmas 7 through 11 into the earlier iteration and reuses quantities computed in previous iteration steps.

> **program** Conjugate-Gradient
> **declare**
> $\quad k$: integer ;
> $\quad \xi, \beta, \pi_{rr}^0, \pi_{rr}^1, \pi_{pw}$: real ;
> $\quad \vec{p}, \vec{r}, \vec{w}, \vec{x}$: vector
> **assign**
> $\quad \vec{r} := \vec{b} - A\vec{x}$;
> $\quad \vec{p} := \vec{r}$;
> $\quad \pi_{rr}^0 := \vec{r}^T \vec{r}$;
> $\quad \langle \; ; \; k \; : \; 0 \leq k < K \; ::$
> $\qquad \vec{w} := A\vec{p}$;
> $\qquad \pi_{pw} := \vec{p}^T \vec{w}$;
> $\qquad \xi := \pi_{rr}^0 / \pi_{pw}$;
> $\qquad \vec{x} := \vec{x} + \xi\vec{p}$;
> $\qquad \vec{r} := \vec{r} - \xi\vec{w}$;
> $\qquad \pi_{rr}^1 := \vec{r}^T \vec{r}$;
> $\qquad \beta := \pi_{rr}^1 / \pi_{rr}^0$;
> $\qquad \vec{p} := \vec{r} + \beta\vec{p}$;
> $\qquad \pi_{rr}^0 := \pi_{rr}^1$
> $\quad \rangle$
> **end**

The scaling of the search direction $\vec{p}$ is omitted, because Lemma 9 fixed the scaling factor. One iteration step of program Conjugate-Gradient requires one matrix–vector product, two inner products, three vector sums, and one scalar-times-vector operation. The multicomputer implementation hinges on the representation of the matrix A, and the discussion of Section 3.1.2 applies here as well.

3.3.5   Preconditioning

Given an arbitrary initial guess $\vec{x}_0$, one line-minimization step in direction $A^{-1}\vec{r}_0$ finds the exact minimum $\vec{x}^*$ of $\mathcal{F}(\vec{x})$ over $\mathbb{R}^M$! Of course, if we could compute that direction, we could also find $\vec{x}^*$ by simply solving the system $A\vec{x} = \vec{b}$. Suppose, however, that there exists an easily inverted matrix C approximating A in a manner to be made precise later. In this case, a modified steepest-descent method performs the k-th line-minimization step not along the steepest-descent direction $\vec{r}_k$, but along $C^{-1}\vec{r}_k$. From this modified steepest-descent method, we now derive the *preconditioned conjugate-gradient method*.

Instead of Equations 3.31 and 3.32, the search directions now satisfy

$$\vec{p}_k = p_k(C^{-1}A)\vec{p}_0 \in \mathcal{K}_k(\vec{p}_0, C^{-1}A),$$

where $\vec{p}_0 = C^{-1}\vec{r}_0$ and $p_k(t)$ is a polynomial of strict degree k. The minimum $\vec{x}_n$ of $\mathcal{F}(\vec{x})$ over the set $\vec{x}_0 + \mathcal{K}_{n-1}(\vec{p}_0, C^{-1}A)$ has a residual vector

$$\vec{r}_n \in \mathcal{K}_n(\vec{r}_0, AC^{-1}).$$

The subspace minimization is reduced to solving the linear system:

$$\forall k \in 0..n-1 : \quad \sum_{j=0}^{n-1}((p_k(C^{-1}A)\vec{p}_0)^T Ap_j(C^{-1}A)\vec{p}_0)\xi_j = \vec{p}_k^T\vec{r}_0, \qquad (3.48)$$

which is analogous to Equations 3.30 and 3.36. The coefficient matrix of this system is reduced to a diagonal matrix by choosing the polynomials $p_k(t)$ orthogonal with respect to the inner product

$$(p(t), q(t)) = (p(C^{-1}A)\vec{p}_0)^T Aq(C^{-1}A)\vec{p}_0. \qquad (3.49)$$

(This expression defines an inner product on $\mathcal{P}_n$ provided A and C are symmetric positive-definite matrices and $\vec{p}_0$ satisfies some technical restrictions.) With the inner product known, the three-term recursion relation for the vectors $\vec{p}_k$ is established, and the following iteration is obtained.

$\vec{r}_0 := \vec{b} - A\vec{x}_0$;
$\vec{p}_0 := C^{-1}\vec{r}_0$;
$\vec{p}_{-1} := \vec{0}$;
$\langle\ ;\ k\ :\ 0 \le k < K\ ::$
   $\vec{w}_k := A\vec{p}_k$;
   $\xi_k := \vec{p}_k^T\vec{r}_0/\vec{w}_k^T\vec{p}_k$;
   $\vec{x}_{k+1} := \vec{x}_k + \xi_k\vec{p}_k$;
   $\vec{r}_{k+1} := \vec{r}_k - \xi_k\vec{w}_k$;
   $\mu_k := \vec{w}_k^T C^{-1}\vec{w}_k/\vec{w}_k^T\vec{p}_k$;
   $\nu_k := \vec{w}_k^T C^{-1}\vec{w}_{k-1}/\vec{w}_{k-1}^T\vec{p}_{k-1}$;
   $\vec{p}_{k+1} := C^{-1}\vec{w}_k - \mu_k\vec{p}_k - \nu_k\vec{p}_{k-1}$;
   scale $\vec{p}_{k+1}$ if needed
$\rangle$

The following lemmas, given without proof, are analogous to Lemmas 8 through 11. They simplify and reduce the operation count of the above iteration.

Lemma 12 $\forall k \geq 0 : \vec{p}_k^T \vec{r}_0 = \vec{p}_k^T \vec{r}_k$.

Lemma 13 $\forall k \geq 0 : \vec{p}_{k+1} = C^{-1}\vec{r}_{k+1} + \beta_k \vec{p}_k$, with $\beta_k = -\dfrac{(C^{-1}A\vec{p}_k)^T \vec{r}_{k+1}}{\vec{p}_k^T A \vec{p}_k}$.

Lemma 14 $\forall k \geq 0 : \vec{p}_k^T \vec{r}_k = (C^{-1}\vec{r}_k)^T \vec{r}_k$.

Lemma 15 $\forall k \geq 0 : \beta_k = (C^{-1}\vec{r}_{k+1})^T \vec{r}_{k+1} / (C^{-1}\vec{r}_k)^T \vec{r}_k$.

Program Preconditioned-Conjugate-Gradient is obtained after incorporating Lemmas 12 through 15 and reusing previously computed quantities.

```
program Preconditioned-Conjugate-Gradient
declare
    k : integer ;
    ξ, β, π⁰_rz, π¹_rz, π_pw : real ;
    p⃗, r⃗, w⃗, x⃗, z⃗ : vector
assign
    r⃗ := b⃗ - Ax⃗ ;
    z⃗ := C⁻¹r⃗ ;
    p⃗ := z⃗ ;
    π⁰_rz := r⃗ᵀz⃗ ;
    ⟨ ; k : 0 ≤ k < K ::
        w⃗ := Ap⃗ ;
        π_pw := p⃗ᵀw⃗ ;
        ξ := π⁰_rz/π_pw ;
        x⃗ := x⃗ + ξp⃗ ;
        r⃗ := r⃗ - ξw⃗ ;
        z⃗ := C⁻¹r⃗ ;
        π¹_rz := r⃗ᵀz⃗ ;
        β := π¹_rz/π⁰_rz ;
        p⃗ := z⃗ + βp⃗ ;
        π⁰_rz := π¹_rz
    ⟩
end
```

As in the method without preconditioning, the scaling of the search direction $\vec{p}$ is fixed by Lemma 13. This program requires, per iteration step, one matrix–vector product, three vector sums, one scalar-times-vector operation, two inner products, and a solution procedure for the system $C\vec{z} = \vec{r}$.

3.3.6 Termination and Convergence

The conjugate-gradient method in exact arithmetic, whether or not it is preconditioned, computes the exact solution $\vec{x}^*$ of the quadratic minimization problem in a finite number of steps. Finite termination is an interesting theoretical property. Unfortunately, the iteration step where the procedure terminates usually is prohibitively large. The success of the conjugate-gradient method is not due to finite termination.

Well before the exact solution is found, the iterates are good approximations to the exact solution. Practical implementations of the conjugate-gradient method estimate the accuracy of the computed solution and terminate the iteration when a stopping criterion is satisfied. Quantities that can be used in the convergence test include $\| \xi \vec{p} \|$ and $\| \vec{r} \|$.

The conjugate-gradient method is successful, because it is a good iterative method. Given the condition number of $C^{-1}A$,

$$\kappa = \| C^{-1}A \|_A \| (C^{-1}A)^{-1} \|_A,$$

it can be proven that

$$\| \vec{x}^* - \vec{x}_k \|_A \leq 2 \| \vec{x}^* - \vec{x}_0 \|_A \left[\frac{\sqrt{\kappa} - 1}{\sqrt{\kappa} + 1} \right]^k, \tag{3.50}$$

where $\| . \|_A$ is the norm consistent with the inner product $(\vec{u}, \vec{v}) = \vec{u}^T A \vec{v}$ for $\vec{u}, \vec{v} \in \mathbb{R}^M$. Although Equation 3.50 is a pessimistic upper bound, it confirms practical experience that $\kappa \approx 1$ is crucial for fast convergence. The condition number κ is a quantifiable measure of the earlier statement that C must be a good approximation to A.

The construction of a preconditioner is the main road block encountered when developing a fast solution procedure based on the conjugate-gradient method. Unfortunately, no black-box preconditioners exist: every preconditioner has a limited application range. For many applications, symmetric successive over-relaxation provides a good preconditioner. Note that most relaxation methods cannot be used as preconditioners, because they do not have a symmetric positive-definite iteration matrix. Finding a good preconditioner usually requires a separate analysis for each application.

In addition to the mathematical problem of finding a good preconditioner, there is also an algorithmic problem. Often, the preconditioner is a complicated algorithm and requires a specialized data distribution for its multicomputer implementation. The data distributions of the conjugate-gradient procedure, the matrix–vector product, and the preconditioner must all be compatible with one another. In addition, the data distribution(s) must achieve a reasonable global efficiency.

3.4   The Quasi-Minimal-Residual Method

For general matrices $A \in \mathbb{R}^{M \times M}$ (as opposed to symmetric positive-definite matrices), there is no longer a unique correspondence between solving the linear system $A\vec{x} = \vec{b}$ and minimizing quadratic functionals of the form given by Equation 3.24.

It is impossible to do justice to all proposed general Krylov-space methods. Instead, we shall explore some of the difficulties encountered with general systems by studying one new promising approach. This study will require familiarity with least-squares minimization and Givens rotations. For an introduction to these topics, see Chapter 5.

3.4.1   Krylov-Space Bases

As in the symmetric positive-definite case, we shall need basis vectors $\vec{v}_k$ of Krylov spaces $\mathcal{K}_{n-1}(\vec{v}_0, A)$. Given an inner product on $\mathbb{R}^M$, it is possible to construct an orthogonal basis for $\mathcal{K}_{n-1}(\vec{v}_0, A)$. For example, Gram–Schmidt orthogonalization of Section 5.1.2 leads to a recursion of the form

$$\vec{v}_{k+1} = A\vec{v}_k - \sum_{\ell=0}^{k} \alpha_\ell^{(k)} \vec{v}_\ell.$$

Such a recursion requires an unacceptable amount of memory, because all previously computed basis vectors must remain stored to compute $\vec{v}_{k+1}$.

To generate basis vectors connected by means of a three-term recursion relation, we give up on orthogonality. Assume that there exists a nonsingular matrix

$$V = [\vec{v}_0 \vec{v}_1 \dots \vec{v}_{M-1}], \tag{3.51}$$

such that

$$H = V^{-1}AV \tag{3.52}$$

is a tridiagonal matrix, whose elements are given by:

$$H = \begin{bmatrix} \alpha_0 & \beta_1 & 0 & \dots & & 0 \\ \gamma_0 & \alpha_1 & \beta_2 & & & \\ 0 & \gamma_1 & \alpha_2 & \ddots & & \\ \vdots & & \ddots & \ddots & & \beta_{M-1} \\ 0 & & & & \gamma_{M-2} & \alpha_{M-1} \end{bmatrix}. \tag{3.53}$$

We wish to compute V and H or, equivalently, the vectors $\vec{v}_k$ and the coefficients α_k, β_k, and γ_k. We shall also use the matrix

$$W = (V^{-1})^T = [\vec{w}_0 \vec{w}_1 \dots \vec{w}_{M-1}],$$

which exists because V is nonsingular. It follows that

$$W^T V \;=\; I \tag{3.54}$$
$$AV \;=\; VH \tag{3.55}$$
$$W^T A \;=\; HW^T. \tag{3.56}$$

These matrix equations reduce to sets of scalar and vector equations. Equation 3.57 is the element in row k and column ℓ of Equation 3.54, Equation 3.58 is column k of Equation 3.55, and Equation 3.59 is row k of Equation 3.56:

$$\vec{w}_k^T \vec{v}_\ell \;=\; \delta_{k,\ell} \tag{3.57}$$
$$A\vec{v}_k \;=\; \beta_k \vec{v}_{k-1} + \alpha_k \vec{v}_k + \gamma_k \vec{v}_{k+1} \tag{3.58}$$
$$\vec{w}_k^T A \;=\; \gamma_{k-1}\vec{w}_{k-1}^T + \alpha_k \vec{w}_k^T + \beta_{k+1}\vec{w}_{k+1}^T. \tag{3.59}$$

The symbol $\delta_{k,\ell}$ in Equation 3.57 is the Kronecker delta. Formulas for the coefficients α_k and β_k are obtained by applying Equation 3.57 to Equations 3.58 and 3.59 for appropriate values of the indices k and ℓ. The coefficients γ_k are scaling factors and are obtained somewhat arbitrarily by scaling the vectors $\vec{v}_{k+1}$ to unit length.

Given a vector $\vec{v}_0$ and setting $\vec{w}_0 = \vec{v}_0$, Equations 3.58 and 3.59 define a three-term recursion for each of the vector sequences $\vec{v}_k$ and $\vec{w}_k$. Because of Equation 3.57, the two sequences $\vec{v}_k$ and $\vec{w}_k$ are said to be *bi-orthogonal* with respect to one another. Bi-orthogonal sequences can be generated by means of the *nonsymmetric Lanczos procedure*, which is derived from Equations 3.57 through 3.59. Its details are given by the following program segment. As can be seen, an arbitrary starting vector $\vec{v}_0$ is required.

$\beta_0 := 0$;
$\gamma_{-1} := \| \vec{v}_0 \|$;
$\vec{w}_0 := \vec{v}_0 / \gamma_{-1}$;
$\langle \; ; \; k \; : \; 0 \le k < M - 1 \; :: $
$\qquad \vec{v}_k := \vec{v}_k / \gamma_{k-1}$;
$\qquad \alpha_k := \vec{w}_k^T A\vec{v}_k$;
$\qquad \vec{v}_{k+1} := A\vec{v}_k - \alpha_k \vec{v}_k - \beta_k \vec{v}_{k-1}$;
$\qquad \gamma_k := \| \vec{v}_{k+1} \|$;
$\qquad \beta_{k+1} := \vec{w}_k^T A\vec{v}_{k+1} / \gamma_k$;
$\qquad \vec{w}_{k+1} := (A^T \vec{w}_k - \alpha_k \vec{w}_k - \gamma_{k-1}\vec{w}_{k-1}) / \beta_{k+1}$
$\rangle$

The nonsymmetric Lanczos procedure breaks down if the assumption that the matrix V exists and is nonsingular does not hold. The symptom of break down is that $\gamma_k = 0$ or $\beta_{k+1} = 0$ for some value of k. In floating-point arithmetic, exact break down rarely occurs. However, near break down, $\gamma_k \approx 0$ or $\beta_{k+1} \approx 0$, can lead to numerical instability. The *look-ahead Lanczos procedure* avoids most occurrences of break down and near break down by constructing a matrix V such that $V^{-1}AV$ is a block-tridiagonal matrix. We shall ignore break down and pretend that the nonsymmetric Lanczos procedure always generates basis vectors $\vec{v}_k$ connected by a three-term recursion.

3.4.2 Quasi-Minimization in a Subspace

Let $\vec{x}_0$ be an initial guess for the solution of the system $A\vec{x} = \vec{b}$. Let $\vec{r}_0 = \vec{b} - A\vec{x}_0$, $\epsilon_0 = \|\vec{r}_0\|$, and $\vec{v}_0 = \vec{r}_0/\epsilon_0$. The matrix V and its column vectors $\vec{v}_k = V\vec{e}_k$ are defined by the nonsymmetric Lanczos procedure.

As in the derivation of the conjugate-gradient method, the problem in $\mathbb{R}^M$ is approximated by a problem in $\vec{x}_0 + \mathcal{K}_{n-1}(\vec{v}_0, A)$, where $1 \leq n \leq M$. Because the vectors $\vec{v}_0, \vec{v}_1, \ldots, \vec{v}_{n-1}$ form a basis of $\mathcal{K}_{n-1}(\vec{v}_0, A)$, all vectors $\vec{x} \in \vec{x}_0 + \mathcal{K}_{n-1}(\vec{v}_0, A)$ can be decomposed in the form

$$\vec{x} = \vec{x}_0 + VI_{M,n}\vec{z}.$$

The matrix $I_{M,n}$ is the $M \times n$ identity matrix, which consists of the first n columns of the $M \times M$ identity matrix. The vector $\vec{z} \in \mathbb{R}^n$. The residual vector corresponding to $\vec{x}$ is given by

$$\vec{r} = V(\epsilon_0\vec{e}_0 - H_n\vec{z}) = V\vec{s}.$$

The matrix $H_n = HI_{M,n}$ is the $M \times n$ matrix defined by the first n columns of the matrix H (see Equations 3.52, 3.53, and 3.55).

We now construct a particular vector $\vec{x}_n \in \vec{x}_0 + \mathcal{K}_{n-1}(\vec{v}_0, A)$ such that the norm of the vector $\vec{s}_n$ associated with $\vec{x}_n$ is minimal. There is no a priori reason to construct $\vec{x}_n$ in this fashion. For example, we could compute $\vec{x}_n$ such that the associated residual vector $\vec{r}_n = V\vec{s}_n$ has minimal norm. This would lead to a *minimal-residual method*. However, the route followed here avoids some problems associated with minimal-residual methods.

The least-squares minimization problem

$$\min_{\vec{z} \in \mathbb{R}^n} \| \epsilon_0\vec{e}_0 - H_n\vec{z} \|$$

solves the overdetermined system

$$H_n\vec{z} = \epsilon_0\vec{e}_0.$$

Because H_n is an $M \times n$ tridiagonal matrix, the Givens QR-decomposition of Section 5.3 is a particularly appropriate solution procedure. The Givens QR-decomposition applied to H_n computes the $M \times M$ orthogonal matrix Q_n and the $M \times n$ upper-triangular matrix R_n such that

$$H_n = Q_nR_n.$$

The matrix $Q_n^T = Q_n^{-1}$ is the product of n Givens rotations:

$$Q_n^T = G_{n-1}G_{n-2}\ldots G_0. \tag{3.60}$$

When Givens rotation G_k is applied to the matrix $G_{k-1}\ldots G_0 H_n$, it combines rows k and $k+1$ of this matrix to zero out the element in row $k+1$ and column k, the element that previously contained the value γ_k; see Equation 3.53. (In the notation of Section 5.3.1, the Givens rotation G_k would be denoted by $G_{k,k+1}$.) The properties of Givens rotations imply that the upper-triangular matrix R_n has at most three non-zero diagonals:

$$R_n = \begin{bmatrix} \rho_0 & \sigma_1 & \tau_2 & & 0 & 0 \\ 0 & \rho_1 & \sigma_2 & \ddots & & \vdots \\ 0 & 0 & \rho_2 & \ddots & \tau_{n-2} & 0 \\ \vdots & \vdots & & \ddots & \sigma_{n-2} & \tau_{n-1} \\ 0 & 0 & & & \rho_{n-2} & \sigma_{n-1} \\ 0 & 0 & \ldots & & 0 & \rho_{n-1} \\ 0 & 0 & \ldots & & 0 & 0 \\ \vdots & \vdots & & & \vdots & \vdots \\ 0 & 0 & \ldots & & 0 & 0 \end{bmatrix}. \tag{3.61}$$

Given the QR-decomposition of H_n, the optimal vector $\vec{z}_n \in \mathbb{R}^n$ is the solution of the $n \times n$ linear system defined by the first n rows of

$$R_n\vec{z} = \epsilon_0 Q_n^T \vec{e}_0.$$

Equation 3.60 and the definition of the Givens rotations imply that components $n+1$ through $M-1$ of $Q_n^T \vec{e}_0$ vanish. We also know that rows n through $M-1$ of R_n are equal to zero. Components $n+1$ through $M-1$ of the residual of the overdetermined system must, therefore, vanish for any vector $\vec{z} \in \mathbb{R}^n$. For the optimal vector $\vec{z}_n$, components 0 through $n-1$ of the residual also vanish. This residual must be proportional to $\vec{e}_n$, and the vector $\vec{s}_n$ associated with $\vec{z}_n$ must satisfy

$$Q_n^T \vec{s}_n = G_{n-1}G_{n-2}\ldots G_0\vec{s}_n = \epsilon_n\vec{e}_n. \tag{3.62}$$

3.4.3 Updating

To obtain a viable procedure, the vector $\vec{x}_{n+1}$ must be computed as an update of $\vec{x}_n$. Because $\vec{x}_n \in \vec{x}_0 + \mathcal{K}_{n-1}(\vec{x}_0, A)$, $\vec{x}_{n+1} \in \vec{x}_0 + \mathcal{K}_n(\vec{x}_0, A)$, and because the vectors $\vec{v}_0, \ldots, \vec{v}_n$ form a basis of $\mathcal{K}_n(\vec{x}_0, A)$, there exists a vector $\vec{y}_{n+1} \in \mathbb{R}^{n+1}$ such that

$$\vec{x}_{n+1} = \vec{x}_n + VI_{M,n+1}\vec{y}_{n+1}. \tag{3.63}$$

With $H_{n+1} = HI_{M,n+1}$, it is easily verified that

$$\vec{s}_{n+1} = \vec{s}_n - H_{n+1}\vec{y}_{n+1}. \tag{3.64}$$

Equation 3.53 and the definitions of H_n, H_{n+1}, R_n, and R_{n+1} imply that

$$
\begin{aligned}
R_{n+1} &= G_n \ldots G_0 H_{n+1} \\
&= G_n \ldots G_0 \left[H_n \mid \beta_n \vec{e}_{n-1} + \alpha_n \vec{e}_n + \gamma_n \vec{e}_{n+1} \right] \\
&= \left[G_n R_n \mid G_n \ldots G_0 (\beta_n \vec{e}_{n-1} + \alpha_n \vec{e}_n + \gamma_n \vec{e}_{n+1}) \right].
\end{aligned}
$$

Because rows n and $n+1$ of R_n vanish and G_n acts only on these two rows, we have that $G_n R_n = R_n$. Similarly, Givens rotations G_{k-2} through G_0 have no effect on $\vec{e}_k$, because they act only on its zero components. These observations further simplify R_{n+1} to:

$$R_{n+1} = \left[R_n \mid G_n(G_{n-1}(\beta_n G_{n-2}\vec{e}_{n-1} + \alpha_n \vec{e}_n) + \gamma_n \vec{e}_{n+1}) \right]. \tag{3.65}$$

The matrix R_{n+1} is thus obtained by attaching an extra column to R_n. This extra column is defined by the coefficients ρ_n, σ_n, and τ_n (apply Equation 3.61 for $n + 1$). Formulas for these coefficients are obtained by multiplying through G_{n-2}, G_{n-1}, and G_n in Equation 3.65. While G_{n-2} and G_{n-1} are known from previous steps, Givens rotation G_n is computed to zero the matrix element containing γ_n. A straightforward calculation leads to the following formulas:

$$
\begin{aligned}
\tau_n &= \beta_n d_{n-2} & (3.66) \\
\sigma_n &= \beta_n c_{n-2} c_{n-1} + \alpha_n d_{n-1} & (3.67) \\
\rho_n &= \sqrt{\gamma_n^2 + (-\beta_n c_{n-2} d_{n-1} + \alpha_n c_{n-1})^2} & (3.68) \\
c_n &= (-\beta_n c_{n-2} d_{n-1} + \alpha_n c_{n-1})/\rho_n & (3.69) \\
d_n &= \gamma_n/\rho_n. & (3.70)
\end{aligned}
$$

The coefficients c_n and d_n define the Givens rotation G_n. (In the notation of Section 5.3.1, G_n would be denoted by $G_{n,n+1}$, c_n by c, and d_n by s.)

Because of Equation 3.64, the vector $\vec{y}_{n+1} \in \mathbb{R}^{n+1}$ that leads to a vector $\vec{s}_{n+1}$ with minimal norm is the solution of the least-squares minimization problem given by:

$$\min_{\vec{y} \in \mathbb{R}^{n+1}} \| \vec{s}_n - H_{n+1}\vec{y} \|.$$

The QR-decomposition of H_{n+1} implies that $\vec{y}_{n+1}$ is the solution of the $(n + 1) \times (n + 1)$ system defined by the first $n + 1$ rows of

$$R_{n+1}\vec{y} = Q_{n+1}^T \vec{s}_n. \tag{3.71}$$

Applying Equation 3.62 to the right-hand side, we obtain the overdetermined system

$$R_{n+1}\vec{y} = \epsilon_n G_n \vec{e}_n = \epsilon_n(c_n \vec{e}_n - d_n \vec{e}_{n+1}).$$

Because rows $n + 1$ through $M - 1$ of R_{n+1} vanish, it is impossible to satisfy equation number $n+1$ of this overdetermined system. On the other hand, rows 0 through n of R_{n+1} form a nonsingular matrix, and equations 0 through n of this system can be satisfied. It follows that the optimal vector $\vec{y}_{n+1}$ satisfies

$$R_{n+1}\vec{y}_{n+1} = c_n \epsilon_n \vec{e}_n \qquad (3.72)$$

exactly, and the residual vector of Equation 3.71 corresponding to $\vec{y}_{n+1}$ is

$$Q_{n+1}^T \vec{s}_{n+1} = -d_n \epsilon_n \vec{e}_{n+1}.$$

Comparison with Equation 3.62 yields that

$$\epsilon_{n+1} = -d_n \epsilon_n. \qquad (3.73)$$

(Note that $|\epsilon_{n+1}| \leq |\epsilon_n|$, because $|d_n| \leq 1$.)

We now construct an update formula for the solution of the system defined by Equation 3.72. The matrix R_n consists of the first n columns of R, where R is the triangular factor of the QR-decomposition of H. We may assume that R is nonsingular because of our earlier assumption that H is nonsingular. Because

$$\rho_n \vec{e}_n + \sigma_n \vec{e}_{n-1} + \tau_n \vec{e}_{n-2}$$

is column vector number n of R, we have that

$$R^{-1}(\rho_n \vec{e}_n + \sigma_n \vec{e}_{n-1} + \tau_n \vec{e}_{n-2}) = \vec{e}_n.$$

This immediately implies that the column vectors of R^{-1}, which are given by $\vec{t}_n = R^{-1}\vec{e}_n$, satisfy the three-term recursion relation

$$\rho_n \vec{t}_n + \sigma_n \vec{t}_{n-1} + \tau_n \vec{t}_{n-2} = \vec{e}_n.$$

This recursion is initialized by setting $\vec{t}_{-2} = \vec{t}_{-1} = \vec{0}$. Because R^{-1} is an upper-triangular matrix, components $n + 1$ through $M - 1$ of $\vec{t}_n$ vanish, and

$$R\vec{t}_n = [R_{n+1} \mid 0]\, \vec{t}_n = \vec{e}_n.$$

Together with Equation 3.72, this implies that

$$I_{M,n+1}\vec{y}_{n+1} = \begin{bmatrix} \vec{y}_{n+1} \\ \vec{0} \end{bmatrix} = c_n \epsilon_n \vec{t}_n$$

and, with Equation 3.63, that

$$\vec{x}_{n+1} = \vec{x}_n + c_n \epsilon_n V \vec{t}_n \qquad (3.74)$$
$$\vec{r}_{n+1} = \vec{r}_n - c_n \epsilon_n A V \vec{t}_n. \qquad (3.75)$$

The recursion for the vectors $\vec{t}_n$ also implies recursions for the vectors $\vec{p}_n = V\vec{t}_n$ and $\vec{q}_n = AV\vec{t}_n$:

$$\vec{p}_n = (\vec{v}_n - \sigma_n \vec{p}_{n-1} + \tau_n \vec{p}_{n-2})/\rho_n \qquad (3.76)$$
$$\vec{q}_n = (A\vec{v}_n - \sigma_n \vec{q}_{n-1} + \tau_n \vec{q}_{n-2})/\rho_n, \qquad (3.77)$$

which are initialized by $\vec{p}_{-1} = \vec{q}_{-1} = \vec{p}_0 = \vec{q}_0 = \vec{0}$.

3.4.4   The Program

Equations 3.66 through 3.77 are assembled into the quasi-minimal-residual iteration. The nonsymmetric Lanczos procedure of Section 3.4.1 generates the coefficients α_k and β_k. The coefficients γ_k are found by scaling the vectors $\vec{v}_{k+1}$ to unit length. The computation of the basis vectors breaks down if β_k or γ_k vanish. It is then necessary to restart the iteration. The recursion relation of Equation 3.77 reduces the number of matrix–vector products per iteration step to two. To obtain a program, quantities computed in previous iterations must be reused. This is done in program Quasi-Minimal-Residual.

program Quasi-Minimal-Residual
declare
  k : integer ;
  $\vec{x}, \vec{b}, \vec{r}, \vec{t}, \vec{u}$: vector ;
  $\vec{v}_0, \vec{w}_0, \vec{p}_0, \vec{q}_0$: vector ;
  $\vec{v}_1, \vec{w}_1, \vec{p}_1, \vec{q}_1$: vector ;
  $\alpha, \beta_0, \beta_1, \gamma_0, \gamma_1, \epsilon, \rho, \sigma, \tau, \theta, c_0, c_1, d_0, d_1$: real
assign
  $\vec{r} := \vec{b} - A\vec{x}$;
  $\epsilon := \| \vec{r} \|$;
  $\vec{v}_1, \vec{w}_1 := \vec{r}, \vec{r}/\epsilon$;
  $\gamma_0, \beta_0 := \epsilon, 0$;
  $\vec{p}_0, \vec{q}_0, \vec{v}_0, \vec{w}_0, \vec{p}_1, \vec{q}_1 := \vec{0}, \vec{0}, \vec{0}, \vec{0}, \vec{0}, \vec{0}$;
  $c_0, c_1, d_0, d_1 := 1, 1, 0, 0$;
  $\langle$; k : $0 \leq k < K$ ::
      $\{$ Compute new $\vec{v}_0$ and $\vec{w}_0$.$\}$
      $\vec{v}_1 := \vec{v}_1/\gamma_0$;
      $\vec{t} := A\vec{v}_1$;
      $\alpha := \vec{w}_1^T \vec{t}$;
      $\vec{v}_0 := \vec{t} - \alpha\vec{v}_1 - \beta_0\vec{v}_0$;
      $\gamma_1 := \| \vec{v}_0 \|$;
      if $\gamma_1 = 0$ **then quit** ;
      $\vec{u} := A^T \vec{w}_1$;
      $\beta_1 := \vec{u}^T \vec{v}_0/\gamma_1$;
      if $\beta_1 = 0$ **then quit** ;
      $\vec{w}_0 := (\vec{u} - \alpha\vec{w}_1 - \gamma_0\vec{w}_0)/\beta_1$;
      $\{$ Compute ρ, σ, and τ.$\}$
      $\tau := \beta_0 d_0$;
      $\sigma := \beta_0 c_0 c_1 + \alpha d_1$;
      $\theta := -\beta_0 c_0 d_1 + \alpha c_1$;
      $\rho := \sqrt{\theta^2 + \gamma_1^2}$;
      $c_1, c_0 := c_0, \theta/\rho$;
      $d_1, d_0 := d_0, \gamma_1/\rho$;

{ Compute new $\vec{p}_0$, $\vec{q}_0$, $\vec{x}$, $\vec{r}$.}
$\vec{p}_0 := (\vec{v}_1 - \sigma \vec{p}_1 - \tau \vec{p}_0)/\rho$;
$\vec{q}_0 := (\vec{t} - \sigma \vec{q}_1 - \tau \vec{q}_0)/\rho$;
$\vec{x} := \vec{x} + c_0 \epsilon \vec{p}_0$;
$\vec{r} := \vec{r} - c_0 \epsilon \vec{q}_0$;
$\epsilon := -d_0 \epsilon$;
{ Switch values.}
$\vec{v}_0, \vec{v}_1 := \vec{v}_1, \vec{v}_0$;
$\vec{w}_0, \vec{w}_1 := \vec{w}_1, \vec{w}_0$;
$\vec{p}_0, \vec{p}_1 := \vec{p}_1, \vec{p}_0$;
$\vec{q}_0, \vec{q}_1 := \vec{q}_1, \vec{q}_0$;
$\beta_0, \gamma_0 := \beta_1, \gamma_1$

$\rangle$

end

Although rarely applied to problems with full matrices, this program is an example of the situation mentioned in Section 2.1.3. If optimal efficiency of each individual operation were the sole criterion, each of the assignments

$\vec{t} := A \vec{v}_1$;
$\vec{u} := A^T \vec{w}_1$;

would require a different data distribution for the matrix A. Because both assignments occur within the same program, a compromise distribution achieves the best global performance.

Solving general systems of equations carries a significant memory expense over solving symmetric positive-definite systems. It suffices to compare program Quasi-Minimal-Residual, which needs 13 vectors, and program Preconditioned-Conjugate-Gradient, which needs 5 vectors.

As for the conjugate-gradient method, the fixed number of iteration steps must be replaced by a stopping criterion. This may be based on the value ϵ_{k+1}, on the norm of the residual vector $\vec{r}_{k+1}$, on the norm of the correction $c_k \epsilon_k \vec{p}_{k+1}$, or on a combination of these factors.

3.5   Notes and References

The inclusion of the complex case in the power method follows the development of Franklin [32]. The power method can be generalized in several ways. The *inverse-power method* uses the iteration for A^{-1} to compute the smallest eigenvalues of A; it requires solving a system of the form $A \vec{x} = \vec{b}$ in every iteration step. The *shifted-inverse-power method* computes eigenvalues near a given value λ by applying the inverse-power method to $(A - \lambda I)$. To compute several eigenvalues, one can use simultaneous iteration: the power method is applied to a set of vectors, which are kept linearly independent by orthogonalization. The QR-algorithm uses this idea to compute all eigenvalues and eigenvectors. The classic reference for numerical methods to compute eigenvalues and eigenvectors is Wilkinson [85]. For a more recent introduction to the subject, see Golub and Van Loan [40]. Sequential software for the algebraic eigenvalue problem is available through LAPACK [21]. It is a challenge to obtain efficient multicomputer implementations of existing robust eigensolvers, because they involve data-distribution issues of matrix–vector multiplication, matrix–matrix multiplication, and LU-decomposition and load-balancing issues similar to those of particle methods. Although solutions exist to address each of these issues, the fact that they occur in combination complicates matters considerably. Not surprisingly, research in numerical methods for eigenvalue problems that are better suited for concurrent computing remains very active. Modi [66] and Shroff [75] use the Jacobi method to develop a concurrent method; Kim and Chronopoulos [57] use the Arnoldi iteration.

The classic reference on relaxation methods is Young [88], but most basic numerical-analysis texts cover relaxation methods in some detail. The implementation aspects of concurrent-relaxation methods are best postponed until Chapter 8, where grid-oriented formulations of relaxation methods are developed. The matrix formulation is used mainly for analysis purposes. O'Leary and White [67] introduced *multisplittings*; some of these lead to relaxation methods with desirable properties for concurrent implementation. Papatheodorou and Saridakis [70] investigate the concurrent implementation of relaxation methods based on multisplitting. Chazan and Miranker [16] and Baudet [6] examine asynchronous relaxation methods to reduce communication overhead by using available values instead of waiting for new values to arrive.

The conjugate-gradient method was originally proposed by Hestenes and Stiefel [49]. For a history of the subsequent development of the method, see Golub and O'Leary [39]. The theory of orthogonal polynomials, which we used in our derivation of the conjugate-gradient method, is found in Szegö [78]. Implementations of the conjugate-gradient method for particular applications usually exploit special-purpose matrix–vector procedures that use the structure of the underlying problem; see Section 8.6 and Chapter 10. For a survey of preconditioned iterative methods, see Axelsson [4].

The quasi-minimal-residual method was proposed by Freund and Nachtigal [36] and Freund, Gutknecht, and Nachtigal [35]. The procedure used to compute basis vectors is due to Lanczos [59], and it is a crucial technique of many iterative methods. The quasi-minimal-residual method of Section 3.4 is a simplification of the original. To avoid break down, the method of Freund and Nachtigal is based on a block-Lanczos procedure, known as the look-ahead Lanczos procedure. The nonsymmetric Lanczos procedure, the basis of our simplified quasi-minimal-residual method, can break down. When break down occurs, the computed solution vector remains a valid estimate, but it is impossible to compute the next basis vector via the

Lanczos procedure. A simple remedy is to restart the iteration from the current solution estimate.

The transpose-matrix–vector product is often the cause of some implementation difficulties. Freund [34] suggested a transpose-free version of the quasi-minimal-residual method to avoid the problem.

As for the conjugate-gradient method, one can include preconditioners in the quasi-minimal-residual method. Because the basic algorithm is applicable to general coefficient matrices, preconditioners for the quasi-minimal-residual method need not be symmetric positive-definite matrices. In particular, all relaxation methods may be used.

The quasi-minimal-residual method is only one of a long list of generalizations of the conjugate-gradient method. Although it is not clear which will emerge as the best method(s), the importance of the Krylov-space methodology to concurrent scientific computing cannot be disputed. Besides being almost ideally suited for concurrent implementation, Krylov-space methods with proper preconditioning have a remarkable convergence rate.

Exercises

Exercise 11 *Implement the power method, the conjugate-gradient method, or the quasi-minimal-residual method on your favorite multicomputer.*

Exercise 12 *Prove that the Jacobi relaxation converges for systems with strictly diagonally dominant coefficient matrices.*

Exercise 13 *The monomials $1, t, t^2, \ldots,$ are an obvious basis of $\mathcal{P}_n$. Use Gram–Schmidt orthogonalization of Section 5.1.2 and Exercise 4 of Chapter 1 to find the first few orthogonal polynomials with respect to the inner product:*

$$(p(t), q(t)) = \int_0^1 p(t)q(t)dt.$$

Exercise 14 *Verify that the three-term recursion relation given by Equations 3.38 through 3.40 indeed generates a sequence of orthogonal polynomials. Apply the recursion to $\mathcal{P}_n$ with the same inner product as in Exercise 13.*

Exercise 15 *Prove that the search directions occurring in the steepest-descent method satisfy Equation 3.31. Prove that there exists a polynomial $p_k(t)$ of degree less than or equal to k such that $\vec{p}_k = p_k(A)\vec{r}_0$.*

Exercise 16 *Prove that Equation 3.37 defines a valid inner product on $\mathcal{P}_n$, provided that $A \in \mathbb{R}^{M \times M}$ is a symmetric positive-definite matrix and that the vector $\vec{r}_0$ has a nonvanishing component along every eigenvector of A.*

Exercise 17 *Show that the conjugate-gradient method finds the exact solution in one iteration step if $\vec{r}_0$ happens to be an eigenvector of A.*

Exercise 18 *Prove that the inner product defined in Equation 3.49 is indeed an inner product on $\mathcal{P}_n$, provided both A and C are symmetric positive-definite matrices and $\vec{p}_0$ has a nonzero component along every eigenvector of $C^{-1/2}AC^{-1/2}$. Note: $C^{-1/2}$ exists, because C is a symmetric positive-definite matrix.*

Exercise 19 *Prove Lemmas 12 through 15.*

4

LU-Decomposition

Iterative linear-system solvers can be fast and are usually easy to implement. However, they have a limited application range, as they impose requirements on the coefficient matrix ranging from positive definiteness to special sparsity structure. Direct solvers, on the other hand, are almost always applicable. Although their operation count and execution time can be large, both are very predictable given the problem size. In Chapters 4 and 5, we shall develop two different direct solvers for general linear systems of equations.

In Sections 4.1 and 4.2, the coefficient matrix will be factored into the product of two permuted-triangular matrices. This factorization is known as the LU-decomposition of a matrix. After LU-decomposition of the coefficient matrix, the solution of the system can be computed by solving two linear systems with permuted-triangular coefficient matrices. Triangular solvers will be developed in Section 4.3. Throughout, it will be assumed that the reader is already familiar with the basic theory of LU-decomposition and with the sequential programs implementing this procedure.

4.1 LU-Decomposition Without Pivoting

4.1.1 Specification

In most cases, a matrix can be factored into the product of a unit-lower-triangular and an upper-triangular matrix. More precisely, the following theorem can be established.

Theorem 2 *Let A_k denote the leading $k \times k$ principal submatrix of $A \in \mathbb{R}^{M \times N}$. If A_k is nonsingular for $k \in 1..\min(M - 1, N)$, then there exists a unit-lower-triangular matrix $L \in \mathbb{R}^{M \times M}$ and an upper-triangular matrix $U \in \mathbb{R}^{M \times N}$ such that $A = LU$.*

The (omitted) proof of this theorem establishes the correctness of program LU-1. As is customary, elements of the unit-lower-triangular matrix L are stored in the lower-triangular part of A, and elements of the upper-triangular matrix U are stored in the upper-triangular part of A.

> **program** LU-1
> **declare**
> k, m, n : integer ;
> $\mathcal{M}$, $\mathcal{N}$: set of integer ;
> a : array $[0..M - 1 \times 0..N - 1]$ of real ;
> a_c : array $[0..M - 1]$ of real ;
> a_r : array $[0..N - 1]$ of real ;
> a_{rc} : real
> **initially**
> $\mathcal{M}$, $\mathcal{N} = 0..M - 1$, $0..N - 1$;
> $\langle\ ;\ m, n\ :\ (m, n) \in \mathcal{M} \times \mathcal{N}\ ::\ a[m, n] = \tilde{a}_{m,n}\ \rangle$
> **assign**
> $\langle\ ;\ k\ :\ 0 \leq k < \min(M, N)\ ::$
> $a_{rc} := a[k, k]$;
>   **if** $a_{rc} = 0.0$ **then quit** ;
> $\mathcal{M} := \mathcal{M} \setminus \{k\}$;
> $\mathcal{N} := \mathcal{N} \setminus \{k\}$;
>
>   {Copy the Pivot Row}
> $\langle\ \|\ n\ :\ n \in \mathcal{N}\ ::\ a_r[n] := a[k, n]\ \rangle$;
>
>   {Compute the Multiplier Column}
> $\langle\ \|\ m\ :\ m \in \mathcal{M}\ ::\ a_c[m] := a[m, k] := a[m, k]/a_{rc}\ \rangle$;
>
>   {Elimination}
> $\langle\ \|\ m, n\ :\ (m, n) \in \mathcal{M} \times \mathcal{N}\ ::$
> $a[m, n] := a[m, n] - a_c[m]a_r[n]$
> $\rangle$
> $\rangle$
> **end**

In step k, a multiple of row k is subtracted from rows $k + 1$ through $M - 1$. The multiplier used when subtracting row k from row m is stored in entry $a[m, k]$. Row k is called *the pivot row*, column k is called *the multiplier column*, and entry $a[k, k]$ is called the *pivot* of step k.

The variables $\mathcal{M}$ and $\mathcal{N}$ are declared as sets of integers. Conventional set operators are defined on set variables. The statements $\mathcal{M} := \mathcal{M} \setminus \{k\}$ and $\mathcal{N} := \mathcal{N} \setminus \{k\}$ deactivate the pivot row and the multiplier column, respectively. Once a row or column has been deactivated, its entries are left unchanged by the remaining steps of the program.

Some apparently superfluous work is done in program LU-1: the pivot is copied into the temporary variable a_{rc}, the pivot row into the array a_r, and the multiplier column into the array a_c. These copies will simplify the development of the multicomputer program.

4.1.2 Implementation

To transform program LU-1 into a multicomputer program, we apply the duplication and selection technique of Section 1.4 with two-dimensional data distributions like those of Section 2.1. The array a_r, which contains a copy of the pivot row, is transformed into a column-distributed vector, and the array a_c, which contains a copy of the multiplier column, is transformed into a row-distributed vector. The scalar variable a_{rc} remains duplicated in every process.

Given the one-dimensional data distributions μ and ν, the pivot of step k is mapped to process $(\hat{p}, \hat{q})$, where $\mu(k) = (\hat{p}, \hat{\imath})$ and $\nu(k) = (\hat{q}, \hat{\jmath})$. Because a_{rc} is duplicated in every process, the assignment $a_{rc} := a[k, k]$ of program LU-1 is transformed into a broadcast from process $(\hat{p}, \hat{q})$ to all other processes. Similarly, the quantification that copies the pivot row into array a_r is also transformed into a broadcast. However, because a_r is a column-distributed vector, it suffices to broadcast the vector from one process row to all other process rows. After this partial broadcast, the vector is duplicated in every process row. Analogously, the quantification that copies the multiplier column into array a_c is transformed into a broadcast from one process column to all other process columns.

It is also necessary to distribute the index sets $\mathcal{M}$ and $\mathcal{N}$. As discussed in Section 1.3, the row distribution μ defines a partition of $0..M - 1$ that consists of P subsets $\mathcal{M}_p$. After duplication and selection, the index-set variable $\mathcal{M}$ of process (p, q) is initialized to $\mathcal{M}_p$. It follows that the pivot row is only deactivated in the process row containing row k of the coefficient matrix. The index set $\mathcal{N}$ is similarly distributed over the process columns. Although the index-set variables remain duplicated in every process, the maximum number of values in the index sets is decreased; this reduces the memory requirements to represent the index sets.

When converting global into local indices, global-index sets $\mathcal{M}$ and $\mathcal{N}$ are replaced by local-index sets $\mathcal{I}$ and $\mathcal{J}$, respectively. Initially, all rows are active, and $\mathcal{I}$ is initialized to local-index set $\mathcal{I}_p$, which is defined by the row distribution μ. In the transformed program, the statement

if $p = \hat{p}$ **then** $\mathcal{I} := \mathcal{I} \setminus \{\hat{\imath}\}$;

deactivates a pivot row; it replaces the statement $\mathcal{M} := \mathcal{M} \setminus \{k\}$. Deactivation of a multiplier column is handled analogously. Program LU-2 is the result of applying duplication, selection, and conversion to local indices to program LU-1

$0..P - 1 \times 0..Q - 1 \parallel (p,q)$ **program** LU-2
declare
      $k, i, j, \hat{p}, \hat{q}, \hat{\imath}, \hat{\jmath}$: integer ;
      $\mathcal{I}, \mathcal{J}$: set of integer ;
      a : array $[\mathcal{I}_p \times \mathcal{J}_q]$ of real ;
      a_c : array $[\mathcal{I}_p]$ of real ;
      a_r : array $[\mathcal{J}_q]$ of real ;
      a_{rc} : real
initially
      $\mathcal{I}, \mathcal{J} = \mathcal{I}_p, \mathcal{J}_q$;
      $\langle\ ;\ i,j\ :\ (i,j) \in \mathcal{I} \times \mathcal{J}\ ::\ a[i,j] = \tilde{a}_{\mu^{-1}(p,i),\nu^{-1}(q,j)}\ \rangle$
assign
      $\langle\ ;\ k\ :\ 0 \le k < \min(M,N)\ ::$
         $\hat{p}, \hat{\imath} := \mu(k)$;
         $\hat{q}, \hat{\jmath} := \nu(k)$;
         if $p = \hat{p}$ **and** $q = \hat{q}$ **then begin**
            $a_{rc} := a[\hat{\imath}, \hat{\jmath}]$;
            send a_{rc} **to all**
         end else receive a_{rc} **from** $(\hat{p}, \hat{q})$;
         if $a_{rc} = 0.0$ **then quit** ;
         if $p = \hat{p}$ **then** $\mathcal{I} := \mathcal{I} \setminus \{\hat{\imath}\}$;
         if $q = \hat{q}$ **then** $\mathcal{J} := \mathcal{J} \setminus \{\hat{\jmath}\}$;

         {Broadcast the Pivot Row}
         if $p = \hat{p}$ **then begin**
            $\langle\ ;\ j\ :\ j \in \mathcal{J}\ ::\ a_r[j] := a[\hat{\imath}, j]\ \rangle$;
            send $\{a_r[j] : j \in \mathcal{J}\}$ **to** $(\textbf{all}, q)$
         end else receive $\{a_r[j] : j \in \mathcal{J}\}$ **from** $(\hat{p}, q)$;

         {Compute and Broadcast the Multiplier Column}
         if $q = \hat{q}$ **then begin**
            $\langle\ ;\ i\ :\ i \in \mathcal{I}\ ::\ a_c[i] := a[i, \hat{\jmath}] := a[i, \hat{\jmath}]/a_{rc}\ \rangle$;
            send $\{a_c[i] : i \in \mathcal{I}\}$ **to** $(p, \textbf{all})$
         end else receive $\{a_c[i] : i \in \mathcal{I}\}$ **from** $(p, \hat{q})$;

         {Elimination}
         $\langle\ ;\ i,j\ :\ (i,j) \in \mathcal{I} \times \mathcal{J}\ ::\ a[i,j] := a[i,j] - a_c[i]a_r[j]\ \rangle$
      $\rangle$
end

4.1.3 Performance Analysis

Our performance analysis relies on the assumptions of the multicomputer-performance model of Section 1.6.4. One step of program LU-2 consists of four computational parts: the pivot broadcast, the pivot-row broadcast, the multiplier-column calculation and broadcast, and the elimination. Each part can be analyzed separately.

Almost all floating-point operations are performed in the elimination step, where one addition and one multiplication is performed for each active matrix entry. The number of floating-point operations in the elimination step of process (p, q) is, therefore, given by

$$2 \times \#\mathcal{I}[p, q] \times \#\mathcal{J}[p, q],$$

where $[p, q]$ is the process identifier of the index-set variables. This number of operations depends on k, p, q, and the data distribution $\mu \times \nu$. For optimal performance, the elimination step must be load balanced, which is achieved only if the number of operations is about equal in every process. Even if μ and ν are perfectly load-balanced distributions, we must expect some load imbalance caused by the deactivation of a row and a column in every step. Moreover, as the number of active rows and columns decrease to $O(P)$ and $O(Q)$, respectively, fine-granularity effects become important. The inefficiency due to fine granularity can be ignored only if M and N are sufficiently large. Then, the execution time of the last $O(P + Q)$ steps is negligible compared with the total execution time.

Using the scatter distribution of Section 2.3 for both the row and the column distribution leads to a load-balanced matrix distribution $\mu \times \nu$ and to a load-balanced first step of the multicomputer LU-decomposition. Moreover, every block of P successive steps deactivates exactly one row in every process row, and every block of Q successive steps deactivates exactly one column in every process column. It follows that the scatter distribution preserves the load balance throughout the computation. In this case, we have that for all $(p, q) \in 0..P - 1 \times 0..Q - 1$:

$$\#\mathcal{I}[p, q] \approx (M - k)/P$$
$$\#\mathcal{J}[p, q] \approx (N - k)/Q,$$

and the multicomputer-execution time of the k-th elimination step is approximately

$$2\frac{(M - k)(N - k)}{PQ}\tau_A.$$

Compared with the sequential-execution time of the elimination step, an almost perfect speed-up is achieved.

We estimate the execution time of the broadcast operation as the time of a recursive-doubling operation with the same number of processes. The pivot-broadcast time is then given by

$$\tau_C(1)\log_2(PQ) = (\tau_S + \beta)\log_2(PQ).$$

This is actually an overestimate. Although a broadcast can be implemented by a recursive-doubling procedure, it does not require message exchanges, merely one-way node-to-node communication. Moreover, most multicomputer systems provide an efficient broadcast operation, which is implemented by low-level primitives with significantly-reduced latency. Although inadequate for performance modeling of real multicomputers, the assumption that a broadcast is a recursive-doubling procedure is sufficient to gain some insights into the performance of multicomputer LU-decomposition.

In each process column, a segment of the pivot row is broadcast. The cost of this operation depends on P and the message length $\#\mathcal{J}[p,q]$, which depends on k, q, and ν. For the scatter distribution, the message length is approximately $(N-k)/Q$ floating-point words. Assuming a broadcast by recursive doubling, we obtain the execution time

$$\tau_C((N-k)/Q)\log_2(P) = (\tau_S + \beta(N-k)/Q)\log_2(P)$$

for the pivot-row broadcast. Similarly, the multiplier-column broadcast with the scatter distribution takes a time approximately given by:

$$\tau_C((M-k)/P)\log_2(Q) = (\tau_S + \beta(M-k)/P)\log_2(Q).$$

Other data distributions require more communication time because of load imbalance.

Intuitively, it might seem that the communication time is minimized by setting either $P = 1$ or $Q = 1$. The pivot-row broadcast is entirely eliminated if $P = 1$, and the multiplier-column broadcast is eliminated if $Q = 1$. In fact, it is usually more efficient to choose $P \approx Q$. Using the above communication-time estimates for the scatter distribution, assuming that $M = N$ and that the total number of processes is fixed and equal to $R = PQ$, one should choose P and Q to minimize the sum of the two array-broadcast times (the pivot broadcast is a fixed cost):

$$\tau_S \log_2(R) + \beta(M-k)\frac{P\log_2(P) + Q\log_2(Q)}{R}.$$

The first term in this sum is fixed. The second term is minimized by a process mesh with $P \approx Q$. Choosing $P \approx Q$ also postpones fine-granularity effects until the total number of processes R is comparable with MN. With $P = 1$ (or $Q = 1$), fine granularity is important as soon as R is comparable to M.

4.2 LU-Decomposition With Pivoting

4.2.1 Specification

If the coefficient matrix does not satisfy the requirements of Theorem 2, programs LU-1 and LU-2 break down. In this case, the LU-decomposition cannot be completed, because a zero pivot occurs prematurely. Even if the coefficient matrix satisfies all requirements, it may be near a matrix that does not. In such a case of near break down, one or more pivots are small in absolute value, which can lead to large multipliers and numerical instability. Near break down is actually more serious than break down, because it occurs more frequently and because the numerical instability may remain undetected. Because of the following result, problems of break down and near break down in LU-decomposition can be avoided by applying row and column permutations to the matrix A.

Theorem 3 *For any real $M \times N$ matrix A, there exist permutation matrices $R \in \mathbb{R}^{M \times M}$ and $C \in \mathbb{R}^{N \times N}$, a unit-lower-triangular matrix $\hat{L} \in \mathbb{R}^{M \times N}$, and an upper-triangular matrix $\hat{U} \in \mathbb{R}^{M \times N}$ such that*

$$RAC^T = \hat{L}I_{N,M}\hat{U}. \tag{4.1}$$

The (omitted) proof of Theorem 3 establishes the correctness proof of LU-decomposition with complete pivoting. In preparation for further algorithmic work, Theorem 3 slightly differs from traditional formulations: the lower-triangular matrix $\hat{L}$ has the same dimensions as the matrix A. In Equation 4.1, the matrix $\hat{L}I_{N,M}$ corresponds to the $M \times M$ unit-lower-triangular matrix of traditional formulations. The post-multiplication of $\hat{L}$ by the $N \times M$ identity matrix $I_{N,M}$ cuts the last $N - M$ zero columns of $\hat{L}$ if $M < N$ and adds $M - N$ zero columns to $\hat{L}$ if $M > N$.

The decomposition of Equation 4.1 explicitly permutes the rows and columns of the matrix. Even without developing a multicomputer program, it is obvious that permutations destroy the data distribution of the matrix A. To avoid this, we permute the rows and columns implicitly. Using that the inverse of a permutation matrix is its transpose, we pre-multiply Equation 4.1 by R^T and and post-multiply it by C to obtain that

$$
\begin{aligned}
A &= R^T\hat{L}I_{N,M}\hat{U}C \\
&= R^T\hat{L}CC^TI_{N,M}RR^T\hat{U}C \\
&= LC^TI_{N,M}RU,
\end{aligned}
$$

where

$$
\begin{aligned}
L &= R^T\hat{L}C \tag{4.2} \\
U &= R^T\hat{U}C. \tag{4.3}
\end{aligned}
$$

The matrix L is called a *permuted unit-lower-triangular matrix* and the matrix U a *permuted upper-triangular matrix*. Theorem 4 is the resulting reformulation of Theorem 3.

Theorem 4 *For any real $M \times N$ matrix A, there exist permutation matrices $R \in \mathbb{R}^{M \times M}$ and $C \in \mathbb{R}^{N \times N}$, a permuted unit-lower-triangular matrix $L \in \mathbb{R}^{M \times N}$, and a permuted upper-triangular matrix $U \in \mathbb{R}^{M \times N}$ such that*

$$A = LC^T I_{N,M} RU. \tag{4.4}$$

To derive the specification program that implements Theorem 4, we need a compact representation of permutation matrices. Let R be the permutation matrix that maps row r_m to row m when it pre-multiplies a matrix: row r_m of A is row m of RA. Let C^T be the permutation matrix that maps column c_n to column n when it post-multiplies a matrix: column c_n of A is column n of AC^T. It is easily verified that

$$\begin{aligned}
R^T &= \begin{bmatrix} \vec{e}_{r_0} & \vec{e}_{r_1} & \cdots & \vec{e}_{r_{M-1}} \end{bmatrix} \\
C^T &= \begin{bmatrix} \vec{e}_{c_0} & \vec{e}_{c_1} & \cdots & \vec{e}_{c_{N-1}} \end{bmatrix},
\end{aligned}$$

where $\vec{e}_k$ denotes the k-th unit vector of a dimension determined by its context: column vectors of R^T have dimension M, while those of C^T have dimension N. The permutations are uniquely defined by the arrays r and c: the element in row m and column n of RAC^T is the same as the element in row r_m and column c_n of A.

Program LU-3 is a first step towards implementing Theorem 4. It is obtained from program LU-1 by replacing all references to array entries $a[m,n]$ by references to entries $a[r_m, c_n]$. As a result, program LU-3 performs an LU-decomposition without pivoting of the matrix RAC^T. Note that the matrix RAC^T is represented by the array a and the two permutation arrays r and c.

program LU-3
declare
    k, m, n : integer ;
    $\mathcal{M}$, $\mathcal{N}$: set of integer ;
    r : array $[0..M-1]$ of integer ;
    c : array $[0..N-1]$ of integer ;
    a : array $[0..M-1 \times 0..N-1]$ of real ;
    a_c : array $[0..M-1]$ of real ;
    a_r : array $[0..N-1]$ of real ;
    a_{rc} : real
initially
    $\mathcal{M}$, $\mathcal{N} = 0..M-1$, $0..N-1$;
    $\langle\ ;\ m,n\ :\ (m,n) \in \mathcal{M} \times \mathcal{N}\ ::\ a[m,n] = \tilde{a}_{m,n}\ \rangle$;

$$\langle \; ; \; m \; : \; m \in \mathcal{M} \; :: \; r[m] = \tilde{r}_m \; \rangle \; ;$$
$$\langle \; ; \; n \; : \; n \in \mathcal{N} \; :: \; c[n] = \tilde{c}_n \; \rangle$$
assign
$$\langle \; ; \; k \; : \; 0 \leq k < \min(M, N) \; ::$$
$$a_{rc} := a[r[k], c[k]] \; ;$$
$$\textbf{if } a_{rc} = 0.0 \textbf{ then quit } ;$$
$$\mathcal{M} := \mathcal{M} \setminus \{r[k]\} \; ;$$
$$\mathcal{N} := \mathcal{N} \setminus \{c[k]\} \; ;$$

{Copy the Pivot Row}
$$\langle \; \| \; n \; : \; n \in \mathcal{N} \; :: \; a_r[n] := a[r[k], n] \; \rangle \; ;$$

{Compute the Multiplier Column}
$$\langle \; \| \; m \; : \; m \in \mathcal{M} \; :: \; a_c[m] := a[m, c[k]] := a[m, c[k]]/a_{rc} \; \rangle \; ;$$

{Elimination}
$$\langle \; \| \; m, n \; : \; (m, n) \in \mathcal{M} \times \mathcal{N} \; ::$$
$$a[m, n] := a[m, n] - a_c[m]a_r[n]$$
$$\rangle$$

$$\rangle$$
end

We now adapt program LU-3 to compute the entries of arrays r and c on the fly. The computation of the pivot indices is called the *pivot strategy*. There are many pivot strategies possible, and a detailed discussion of particular pivot strategies and their implementation is postponed until Section 4.2.3. Because $r[k]$ and $c[k]$ are first used in step k of the LU-decomposition, it is appropriate to postpone computing them until the start of step k. At that time, the following information is available: the step k, the index sets $\mathcal{M}$ and $\mathcal{N}$, and the two-dimensional array a. This information is passed to program Pivot-Strategy-4, which is a program implementing an unspecified pivot strategy and is called by program LU-4.

program LU-4
declare
    {See program LU-3}
initially
$$\mathcal{M}, \; \mathcal{N} = 0..M - 1, \; 0..N - 1 \; ;$$
$$\langle \; ; \; m, n \; : \; (m, n) \in \mathcal{M} \times \mathcal{N} \; :: \; a[m, n] = \tilde{a}_{m,n} \; \rangle$$
assign
$$\langle \; ; \; k \; : \; 0 \leq k < \min(M, N) \; ::$$
$$\text{Pivot-Strategy-4}(k, \mathcal{M}, \mathcal{N}, a, a_{rc}, r[k], c[k]) \; ;$$
$$\{\text{See program LU-3}\}$$
$$\rangle$$
end

In this program, a new notational feature was introduced: the ability to call another program. The calling program (program LU-4) and the called program (program Pivot-Strategy-4) exchange information by means of a parameter list. By convention, the parameters in this list are passed via a call-by-variable mechanism: the parameters are variables shared between the calling and the called program. This is the FORTRAN mechanism of exchanging parameters between two programs. Computer-language experts agree that this mechanism is seriously flawed, because unwanted side effects may occur in the data space of the calling program. In spite of this (very valid) objection, we use the call-by-variable mechanism to avoid additional notation needed by other mechanisms.

4.2.2 Implementation

Program LU-4 is transformed into a multicomputer program using the same technique as before. Program LU-4 is duplicated over a $P \times Q$ process mesh. Subsequently, the selection step introduces a two-dimensional data distribution $\mu \times \nu$ of the matrix A. Finally, global indices are converted into local indices. Again, the array a_r is transformed into a column-distributed vector and the array a_c into a row-distributed vector. The scalar variable a_{rc} remains duplicated in every process. The details of the transformation are left as an exercise. The result is given by program LU-5.

$0..P - 1 \times 0..Q - 1 \parallel (p, q)$ **program** LU-5
declare
$\quad k,\ i,\ j,\ \hat{p},\ \hat{q},\ \hat{\imath},\ \hat{\jmath}\ :\ $ integer ;
$\quad \mathcal{I},\ \mathcal{J}\ :\ $ set of integer ;
$\quad r\ :\ $ array $[0..M - 1]$ of integer ;
$\quad c\ :\ $ array $[0..N - 1]$ of integer ;
$\quad a\ :\ $ array $[\mathcal{I}_p \times \mathcal{J}_q]$ of real ;
$\quad a_c\ :\ $ array $[\mathcal{I}_p]$ of real ;
$\quad a_r\ :\ $ array $[\mathcal{J}_q]$ of real ;
$\quad a_{rc}\ :\ $ real
initially
$\quad \mathcal{I},\ \mathcal{J} = \mathcal{I}_p,\ \mathcal{J}_q\ ;$
$\quad \langle\ ;\ i, j\ :\ (i, j) \in \mathcal{I} \times \mathcal{J}\ ::\ a[i, j] = \tilde{a}_{\mu^{-1}(p,i), \mu^{-1}(q,j)}\ \rangle$
assign
$\quad \langle\ ;\ k\ :\ 0 \le k < \min(M, N)\ ::$

$\qquad$ Pivot-Strategy-5$(k, \mathcal{I}, \mathcal{J}, a, a_{rc}, \hat{p}, \hat{\imath}, \hat{q}, \hat{\jmath})\ ;$

$\qquad$ **if** $a_{rc} = 0.0$ **then quit** ;
$\qquad$ **if** $p = \hat{p}$ **then** $\mathcal{I} := \mathcal{I} \setminus \{\hat{\imath}\}$;
$\qquad$ **if** $q = \hat{q}$ **then** $\mathcal{J} := \mathcal{J} \setminus \{\hat{\jmath}\}$;

$r[k] := \mu^{-1}(\hat{p}, \hat{\imath})$;
$c[k] := \nu^{-1}(\hat{q}, \hat{\jmath})$;

{Broadcast the Pivot Row}
if $p = \hat{p}$ then begin
$\quad \langle \; ; \; j \; : \; j \in \mathcal{J} \; :: \; a_r[j] := a[\hat{\imath}, j] \; \rangle$;
$\quad$ send $\{a_r[j] : j \in \mathcal{J}\}$ to $(\textbf{all}, q)$
end else receive $\{a_r[j] : j \in \mathcal{J}\}$ from $(\hat{p}, q)$;

{Compute and Broadcast the Multiplier Column}
if $q = \hat{q}$ then begin
$\quad \langle \; ; \; i \; : \; i \in \mathcal{I} \; :: \; a_c[i] := a[i, \hat{\jmath}] := a[i, \hat{\jmath}]/a_{rc} \; \rangle$;
$\quad$ send $\{a_c[i] : i \in \mathcal{I}\}$ to $(p, \textbf{all})$
end else receive $\{a_c[i] : i \in \mathcal{I}\}$ from $(p, \hat{q})$;

{Elimination}
$\langle \; \| \; i,j \; : \; (i,j) \in \mathcal{I} \times \mathcal{J} \; :: \; a[i,j] := a[i,j] - a_c[i]a_r[j] \; \rangle$
$\rangle$
end

In program LU-5, the permutation arrays r and c are kept duplicated in every process. These arrays are not required anywhere in the multicomputer LU-decomposition. However, they will be used in Section 4.3 to solve the triangular systems. In Section 4.4, both the LU-decomposition and the triangular-solve programs will be adapted to distribute the permutation arrays over the processes.

4.2.3   Pivot Strategies

The pivot strategy determines the permutations necessary to complete the LU-decomposition. The proofs of Theorems 3 and 4 require that one use *complete pivoting*: in every step, the active matrix element with maximum absolute value among all active matrix elements is chosen as the pivot. In exact arithmetic, the LU-decomposition with complete pivoting terminates in step K, where K is the rank of the matrix A. This desirable property is lost when the algorithm is performed in floating-point arithmetic, and it cannot be guaranteed that the computed rank equals the actual rank of the matrix. However, it can be proven that, under reasonable assumptions, LU-decomposition with complete pivoting is numerically stable for the purpose of solving systems of linear equations.

Given the distributed matrix and the sets of active rows and columns, program Complete-Pivoting computes the complete pivot. This concurrent program is one possible specification of program Pivot-Strategy-5, which is called by program LU-5.

$0..P - 1 \times 0..Q - 1 \parallel (p, q)$
    program Complete-Pivoting$(k, \mathcal{I}, \mathcal{J}, a, a_{rc}, \hat{p}, \hat{\imath}, \hat{q}, \hat{\jmath})$
declare
    k, d, u, i, j, $\hat{\imath}$, $\hat{\jmath}$, i_0, j_0, $\hat{p}$, $\hat{q}$, p_0, q_0, p_u, q_u : integer ;
    $\mathcal{I}$, $\mathcal{J}$: set of integer ;
    a : array $[\mathcal{I}_p \times \mathcal{J}_q]$ of real ;
    a_{rc}, h : real
assign
    $a_{rc}, \hat{p}, \hat{\imath}, \hat{q}, \hat{\jmath} := 0.0, p, -1, q, -1$;
    $\langle\ ;\ i, j\ :\ (i, j) \in \mathcal{I} \times \mathcal{J}\ ::$
        if $|a[i, j]| > |a_{rc}|$ **then** $a_{rc}, \hat{\imath}, \hat{\jmath} := a[i, j], i, j$
    $\rangle$;
    $\langle\ ;\ d\ :\ 0 \leq d < D\ ::$
        $u := (pQ + q) \bar{\vee} 2^d$;
        $p_u, q_u := \lfloor u/Q \rfloor, u \bmod Q$;
        send $a_{rc}, \hat{p}, \hat{\imath}, \hat{q}, \hat{\jmath}$ **to** (p_u, q_u) ;
        receive h, p_0, i_0, q_0, j_0 **from** (p_u, q_u) ;
        if $|h| > |a_{rc}|$ **or**
              $|h| = |a_{rc}|$ **and** $p_0 < \hat{p}$ **or**
              $|h| = |a_{rc}|$ **and** $p_0 = \hat{p}$ **and** $q_0 < \hat{q}$
        then $a_{rc}, \hat{p}, \hat{\imath}, \hat{q}, \hat{\jmath} := h, p_0, i_0, q_0, j_0$

    $\rangle$
    end

The first quantification of this program computes the maximum element in absolute value among all active local-matrix elements. Subsequently, a recursive-doubling procedure computes the global maximum and its row and column indices. The recursive-doubling procedure over a $P \times Q$ process mesh is similar to the procedure of program Inner-Product-6 in Section 2.2.2 and assumes that $D = \log_2(PQ)$. The comparison test ensures that all processes choose the same active matrix element as pivot, even if two or more matrix elements are equal in absolute value. In this case, the components of the process identifier (p, q) are minimized.

Without pivoting, the two-dimensional scatter distribution ensures that all steps of the LU-decomposition remain load balanced. With complete pivoting, the rows and columns are deactivated in an unpredictable order. It is, therefore, impossible to find the best data distribution. On the other hand, the randomness of the pivot locations guarantees that almost all load-balanced data distributions lead to nearly load-balanced computations. A long string of consecutive pivots all from the same process occurs only if one is extremely unlucky.

The main disadvantage of complete pivoting is that the search is expensive: the number of logical operations for one pivot search is of the same order of magnitude as the number of arithmetic operations in the elimination step that precedes the pivot search. To reduce the pivot-search cost, several alternatives to complete pivoting have been suggested. All methods restrict the pivot search to a certain subset of the set of all active matrix elements. For sequential computations, the most commonly used strategy is *row pivoting*: in step k, the pivot search is limited to the active elements of column number k. The multicomputer implementation of row pivoting is given by program Row-Pivoting.

$$0..P - 1 \times 0..Q - 1 \parallel (p, q)$$
$$\textbf{program Row-Pivoting}(k, \mathcal{I}, \mathcal{J}, a, a_{rc}, \hat{p}, \hat{\imath}, \hat{q}, \hat{\jmath})$$
declare
  k, d, u, i, j, $\hat{\imath}$, $\hat{\jmath}$, i_0, $\hat{p}$, $\hat{q}$, p_0, p_u, q_u : integer ;
  $\mathcal{I}$, $\mathcal{J}$: set of integer ;
  a : array $[\mathcal{I}_p \times \mathcal{J}_q]$ of real ;
  a_{rc}, h : real
assign
  $\hat{q}, \hat{\jmath} := \nu(k)$;
  $a_{rc}, \hat{p}, \hat{\imath} := 0.0, p, -1$;
  if $q = \hat{q}$ **then**
    $\langle$; i : $i \in \mathcal{I}$ ::
        if $|a[i, \hat{\jmath}]| > |a_{rc}|$ **then** $a_{rc}, \hat{\imath} := a[i, j], i$
    $\rangle$;
  $\langle$; d : $0 \le d < D$ ::
      $u := (pQ + q) \bar{\vee} 2^d$;
      $p_u, q_u := \lfloor u/Q \rfloor, u \bmod Q$;
      send $a_{rc}, \hat{p}, \hat{\imath}$ **to** (p_u, q_u) ;
      receive h, p_0, i_0 **from** (p_u, q_u) ;
      if $|h| > |a_{rc}|$ **or**
          $|h| = |a_{rc}|$ **and** $p_0 < \hat{p}$
      then $a_{rc}, \hat{p}, \hat{\imath} := h, p_0, i_0$
  $\rangle$
end

Using row pivoting and assuming load balance, the number of logical operations in the local search is reduced to $O((M - k)/P)$. While matrix rows are deactivated in an unpredictable order, columns become inactive in the order dictated by their global index. This suggests using the scatter distribution as the column distribution of the matrix.

Row pivoting restricts the pivot-search set. Unfortunately, it also restricts our choice of data distributions by making the pivot locations less random. Almost any load-balanced data distribution leads to a load-balanced computation, provided the pivot strategy randomizes the pivot locations. This property of dynamic pivoting can be used to our advantage. By generating pivot locations that are as random as possible, load balance is automatically achieved for a wide range of data distributions. This is important, particularly when developing a software library: the user of the library, not the writer, will choose a data distribution. The more restrictions are imposed on this choice, the more difficult it will be to incorporate the library routines in an application.

To randomize the locations of the pivot columns, while keeping the search cost down to that of row pivoting, we change the row-pivoting strategy somewhat. The pivot-search set can be expanded without penalty by searching through one active column in each process column. This expansion does not increase the pivot-search cost, because the local searches can be done concurrently and the recursive-doubling procedure remains essentially unchanged. We call this alternative *multirow pivoting*, and its multicomputer implementation is given by program Multirow-Pivoting. The randomization of the pivot columns leads to better load balance and higher performance. In fact, LU-decomposition with multirow pivoting often outperforms LU-decomposition without pivoting and LU-decomposition with row pivoting.

$$0..P - 1 \times 0..Q - 1 \parallel (p, q)$$
$\quad$ **program** Multirow-Pivoting$(k, \mathcal{I}, \mathcal{J}, a, a_{rc}, \hat{p}, \hat{i}, \hat{q}, \hat{j})$
declare
$\quad k, d, u, i, j, \hat{i}, \hat{j}, i_0, j_0, \hat{p}, \hat{q}, p_0, q_0, p_u, q_u$: integer ;
$\quad \mathcal{I}, \mathcal{J}$: set of integer ;
$\quad a$: array $[\mathcal{I}_p \times \mathcal{J}_q]$ of real ;
$\quad a_{rc}, h$: real
assign
$\quad a_{rc}, \hat{i}, \hat{j} := 0.0, -1, -1$;
$\quad$ **if** $\mathcal{J} \neq \emptyset$ **then begin**
$\quad\quad \hat{j} := \mathbf{any}(\mathcal{J})$;
$\quad\quad \langle \, ; \, i \, : \, i \in \mathcal{I} \, ::$
$\quad\quad\quad$ **if** $|a[i, \hat{j}]| > |a_{rc}|$ **then** $a_{rc}, \hat{i} := a[i, \hat{j}], i$
$\quad\quad \rangle$
$\quad$ **end** ;
$\quad \hat{p}, \hat{q} := p, q$;
$\quad \langle \, ; \, d \, : \, 0 \leq d < D \, ::$
$\quad\quad u := (pQ + q) \bar{\vee} 2^d$;
$\quad\quad p_u, q_u := \lfloor u/Q \rfloor, u \bmod Q$;
$\quad\quad$ **send** $a_{rc}, \hat{p}, \hat{i}, \hat{q}, \hat{j}$ **to** (p_u, q_u) ;
$\quad\quad$ **receive** h, p_0, i_0, q_0, j_0 **from** (p_u, q_u) ;
$\quad\quad$ **if** $\quad |h| > |a_{rc}|$ **or**
$\quad\quad\quad\quad |h| = |a_{rc}|$ **and** $p_0 < \hat{p}$ **or**
$\quad\quad\quad\quad |h| = |a_{rc}|$ **and** $p_0 = \hat{p}$ **and** $q_0 < \hat{q}$
$\quad\quad$ **then** $a_{rc}, \hat{p}, \hat{i}, \hat{q}, \hat{j} := h, p_0, i_0, q_0, j_0$
$\quad \rangle$
end

The operator **any** is defined on variables of type set and returns an arbitrary element of the set.

Of course, row pivoting and multirow pivoting have obvious column counterparts: *column pivoting* and *multicolumn pivoting*.

Individual nodes of a multicomputer often operate near peak performance only on contiguously stored data. Most dynamic-pivoting strategies fragment the active data within a process, because they deactivate rows and columns in an unpredictable order. The *one-per-process pivot strategy* eliminates the local pivot search and submits the matrix element with minimal *local* row and *local* column index to the global pivot search. The local index sets $\mathcal{I}$ and $\mathcal{J}$ remain index ranges throughout the computation. Hence, the data remain in contiguous blocks. Although matrix rows and columns become inactive in a more predictable order, the process rows and columns in which they become inactive are sufficiently random to ensure load balance. The *local-corner pivot strategy* is a slight generalization and achieves the same goal as the one-per-process pivot strategy: it submits the four corners of the local matrix to the pivot search. By expanding the pivot-search set, one hopes to increase numerical stability.

Finally, let us compare the different pivoting strategies from the perspective of numerical stability. A rigorous proof of numerical stability exists only for LU-decomposition with complete pivoting. Moreover, for any partial-pivoting strategy, it is always possible to find some example for which the LU-decomposition breaks down. Row and column pivoting are special, however. In exact arithmetic, LU-decomposition with row or column pivoting breaks down only if the matrix does not have full rank. Multirow and multicolumn pivoting retain this property. One-per-process pivoting and local-corner pivoting, however, can break down in exact arithmetic for matrices of full rank. Of course, break down in exact arithmetic can lead to numerical instability in floating-point arithmetic. From a fundamental point of view, the numerical stability of partial-pivoting techniques is far from settled. The pragmatic point of view, on the other hand, holds that the probability of LU-decomposition breaking down or being numerically unstable is virtually negligible if the pivot-search set is sufficiently large. When reasonable partial-pivoting strategies fail, one should consider switching to other solution methods like, for example, the more expensive but guaranteed stable QR-decomposition of Chapter 5.

4.3   Triangular Systems

Given the LU-decomposition of the coefficient matrix A, the system of linear equations $A\vec{x} = \vec{b}$ is reduced to two systems with permuted-triangular coefficient matrices. By carrying out the LU-decomposition on the matrix A extended with the extra column vector $\vec{b}$, one triangular system can be avoided. We concentrate, therefore, on solving the permuted upper-triangular system

$$U\vec{x} = \vec{b}.$$

Algorithms for permuted lower-triangular systems are similar, however, and a complete multicomputer solver for linear systems will be given in Section 4.4.

4.3.1   Specification

Program Back-Solve-1 solves the upper-triangular system obtained after applying an LU-decomposition without pivoting (program LU-1) to the coefficient matrix A extended with the right-hand-side vector $\vec{b}$.

> **program** Back-Solve-1
> **declare**
> k, m, n : integer ;
> $\mathcal{M}$: set of integer ;
> a : array $[0..M-1 \times 0..N-1]$ of real ;
> b : array $[0..M-1]$ of real ;
> x : array $[0..N-1]$ of real
> **assign**
> $\mathcal{M} := 0..K-1$;
> $\langle\; ;\; k\; :\; K > k \geq 0\; ::$
> $\mathcal{M} := \mathcal{M} \setminus \{k\}$;
> $x[k] := b[k]/a[k,k]$;
> $\langle\; \|\;\; m\; :\; m \in \mathcal{M}\; ::\; b[m] := b[m] - a[m,k]x[k] \rangle$
> $\rangle$
> **end**

In program Back-Solve-1, K is the computed rank of the matrix A: program LU-1 terminated in step K because the pivot vanished. If $M = N = K$, this corresponds to solving the complete system. If $K \leq \min(M, N)$, the $K \times K$ principal subsystem is solved. Program Back-Solve-1 traverses columns 0 through $K-1$ in reverse order. In step k, the value of $x[k]$ is computed, and $x[k]$ times column k of U is subtracted from the right-hand side to obtain a reduced upper-triangular system of dimension $k \times k$ in the unknowns $x[0], \ldots, x[k-1]$. The undeclared value K, like M and N, is an assumed-known constant. In practice, the value of K is passed from the LU-decomposition to the back-solve program. It is assumed that the declared arrays a and b are already appropriately initialized. In particular, the entries of array a that represent the upper-triangular part of A contain the elements of U.

The permuted upper-triangular system obtained program LU-4 (LU-decomposition with implicit pivoting) can be reduced to the non-permuted case using the results of Section 4.2.1. According to Theorem 4, the permuted triangular matrices L and U satisfy Equation 4.4. On the other hand, we know that the LU-decomposition without pivoting applied to RAC^T would result in the decomposition given by Equation 4.1. Moreover, these two factorizations are related by Equations 4.2 and 4.3. It follows that solving $U\vec{x} = \vec{b}$ is equivalent to solving:

$$\hat{U}C\vec{x} = R\vec{b}.$$

Because $\hat{U}$ is an upper-triangular matrix, this suggests to adapt program Back-Solve-1 by replacing every access to $x[n]$ by $x[c[n]]$ and every access to $b[m]$ by $b[r[m]]$. Instead of using elements of the upper-triangular coefficient matrix $\hat{U}$, we may use elements of U, because

$$\hat{u}_{m,n} = u_{r_m,c_n}.$$

Note that matrix element u_{r_m,c_n} is represented by array entry $a[r[m], c[n]]$. After carrying out these substitutions in program Back-Solve-1, we obtain program Back-Solve-4.

> **program** Back-Solve-4
> **declare**
> $k,\ m,\ n$: integer ;
> r : array $[0..M-1]$ of integer ;
> c : array $[0..N-1]$ of integer ;
> $\mathcal{M}$: set of integer ;
> a : array $[0..M-1 \times 0..N-1]$ of real ;
> b : array $[0..M-1]$ of real ;
> x : array $[0..N-1]$ of real
> **assign**
> $\mathcal{M} := \langle\, \bigcup k\ :\ 0 \le k < K\ ::\ \{r[k]\}\, \rangle$;
> $\langle\, ;\ k\ :\ K > k \ge 0\ ::$
> $\mathcal{M} := \mathcal{M} \setminus \{r[k]\}$;
> $x[c[k]] := b[r[k]]/a[r[k], c[k]]$;
> $\langle\, \|\ m\ :\ m \in \mathcal{M}\ ::\ b[m] := b[m] - a[m, c[k]]x[c[k]]\, \rangle$
> $\rangle$
> **end**

The initial assignment to the index-set variable $\mathcal{M}$ activates all rows that occurred as a pivot row in the first K steps of the LU-decomposition. Given the computed rank K of the coefficient matrix, the $K \times K$ subsystem formed by the first K pivot rows and columns is solved. It is assumed that the arrays a, b, r, and c were initialized appropriately by a preceding LU-decomposition.

4.3.2 Implementation

The data distribution for the multicomputer back-solve program is already determined by the multicomputer LU-decomposition. The distribution of the coefficient matrix is given, the right-hand side is transformed into a row-distributed vector, and the solution vector into a column-distributed vector.

Just one matrix column is active in each step of program Back-Solve-4. In follows that, in the multicomputer version, only one process column can be active in each step. The active process column computes the next component of the solution, updates the right-hand side, and sends the updated right-hand side to the next active process column. Program Back-Solve-5 fills in the details.

$0..P - 1 \times 0..Q - 1 \parallel (p, q)$ **program** Back-Solve-5
declare
        $k, i, \hat{\imath}, \hat{\jmath}, \hat{\jmath}_-, \hat{\jmath}_+, \hat{p}, \hat{q}, \hat{q}_-, \hat{q}_+$: integer ;
        r : array $[0..M - 1]$ of integer ;
        c : array $[0..N - 1]$ of integer ;
        $\mathcal{I}$: set of integer ;
        a : array $[\mathcal{I}_p \times \mathcal{J}_q]$ of real ;
        b : array $[\mathcal{I}_p]$ of real ;
        x : array $[\mathcal{J}_q]$ of real
assign
        $\mathcal{I} := \emptyset$;
        $\langle\ ;\ k\ :\ 0 \le k < K\ ::$
                $\hat{p}, \hat{\imath} := \mu(r[k])$;
                if $p = \hat{p}$ **then** $\mathcal{I} := \mathcal{I} \bigcup \{\hat{\imath}\}$
        $\rangle$;
        $\langle\ ;\ k\ :\ K > k \ge 0\ ::$
                if $k < K - 1$ **then** $\hat{q}_+, \hat{\jmath}_+ := \nu(c[k + 1])$;
                $\hat{q}, \hat{\jmath} := \nu(c[k])$;
                if $k > 0$ **then** $\hat{q}_-, \hat{\jmath}_- := \nu(c[k - 1])$;
                $\hat{p}, \hat{\imath} := \mu(r[k])$;
                if $q = \hat{q}$ **then begin**
                        if $k < K - 1$ **and** $q \ne \hat{q}_+$ **then**
                                receive $\mathcal{I},\ \{b[i] : i \in \mathcal{I}\}$ **from** $(p, \hat{q}_+)$;
                        if $p = \hat{p}$ **then begin**
                                $\mathcal{I} := \mathcal{I} \setminus \{\hat{\imath}\}$;
                                $x[\hat{\jmath}] := b[\hat{\imath}]/a[\hat{\imath}, \hat{\jmath}]$;
                                send $x[\hat{\jmath}]$ **to** (**all**, q)
                        end else receive $x[\hat{\jmath}]$ **from** $(\hat{p}, q)$;
                        $\langle\ ;\ i\ :\ i \in \mathcal{I}\ ::\ b[i] := b[i] - a[i, \hat{\jmath}]x[\hat{\jmath}]\ \rangle$;
                        if $k > 0$ **and** $q \ne \hat{q}_-$ **then**
                                send $\mathcal{I},\ \{b[i] : i \in \mathcal{I}\}$ **to** $(p, \hat{q}_-)$
                end
        $\rangle$
end

The arrays a, b, r, and c and the computed rank K are assumed given by a preceding LU-decomposition. The distributed array b is initialized as a row-distributed vector and, initially, is duplicated in every process column. The index set $\mathcal{I}$ is a distributed representation of the index set $\mathcal{M}$ of program

Back-Solve-4. The first quantification of the **assign** section activates the first K pivot rows.

Because the pivot-index arrays r and c are duplicated in every process, they may require a significant amount of memory. If PQ is comparable to $M + N$, the amount of memory required for the pivot-index arrays is comparable to that required for the coefficient matrix! To avoid such massive memory overhead for medium- and fine-grained computations, we need a distributed representation of the pivot information.

In program Back-Solve-5, the matrix columns become active in the reverse order of their occurrence as pivot columns in the LU-decomposition. From the sequence of column indices $c[0], c[1], \ldots, c[K-1]$, eliminate all columns not mapped to process column q to obtain the subsequence:

$$c[k_0], c[k_1], \ldots, c[k_\ell], \ldots, c[k_{L_q-1}].$$

The corresponding sequence of local column indices

$$\hat{\jmath}[0], \hat{\jmath}[1], \ldots, \hat{\jmath}[\ell], \ldots, \hat{\jmath}[L_q - 1]$$

gives the reverse order in which the matrix columns local to process column q become active. The sequence of row indices corresponding to the above sequence of column indices,

$$r[k_0], r[k_1], \ldots, r[k_\ell], \ldots, r[k_{L_q-1}],$$

is associated with the sequence of process-row identifiers

$$\hat{p}[0], \hat{p}[1], \ldots, \hat{p}[\ell], \ldots, \hat{p}[L_q - 1]$$

and the sequence of local row indices

$$\hat{\imath}[0], \hat{\imath}[1], \ldots, \hat{\imath}[\ell], \ldots, \hat{\imath}[L_q - 1].$$

The step in which matrix column $c[k_\ell]$ is active receives its right-hand side from the process column where matrix column $c[k_\ell + 1]$ was active and sends the updated right-hand side to the process column where matrix column $c[k_\ell - 1]$ will be active. These process-column identifiers are, respectively, kept in arrays $\hat{q}_+$ and $\hat{q}_-$. There are two special cases. First, the step in which matrix column $c[K-1]$ is active need not receive the right-hand side, but may use the locally available row-distributed right-hand-side vector. In the process column q where column $c[K-1]$ is active, it is thus appropriate to set $\hat{q}_+[L_q-1]$ equal to q. Second, the step in which matrix column $c[0]$ is active need not send the right-hand side to the next process column. In process column q where column $c[0]$ is active, it is thus appropriate to set $\hat{q}_-[0]$ equal to q.

Program Back-Solve-6 uses the above representation of the pivot information, but is otherwise equivalent to program Back-Solve-5. It is assumed that the arrays $\hat{p}$, $\hat{q}_-$, $\hat{q}_+$, $\hat{\imath}$, and $\hat{\jmath}$ are initialized during the LU-decomposition; see Section 4.4 for those implementation details.

$0..P - 1 \times 0..Q - 1 \parallel (p, q)$ **program** Back-Solve-6
declare

    i, d, ℓ : integer ;

    $\hat{\imath}$, $\hat{\jmath}$, $\hat{p}$, $\hat{q}_-$, $\hat{q}_+$: array $[0..J_q - 1]$ of integer ;

    $\mathcal{I}$, $\mathcal{H}$: set of integer ;

    a : array $[\mathcal{I}_p \times \mathcal{J}_q]$ of real ;

    b : array $[\mathcal{I}_p]$ of real ;

    x : array $[\mathcal{J}_q]$ of real

assign

    $\mathcal{I} := \langle\, \bigcup \ell \,:\, 0 \leq \ell < L_q \text{ and } p = \hat{p}[\ell] \,::\, \{\hat{\imath}[\ell]\} \,\rangle$;

    $\langle\, ; d \,:\, 0 \leq d < \log_2 Q \,::$

        send $\mathcal{I}$ **to** $(p, q\bar{\vee}2^d)$;

        receive $\mathcal{H}$ **from** $(p, q\bar{\vee}2^d)$;

        $\mathcal{I} := \mathcal{I} \bigcup \mathcal{H}$

    $\rangle$;

    $\langle\, ; \ell \,:\, L_q > \ell \geq 0 \,::$

        if $q \neq \hat{q}_+[\ell]$ **then**

            receive $\mathcal{I}$, $\{b[i] : i \in \mathcal{I}\}$ **from** $(p, \hat{q}_+[\ell])$;

        if $p = \hat{p}[\ell]$ **then begin**

            $\mathcal{I} := \mathcal{I} \setminus \{\hat{\imath}[\ell]\}$;

            $x[\hat{\jmath}[\ell]] := b[\hat{\imath}[\ell]]/a[\hat{\imath}[\ell], \hat{\jmath}[\ell]]$;

            send $x[\hat{\jmath}[\ell]]$ **to** $(\textbf{all}, q)$

        end else receive $x[\hat{\jmath}[\ell]]$ **from** $(\hat{p}[\ell], q)$;

        $\langle\, ; i \,:\, i \in \mathcal{I} \,::\, b[i] := b[i] - a[i, \hat{\jmath}[\ell]]x[\hat{\jmath}[\ell]] \,\rangle$;

        if $q \neq \hat{q}_-[\ell]$ **then**

            send $\mathcal{I}$, $\{b[i] : i \in \mathcal{I}\}$ **to** $(p, \hat{q}_-[\ell])$

    $\rangle$

end

Program Back-Solve-6, like program Back-Solve-5, activates the first K pivot rows of the LU-decomposition by initializing the index set $\mathcal{I}$. The new representation of the pivot information complicates this initialization. The first quantification activates those pivot rows in process (p, q) that correspond to pivots in process column q. The recursive-doubling procedure computes the union of all the sets of process row p.

The quantification over ℓ is the actual back-solve procedure. There is a one-to-one correspondence between the body of this quantification and the statement **if** $q = \hat{q}$ **then begin** ... **end** in program Back-Solve-5. In fact, the major difference is the absence of the test itself. In program Back-Solve-6, this test is superfluous, because a process column is activated only when it receives the updated right-hand side. Once the next solution entry has been computed and broadcast, the updated right-hand side is sent to the next process column, thereby activating it. Of course, if the current active process column is also the next active one, there is no need for communication. This explains the **if** statements in front of the communication instructions for the right-hand-side. These conditions also serve to initiate and terminate the back-solve procedure. It is initiated in the process column where $q = \hat{q}_+[L_q - 1]$. The initialization of the array $\hat{q}_+$ guarantees that this holds only in the process column where matrix column $c[K - 1]$ is active. This process column skips the receive instruction for the right-hand side and becomes active immediately. Other process columns execute the receive instruction and wait until an updated right-hand side appears. The back-solve procedure terminates in the process column where $q = \hat{q}_-[0]$. Once again, our initialization guarantees that this holds only in the process column where matrix column $c[0]$ is active. After skipping the send instruction for the right-hand side, this process column terminates its quantification over ℓ. Other process columns have, of course, completed the back-solve procedure earlier.

4.3.3   Performance Analysis

Let us first examine the memory economies of the revised representation of the pivot information. The number of former pivot columns in process column q was denoted by L_q. Because there are $J_q = \#\mathcal{J}_q$ matrix columns in process column q, we have that $L_q \leq J_q$, and inequality occurs only if the computed rank is less than the column dimension of the coefficient matrix. Moreover, assuming a load-balanced column distribution, we have that $J_q \approx N/Q$. It follows that the revised representation requires less memory if $5N/Q < (M+N)$ or, with $M = N$, if $Q > 2.5$. The memory-use ratio, which compares the new and the old representation, is given by

$$\frac{5N/Q}{M+N} = \frac{5}{2Q}$$

if $M = N$. This shows that Q should be as large as possible if memory is a problem.

Under our standard performance-analysis assumptions, at most P processes can be active simultaneously in programs Back-Solve-5 and Back-Solve-6. Step k of program Back-Solve-5 and the corresponding step in program Back-Solve-6 perform $2k - 1$ floating-point operations distributed over P processes. The communication consists of broadcasting $x[\hat{j}]$ to all P processes of the active process column and sending array b to the next active process column (if it differs from the current active process column). The execution time for step k, under the assumption of perfect load balance, is then estimated by

$$\frac{2k - 1}{P}\tau_A + \log_2(P)\tau_C(1) + \tau_C\left(\frac{k}{P}\right).$$

The second term of this expression is a pessimistic execution-time estimate for broadcasting $x[\hat{j}]$ via recursive doubling over P processes.

Given a fixed number of nodes R and one process per node, the execution time of the back-solve procedure is typically minimized by choosing $Q = 1$ and $P = R$. Then, the floating-point operations are performed with maximal concurrency, and sending the right-hand-side vector between process columns is entirely avoided. Only if the broadcast term dominates, is it possible that some $Q > 1$ is a better choice.

In practice, one cannot consider the back-solve algorithm without considering its context. The optimal process mesh really depends on how many triangular solves are performed per LU-decomposition. The larger this number, the more the optimal process mesh will be skewed towards small Q and large P. However, if only a few triangular-solve procedures are performed, optimal performance of the LU-decomposition dictates using two-dimensional process meshes; see Section 4.1.3.

4.4 Linear-System Solver

The combined results of the multicomputer LU-decomposition and back-solve programs lead to a complete linear-system solver. Program LU-7 incorporates the revised representation of the pivot information and shares this information through its argument list, which consists of the coefficient matrix and the pivot information.

$0..P - 1 \times 0..Q - 1 \parallel (p, q)$ **program** LU-7 $(a, L, \hat{\imath}, \hat{\jmath}, \hat{p}, \hat{q}_-, \hat{q}_+)$
declare
    $i,\ j,\ k,\ L,\ \hat{q},\ \hat{q}_o\ :$ integer ;
    $\hat{\imath},\ \hat{\jmath},\ \hat{p},\ \hat{q}_-,\ \hat{q}_+\ :$ array $[0..J_q - 1]$ of integer ;
    $\mathcal{I},\ \mathcal{J}\ :$ set of integer ;
    $a\ :$ array $[\mathcal{I}_p \times \mathcal{J}_q]$ of real ;
    $a_c\ :$ array $[\mathcal{I}_p]$ of real ;
    $a_r\ :$ array $[\mathcal{J}_q]$ of real ;
    $a_{rc}\ :$ real
initially
    $\mathcal{I},\ \mathcal{J} = \mathcal{I}_p,\ \mathcal{J}_q$;
    $L = 0$;
    $\hat{q}_o = q$;
    $\langle\ ;\ k\ :\ 0 \le k < J_q\ ::\ \hat{q}_+[k] = q\ \rangle$
assign
    $\langle\ ;\ k\ :\ 0 \le k < \min(M, N)\ ::$

        Pivot-Strategy-5$(k, \mathcal{I}, \mathcal{J}, a, a_{rc}, \hat{p}[L], \hat{\imath}[L], \hat{q}, \hat{\jmath}[L])$;

        if $a_{rc} = 0.0$ **then quit** ;
        if $p = \hat{p}[L]$ **then** $\mathcal{I} := \mathcal{I} \setminus \{\hat{\imath}[L]\}$;
        if $q = \hat{q}$ **then** $\mathcal{J} := \mathcal{J} \setminus \{\hat{\jmath}[L]\}$;

{Broadcast the Pivot Row}
if $p = \hat{p}[L]$ **then begin**
$\quad \langle\ ; j\ :\ j \in \mathcal{J}\ ::\ a_r[j] := a[\hat{\imath}[L], j]\ \rangle\ ;$
$\quad$ **send** $\{a_r[j] : j \in \mathcal{J}\}$ **to** $(\textbf{all}, q)$
end else receive $\{a_r[j] : j \in \mathcal{J}\}$ **from** $(\hat{p}[L], q)\ ;$

{Compute and Broadcast the Multiplier Column}
if $q = \hat{q}$ **then begin**
$\quad \langle\ ; i\ :\ i \in \mathcal{I}\ ::\ a_c[i] := a[i, \hat{\jmath}[L]] := a[i, \hat{\jmath}[L]]/a_{rc}\ \rangle\ ;$
$\quad$ **send** $\{a_c[i] : i \in \mathcal{I}\}$ **to** $(p, \textbf{all})$
end else receive $\{a_c[i] : i \in \mathcal{I}\}$ **from** $(p, \hat{q})\ ;$

{Elimination}
$\langle\ \|\ i, j\ :\ (i, j) \in \mathcal{I} \times \mathcal{J}\ ::\ a[i, j] := a[i, j] - a_c[i]a_r[j]\ \rangle\ ;$

if $q = \hat{q}$ **then** $\hat{q}_-[L] := \hat{q}_o\ ;$
if $L > 0$ **and** $q = \hat{q}_o$ **then** $\hat{q}_+[L - 1] := \hat{q}\ ;$
if $q = \hat{q}$ **then** $L := L + 1\ ;$
$\hat{q}_o := \hat{q}$

$\quad\rangle$
end

Upon termination of the LU-decomposition, the variable L in process column q will contain the value L_q, which is the number of pivots in process column q as well as the number of initialized entries in the integer arrays $\hat{\imath}$, $\hat{\jmath}$, $\hat{p}$, $\hat{q}_-$, and $\hat{q}_+$. It follows that the variable L is initialized to 0 and incremented by one each time a pivot occurs in the process column.

Throughout step k, the variables $\hat{p}[L]$, $\hat{q}$, $\hat{\imath}[L]$, and $\hat{\jmath}[L]$ contain the process identifiers and local indices of pivot number k. When compared with program LU-5, $\hat{p}[L]$, $\hat{\imath}[L]$, and $\hat{\jmath}[L]$ replace the variables $\hat{p}$, $\hat{\imath}$, and $\hat{\jmath}$. The variable L is incremented in the process column of the pivot, thereby saving this information only in this process column.

The variable $\hat{q}_o$ contains the process-column identifier of the previous pivot and is used in the initialization of arrays $\hat{q}_-$ and $\hat{q}_+$. Initially, we have that $\hat{q}_o = q$ in every process column q. However, this value is used only in the process column of pivot number 0 to initialize $\hat{q}_-[0]$. The motivation for this initialization is discussed in Section 4.3.2. The initialization of entries of $\hat{q}_+$ lags by one step (until the location of the next pivot is known). The last entry of $\hat{q}_+$ in the process column of the last pivot is a special case (see Section 4.3.2), which is taken care of in the **initially** section by initializing the whole array $\hat{q}_+$ to the local process-column identifier.

When the right-hand side is not carried along in the LU-decomposition of the coefficient matrix, the backward solve must be preceded by a forward solve. This is done in program Solve-LU-7.

$0..P - 1 \times 0..Q - 1 \parallel (p, q)$

    program Solve-LU-7 $(a, b, x, L, \hat{\imath}, \hat{\jmath}, \hat{p}, \hat{q}_-, \hat{q}_+)$

declare

    i, ℓ, L : integer ;

    $\hat{\imath}, \hat{\jmath}, \hat{p}, \hat{q}_-, \hat{q}_+$: array $[0..J_q - 1]$ of integer ;

    $\mathcal{I}$: set of integer ;

    a : array $[\mathcal{I}_p \times \mathcal{J}_q]$ of real ;

    b : array $[\mathcal{I}_p]$ of real ;

    x : array $[\mathcal{J}_q]$ of real ;

    t : real

initially

    $\mathcal{I} = \mathcal{I}_p$

assign

    $\langle \; ; \ell \; : \; 0 \leq \ell < L \; ::$

        if $q \neq \hat{q}_-[\ell]$ **then**

            receive $\mathcal{I}, \{b[i] : i \in \mathcal{I}\}$ **from** $(p, \hat{q}_-[\ell])$;

        if $p = \hat{p}[\ell]$ **then begin**

            $\mathcal{I} := \mathcal{I} \setminus \{\hat{\imath}[\ell]\}$;

            send $b[\hat{\imath}[\ell]]$ **to** $(\textbf{all}, q)$

        end else receive t **from** $(\hat{p}[\ell], q)$;

        $\langle \; ; i \; : \; i \in \mathcal{I} \; :: \; b[i] := b[i] - a[i, \hat{\jmath}[\ell]]t \rangle$;

        if $q \neq \hat{q}_+[\ell]$ **then**

            send $\mathcal{I}, \{b[i] : i \in \mathcal{I}\}$ **to** $(p, \hat{q}_+[\ell])$

    $\rangle$;

    $\mathcal{I} := \mathcal{I}_p \setminus \mathcal{I}$;

    $\langle \; ; \ell \; : \; L > \ell \geq 0 \; ::$

        if $q \neq \hat{q}_+[\ell]$ **then**

            receive $\mathcal{I}, \{b[i] : i \in \mathcal{I}\}$ **from** $(p, \hat{q}_+[\ell])$;

        if $p = \hat{p}[\ell]$ **then begin**

            $\mathcal{I} := \mathcal{I} \setminus \{\hat{\imath}[\ell]\}$;

            $x[\hat{\jmath}[\ell]] := b[\hat{\imath}[\ell]]/a[\hat{\imath}[\ell], \hat{\jmath}[\ell]]$;

            send $x[\hat{\jmath}[\ell]]$ **to** $(\textbf{all}, q)$

        end else receive $x[\hat{\jmath}[\ell]]$ **from** $(\hat{p}[\ell], q)$;

        $\langle \; ; i \; : \; i \in \mathcal{I} \; :: \; b[i] := b[i] - a[i, \hat{\jmath}[\ell]]x[\hat{\jmath}[\ell]] \rangle$;

        if $q \neq \hat{q}_-[\ell]$ **then**

            send $\mathcal{I}, \{b[i] : i \in \mathcal{I}\}$ **to** $(p, \hat{q}_-[\ell])$

    $\rangle$

end

The initialization of the index set $\mathcal{I}$ for use in the back-solve procedure (the second quantification over ℓ) is simpler than the corresponding initialization in program Back-Solve-6. Because the forward and backward solves have the same active rows, it suffices to subtract the rows left active by the forward solve from the set of all rows.

4.5   Sparse Systems

Several sparse-matrix representations can be incorporated into the LU-decomposition and triangular-solve programs of previous sections. However, incorporating a sparse-matrix representation into the programs is only a small part of the problem. The LU-decomposition introduces nonzero entries, and the triangular factors of a sparse matrix may have a considerable amount of fill.

In a sparse-matrix context, the pivot strategy must achieve three goals:

- numerical stability,

- limit the amount of fill generated by the LU-decomposition, and

- load balance.

Separately, each of these goals can be accomplished quite well. Standard pivot strategies can achieve numerical stability. Given the fill of a matrix, there exist good methods that preselect a sequence of pivots to minimize the amount of fill generated by LU-decomposition. The generated pivot sequence depends only on the fill structure of the coefficient matrix, not on the values of the matrix elements during the LU-decomposition. The preset pivots of such a *static fill-minimizing pivot strategy* cannot, in general, ensure numerical stability. Load balance requires a data distribution and a pivot strategy that are both tuned to the fill structure of the matrix. This is easily done if the other two goals can be ignored. In practice, the pivot strategy must balance all three goals to achieve an effective and economical compromise.

In sparse LU-decomposition, another source of concurrency becomes important. Elimination steps k and $k + 1$ of the LU-decomposition can be combined and performed simultaneously if

$$a[r[k], c[k + 1]] = a[r[k + 1], c[k]] = 0.$$

To exploit this observation, the pivot strategy must return a set of pivots that mutually satisfy the above condition. This adds a fourth goal to be achieved by the pivot strategy:

- find concurrent pivots.

Concurrent pivots are particularly effective for sparse matrices of special structure that allow the concurrent elimination of whole blocks.

It should come as no surprise that concurrent sparse LU-decomposition is a technology on the front lines of research in concurrent scientific computing. Algorithms, data structures, and adaptive data distributions for sparse matrices and the numerical stability of concurrent pivots are all under active investigation.

4.6   Notes and References

The formulation of Theorems 2 and 3 is from Golub and Van Loan [40]. However, similar formulations and proofs for the theorems can be found in most introductory numerical-analysis texts. Early work on concurrent algorithms for linear algebra is surveyed by Miranker [65] and Heller [47]. More recent work is surveyed by Gallivan, Plemmons, and Sameh [37]. Much of the surveyed work concerns LU-decomposition, but it also includes discussions on elementary matrix operations, iterative methods, and other decomposition methods besides LU-decomposition. Heath, Ng, and Peyton [46] survey concurrent sparse solvers.

Our discussion of pivoting focuses on the algorithmic aspect of implementing certain strategies. One should not assume that all pivoting strategies are equivalent or are numerically stable. The analysis of pivoting strategies has a long and well-documented history, which is not repeated here. Because row pivoting is the most frequently used in sequential computations, most concurrent LU-decompositions are also based on row pivoting. Although there is no developed theory, there are substantial heuristic arguments to use multirow pivoting for concurrent LU-decompositions.

Concurrent LU-decomposition algorithms for full matrices is probably one of the most frequently written concurrent programs. It should be no surprise that the number of papers on the subject is overwhelming. The different algorithms can be categorized by the process mesh, by the distribution of the coefficient matrix over the process mesh, and by the pivoting strategy. Originally used by Van de Velde [82] to study the relation between data distribution, pivoting, and performance, the data-distribution-independent LU-decomposition of this chapter covers most other algorithms. Particular combinations of data distribution and pivot strategy do allow for additional optimization. While such opportunities are lost in the generalization, new opportunities for increased performance arise because of the generalization. For example, data-distribution-independent algorithms make it possible to adapt the distribution dynamically to the computation for increased load balance; see Van de Velde and Lorenz [83].

The Gauss–Jordan algorithm eliminates all pivot-column elements instead of only the subdiagonal elements. As a result, it need not be followed by a triangular-solve step. The multicomputer-program development of the Gauss–Jordan algorithm is virtually identical to that of LU-decomposition. The Cholesky decomposition is applicable to symmetric positive-definite matrices and does not require pivoting. Not surprisingly, the derivation of the concurrent Cholesky-decomposition algorithm is almost identical to that of LU-decomposition without pivoting.

The triangular-solve algorithms can be further optimized. Unfortunately, many optimizations come at the cost of generality and impose a strict set of compatibility criteria on the data distribution. Possible optimizations include: pipelining the communication, overlapping arithmetic and communication, and pipelining multiple right-hand sides.

Exercises

Exercise 20 *Reformulate program LU-2 (LU-decomposition without pivoting on a $P \times Q$ process mesh), such that there is no separate broadcast of the pivot. Can this be generalized to multicomputer LU-decomposition with pivoting?*

Exercise 21 *If the matrix A is symmetric positive definite, it has a decomposition of the form:*

$$A = L^T L,$$

where L is a lower-triangular matrix. This is called the Cholesky decomposition. Write down the specification program and develop it into a multicomputer program.

Exercise 22 *There are other distributed representations of the pivot information. Consider the following alternative: "invert" the arrays r and c. For every row index m, let $k_r[m]$ be the LU-decomposition step in which row m was the pivot row. Similarly, for every column index n, let $k_c[n]$ be the LU-decomposition step in which column n was the multiplier column. When developing the multicomputer program, arrays k_r and k_c are transformed into row- and column-distributed vectors, respectively.*

Adapt the back-solve procedure such that it uses this compact representation of the pivot information?

Exercise 23 *Given the LU-decomposition of the matrix $A = LC^T I_{N,M} RU$ and many right hand sides $B = \begin{bmatrix} \vec{b}_0 & \vec{b}_1 & \dots \end{bmatrix}$, one may, in principle, solve all right-hand sides concurrently. However, the data distribution of the matrix A makes this difficult to achieve in practice. It is possible, however, to pipeline the individual triangular solves. Develop and write the program.*

5
QR-Decomposition

The *QR-decomposition* of a matrix is used to solve nearly singular, overdetermined, and underdetermined systems of linear equations. In specialized form, it is an important component of the *QR-algorithm*, which computes all eigenvalues and eigenvectors of a matrix. The multicomputer program for QR-decomposition is an application of recursive doubling, but several interesting complications arise.

5.1 Introduction

5.1.1 Orthogonal Matrices

A matrix $Q \in \mathbb{R}^{M \times M}$ is an orthogonal matrix if

$$Q^T Q = I_{M,M}, \qquad (5.1)$$

where $I_{M,M}$ is the $M \times M$ identity matrix. Equation 5.1 implies that the set of M column vectors of an orthogonal matrix is an orthonormal basis of $\mathbb{R}^M$. Conversely, any orthonormal basis $\vec{q}_0, \vec{q}_1, \ldots, \vec{q}_{M-1}$ of $\mathbb{R}^M$ defines an $M \times M$ orthogonal matrix $Q = [\vec{q}_0 \ \vec{q}_1 \ \ldots \ \vec{q}_{M-1}]$. The study of orthogonal matrices is, therefore, equivalent to the study of orthonormal bases. The inverse of an orthogonal matrix is its transpose:

$$Q^{-1} = Q^T. \qquad (5.2)$$

In turn, this implies that orthogonal matrices preserve the norm of a vector:

$$\| Q\vec{x} \|^2 = \vec{x}^T Q^T Q\vec{x} = \vec{x}^T \vec{x} = \| \vec{x} \|^2 . \tag{5.3}$$

5.1.2   Gram–Schmidt Orthogonalization

Gram–Schmidt orthogonalization is a well-known procedure to compute an orthonormal basis for a subspace $\mathcal{A}_N = \text{span}(\vec{a}_0, \vec{a}_1, \ldots, \vec{a}_{N-1})$ of $\mathbb{R}^M$. The dimension of $\mathcal{A}_N$ is less than or equal to N, and equality is attained only if the vectors $\vec{a}_n$ are linearly independent.

Gram–Schmidt orthogonalization replaces a sequence of independent vectors $\vec{a}_0, \vec{a}_1, \ldots, \vec{a}_{N-1}$ with orthonormal-basis vectors $\vec{q}_0, \vec{q}_1, \ldots, \vec{q}_{N-1}$. The procedure consists of N steps, and each step computes one basis vector. We make the following induction hypothesis. At the start of step k, the vectors $\vec{a}_0, \vec{a}_1, \ldots, \vec{a}_{k-1}$ already contain $\vec{q}_0, \vec{q}_1, \ldots, \vec{q}_{k-1}$. Moreover, the vectors $\vec{a}_k, \vec{a}_{k+1}, \ldots, \vec{a}_{N-1}$ have no components along $\vec{q}_0, \vec{q}_1, \ldots, \vec{q}_{k-1}$. This induction hypothesis trivially holds at the start of step 0.

By induction, the vector $\vec{a}_k$ is already orthogonal with respect to the previous vectors in the sequence, which contain $\vec{q}_0, \vec{q}_1, \ldots, \vec{q}_{k-1}$. It follows that the next basis vector, $\vec{q}_k$, can be found simply by normalizing $\vec{a}_k$. To establish the induction hypothesis for step $k+1$, we only have to subtract from $\vec{a}_{k+1}, \vec{a}_{k+2}, \ldots, \vec{a}_{N-1}$ their component along $\vec{q}_k$.

The following program segment allows for the possibility that the initial vectors might be linearly dependent. In this case, some of the computed "basis vectors" will vanish.

$$\langle \; ; \; n \; : \; 0 \leq n < N \; ::$$

$$\alpha := (\vec{a}_n, \vec{a}_n) \; ;$$

if $\alpha \neq 0$ **then begin**

$$\vec{a}_n := \vec{a}_n / \sqrt{\alpha} \; ;$$

$$\langle \; \| \; j \; : \; n+1 \leq j < N \; :: \; \vec{a}_j := \vec{a}_j - (\vec{a}_j, \vec{a}_n)\vec{a}_n \; \rangle$$

end

$$\rangle$$

A multicomputer version was developed in Exercise 4 of Chapter 1.

Unfortunately, accumulation of round-off errors causes nonorthogonality of the computed vectors. For large N, we are forced to consider alternative methods that avoid this problem.

5.1.3   QR-Decomposition

Gram–Schmidt orthogonalization in exact arithmetic implies the existence of the QR-decomposition of a matrix, as is shown by the following theorem.

Theorem 5 *For any $M \times N$ matrix A of rank $K \leq \min(M, N)$, there exists an $N \times N$ permutation matrix C, an $M \times M$ orthogonal matrix Q, and an $M \times N$ upper-triangular matrix R such that $AC^T = QR$.*

If the matrix A has rank K, exactly K column vectors of A are linearly independent. Let C be a permutation matrix such that these vectors are the first K columns of the matrix $AC^T = [\vec{a}_0 \vec{a}_1 \ldots \vec{a}_{N-1}]$. It suffices to apply Gram–Schmidt orthogonalization to the first K column vectors of AC^T and to set

$$\hat{Q} = [\vec{q}_0 \vec{q}_1 \ldots \vec{q}_{K-1}].$$

From the observation that

$$\forall n \in 0..K - 1 : \quad \vec{a}_n \in \mathrm{span}(\vec{q}_0, \vec{q}_1, \ldots, \vec{q}_n)$$
$$\forall n \in K..N - 1 : \quad \vec{a}_n \in \mathrm{span}(\vec{q}_0, \vec{q}_1, \ldots, \vec{q}_{K-1}),$$

it follows that $AC^T = \hat{Q}\hat{R}$, with $\hat{R}$ a $K \times N$ upper-triangular matrix. The matrix $\hat{Q}$ is an $M \times K$ matrix with mutually-orthogonal column vectors.

Expand the basis $\vec{q}_0, \vec{q}_1, \ldots, \vec{q}_{K-1}$ with $M - K$ additional vectors to obtain an orthonormal basis for $\mathbb{R}^M$ and the $M \times M$ orthogonal matrix $Q = [\vec{q}_0, \vec{q}_1, \ldots, \vec{q}_{M-1}]$. Then, we have that $AC^T = QR$, where R is the $M \times N$ matrix obtained by expanding $\hat{R}$ with $M - K$ zero rows.   □

5.1.4 QR-Decomposition and Linear Systems

Consider the system of linear equations

$$A\vec{x} = \vec{b}$$

with $A \in \mathbb{R}^{M \times N}$ a matrix of rank K.

If $K = N$, all column vectors of the matrix A are linearly independent, and the proof of Theorem 5 implies that the QR-decomposition of the matrix A does not require column permutations. Setting $C = I_{N,N}$, we have that $A = QR$ and, using Equation 5.3, that

$$\forall \vec{x} \in \mathbb{R}^N : \parallel \vec{b} - A\vec{x} \parallel = \parallel Q^T\vec{b} - R\vec{x} \parallel .$$

The last $M - N$ rows of R vanish, and the vector minimizing the norm of the residual is, therefore, the solution of the $N \times N$ principal subsystem of

$$R\vec{x} = Q^T\vec{b}.$$

If $M = N = K$, this system, its $N \times N$ principal subsystem, and the original system $A\vec{x} = \vec{b}$ are equivalent. The QR-decomposition is useful for regular systems with nearly singular coefficient matrices, because the LU-decomposition of these matrices may fail in floating-point arithmetic. For robustness sake, one might prefer the guaranteed stable, but more expensive, QR-decomposition approach.

The more interesting case occurs when $M > N = K$. Then, the solution of the $N \times N$ principal subsystem is the vector that minimizes the norm of the residual. This proposed approximate-solution method for *overdetermined systems* is called a *least-squares-minimization procedure*.

For *underdetermined systems*, where $K < N$, column pivoting is required to obtain a QR-decomposition. There exist several dynamic pivot strategies that select the most independent column vectors according to some criterion. Although the numerical details of pivoting for QR-decomposition differ from those of pivoting for LU-decomposition, pivoting gives rise to the same algorithmic problems. These can be dealt with as in Chapter 4. In the remainder of this chapter, we shall ignore column pivoting and assume that $K = N$.

5.2 Householder QR-Decomposition

5.2.1 Householder Reflections

Householder reflections are orthogonal matrices of the form:

$$H = I - 2\frac{\vec{u}\vec{u}^T}{\vec{u}^T\vec{u}} = I - \theta\vec{u}\vec{u}^T.$$

In computations, Householder reflections can be used to annihilate vector components. Given a vector $\vec{x} = [\xi_m]_{m=0}^{M-1}$, an index set $\mathcal{K} \subset 0..M-1$, and an index $k \in \mathcal{K} \subset 0..M-1$, construct the vector $\vec{u} = [v_m]_{m=0}^{M-1}$ with

$$v_m = \begin{cases} \xi_m & \textbf{if } m \in \mathcal{K} \setminus \{k\} \\ \xi_k + \alpha & \textbf{if } m = k \\ 0 & \textbf{if } m \notin \mathcal{K} \end{cases}$$

and $\alpha^2 = \sum_{m\in\mathcal{K}} \xi_m^2$. The sign of α is chosen equal to that of ξ_k to avoid summing nearly equal quantities of opposite sign in v_k. The Householder reflection based on this vector $\vec{u}$ is denoted by

$$H_{k,\mathcal{K}} = H_{k,\mathcal{K}}(\vec{x}),$$

and when this transformation is applied to $\vec{x}$, we obtain the vector

$$\vec{y} = [\eta_m]_{m=0}^{M-1} = H_{k,\mathcal{K}}(\vec{x})\vec{x}.$$

An easy calculation shows that

$$\eta_m = \begin{cases} 0 & \textbf{if } m \in \mathcal{K} \setminus \{k\} \\ -\alpha & \textbf{if } m = k \\ \xi_m & \textbf{if } m \notin \mathcal{K}. \end{cases}$$

The Householder reflection $H_{k,\mathcal{K}}$ annihilates components $m \in \mathcal{K} \setminus \{k\}$.

Given a vector $\vec{x}$, an index set $\mathcal{K}$, and an index k, program Householder computes the Householder reflection $H_{k,\mathcal{K}}(\vec{x})$.

> **program** Householder$(k, \mathcal{K}, x, \theta, u)$
> **declare**
> > x, u : array $[0..M-1]$ of real ;
> > θ, α : real ;
> > k, m : integer ;
> > $\mathcal{K}$: set of integer
>
> **assign**
> > $\langle \parallel m : m \in 0..M-1 :: u[m] := 0 \rangle$;
> > $\langle \parallel m : m \in \mathcal{K} :: u[m] := x[m] \rangle$;
> > $\alpha := \langle + m : m \in \mathcal{K} :: x[m]^2 \rangle$;
> > $\alpha := \text{sign}(x[k])\sqrt{\alpha}$;
> > $u[k] := x[k] + \alpha$;
> > $\theta := 1/(\alpha u[k])$
>
> **end**

Computing the vector $\vec{u}$ and the scalar θ of a Householder reflection requires approximately $2(\#\mathcal{K})$ floating-point operations and one square-root evaluation.

Programs that use Householder reflections often evaluate vector assignments of the form

$$\vec{a} := H_{k,\mathcal{K}}\vec{a}.$$

This is implemented by program Householder-Assignment, which is based on the observation that

$$H_{k,\mathcal{K}}\vec{a} = (I - \theta\vec{u}\vec{u}^T)\vec{a} = \vec{a} - \theta(\vec{u}^T\vec{a})\vec{u}.$$

> **program** Householder-Assignment$(\mathcal{K}, \theta, u, a)$
> **declare**
> > a, u : array $[0..M-1]$ of real ;
> > θ, τ : real ;
> > m : integer ;
> > $\mathcal{K}$: set of integer
>
> **assign**
> > $\tau := \langle + m : m \in \mathcal{K} :: u[m]a[m] \rangle$;
> > $\tau := \theta\tau$;
> > $\langle \parallel m : m \in \mathcal{K} :: a[m] := a[m] - \tau u[m] \rangle$
>
> **end**

The product of a Householder reflection and an arbitrary vector requires approximately $4(\#\mathcal{K})$ floating-point operations.

5.2.2   QR-Decomposition

Given a matrix $A \in \mathbb{R}^{M \times N}$, we consider computing its QR-decomposition by incremental updating of the trivial decomposition $A = Q_0 R_0$ with $Q_0 = I_{M,M}$ and $R_0 = A$. In step k, the decomposition $A = Q_k R_k$ is modified into the decomposition $A = Q_{k+1} R_{k+1}$, such that Q_{k+1} remains orthogonal, R_{k+1} is nearer to upper-triangular form, and the new factors can be computed by incremental updates of the old factors. Because the product of two orthogonal matrices is an orthogonal matrix, we choose an orthogonal matrix U_k as incremental update and set

$$
\begin{aligned}
Q_{k+1} &= Q_k U_k^T \\
R_{k+1} &= U_k R_k.
\end{aligned}
$$

It remains to choose the orthogonal incremental updates U_k appropriately.

Given the matrix $A = [\vec{a}_0 \vec{a}_1 \ldots \vec{a}_{N-1}]$, the first incremental update U_0 is the Householder reflection $H_{0,0..M-1}(\vec{a}_0)$, which zeroes components 1 through $M - 1$ of the column vector $\vec{a}_0$. In the following step, components 2 through $M - 1$ of the updated column vector $\vec{a}_1$ are zeroed by a Householder reflection $H_{1,1..M-1}(\vec{a}_1)$. In step k, components $k + 1$ through $M - 1$ of the updated column vector $\vec{a}_k$ are zeroed by a Householder reflection $H_{k,k..M-1}(\vec{a}_n)$. It is easily verified that these Householder reflections do not destroy zeroes created in previous steps and that the QR-decomposition is obtained after N steps. We obtain the program outline:

```
Q := I ;
K := 0..M − 1 ;
〈 ; k : 0 ≤ k < N ::
    U := H_{k,K}(\vec{a}_k) ;
    Q, A := QU^T, UA ;
    K := K \ {k}
〉
```

In step k, the Householder reflection $H_{k,k..M-1}(\vec{a}_k)$ is computed and applied to $(N - k - 1)$ column vectors of the matrix A. Ignoring the computation of Q (in practice, Q is left in a factored representation or is not computed at all), the approximate operation count is given by

$$
\sum_{k=0}^{N-1} \left(2(M - 1 - k) + \sum_{n=k+1}^{N-1} 4(M - 1 - k) \right) \approx 2MN^2. \qquad (5.4)
$$

At this level of specification, there is no concurrency. However, each assignment of the program outline corresponds to an extensive vector and matrix operation with concurrency available by simple data distribution. Further transformation into a multicomputer program is left as an exercise.

5.3   Givens QR-Decomposition

5.3.1   Givens Rotations

Givens rotations are orthogonal matrices of the form

$$
G_{k,j} = \begin{bmatrix}
1 & 0 & \cdots & 0 & \cdots & 0 & \cdots & 0 \\
0 & 1 & & 0 & \cdots & 0 & \cdots & 0 \\
\vdots & & & & & & & \vdots \\
0 & 0 & & c & & s & & 0 \\
\vdots & \vdots & & & & & & \vdots \\
0 & 0 & & -s & & c & & 0 \\
\vdots & \vdots & & & & & & \\
0 & 0 & \cdots & 0 & \cdots & 0 & & 1
\end{bmatrix},
$$

where $c = \cos(\theta)$ and $s = \sin(\theta)$ for some θ. If the coefficients c are found in elements (k,k) and (j,j) of the matrix, then $-s$ and s are found in elements (j,k) and (k,j), respectively.

Givens rotations, like Householder reflections, can be used to annihilate components of a vector. To annihilate component j with component k of the vector $\vec{x}$, choose the Givens rotation $G_{k,j} = G_{k,j}(\vec{x})$ with

$$
c = \xi_k / \sqrt{\xi_k^2 + \xi_j^2} \quad \text{and} \quad s = \xi_j / \sqrt{\xi_k^2 + \xi_j^2}.
$$

Then, we have that

$$
G_{k,j}\vec{x} = \begin{bmatrix}
[\xi_m]_{m=0}^{k-1} \\
\sqrt{\xi_k^2 + \xi_j^2} \\
[\xi_m]_{m=k+1}^{j-1} \\
0 \\
[\xi_m]_{m=j+1}^{M-1}
\end{bmatrix}.
$$

Given a vector $\vec{x}$ and indices k and j, program Givens computes the Givens rotation $G_{k,j}(\vec{x})$.

program Givens(k, j, x, c, s)
declare
    x : array $[0..M-1]$ of real ;
    c, s, t : real ;
    j, k : integer
assign
    $t := \sqrt{x[k]^2 + x[j]^2}$;
    $c, s := x[k]/t, x[j]/t$
end

The vector assignment for arbitrary vectors $\vec{a}$

$$\vec{a} := G_{k,j}\vec{a}$$

reduces to one simultaneous assignment; see program Givens-Assignment.

program Givens-Assignment(k, j, c, s, a)
declare
    a : array $[0..M-1]$ of real ;
    j, k : integer
assign
    $a[k], a[j] := ca[k] + sa[j], -sa[k] + ca[j]$
end

The computation of a Givens rotation takes five floating-point operations and one square-root evaluation. The product of a Givens rotation with an arbitrary vector requires six floating-point operations.

5.3.2   QR-Decomposition

As in Section 5.2.2, we use an incremental-update strategy, but this time we use Givens rotations instead of Householder reflections as incremental updates. Annihilating the subdiagonal elements in column-by-column order, the following program outline results:

$Q := I$;
$\langle\ ;\ k\ :\ 0 \leq k < N\ ::$
    $\langle\ ;\ m\ :\ k < m < M\ ::$
        $U := G_{k,m}(\vec{a}_k)$;
        $Q, A := QU^T, UA$
    $\rangle$
$\rangle$

In step k of the computation, $(M-k-1)$ Givens rotations are computed, each of which is applied to $(N-k-1)$ columns of the matrix A. Ignoring the computation of Q, we obtain an operation count of

$$\sum_{k=0}^{N-1} \sum_{m=k+1}^{M-1} (6 + \sum_{n=k+1}^{N-1} 6) \approx 3MN^2.$$

Here, a square-root evaluation is counted, rather optimistically, as one floating-point operation. Comparison with Equation 5.4 shows that Givens QR-decomposition requires about 50% more floating-point operations than Householder QR-decomposition. Numerically, both procedures are virtually equivalent and, for sequential programs, the Householder approach is preferred. The sequential Givens QR-decomposition is usually reserved for sparse-matrix computations.

For comparison with the multicomputer program, a detailed specification of the above program outline follows. Program QR-Givens-1 does not compute or store the matrix Q. This could be accomplished by performing the QR-decomposition on the extended matrix $[A|I_{M,M}]$, in which case the matrix $[R|Q^T]$ would be obtained upon termination.

> **program** QR-Givens-1
> **declare**
> > m, n, k : integer ;
> > a : array $[0..M-1 \times 0..N-1]$ of real ;
> > c, s, t : real
>
> **assign**
> > $\langle$; k : $0 \le k < N$ ::
> > > $\langle$; m : $k < m < M$ ::
> > > > $t := \sqrt{a[k,k]^2 + a[m,k]^2}$;
> > > > **if** $t \ne 0$ **then begin**
> > > > > $a[k,k]$, c, $s := t$, $a[k,k]/t$, $a[m,k]/t$;
> > > > > $\langle$; n : $k < n < N$ ::
> > > > > > $$\begin{bmatrix} a[k,n] \\ a[m,n] \end{bmatrix} := \begin{bmatrix} c & s \\ -s & c \end{bmatrix} \begin{bmatrix} a[k,n] \\ a[m,n] \end{bmatrix}$$
> > > > > $\rangle$
> > > > **end**
> > > $\rangle$
> > $\rangle$
> **end**

5.4 Multicomputer QR-Decomposition

As in Chapter 4, the matrix A is distributed over a $P \times Q$ process mesh by means of a two-dimensional data distribution $\mu \times \nu$. After applying the recursive-doubling strategy discussed below and after converting global into local indices, we shall obtain multicomputer program QR-Givens-2. In the discussion that follows, we assume that $Q = 1$ and ignore the column distribution ν. The generalization to $Q \ne 1$ need not complicate the text, because it is easily achieved by a simple pivot-column broadcast.

The algorithms for QR- and LU-decomposition are similar. Both deactivate one row and one column of the matrix in each step of the computation. In step k of program QR-Givens-1, matrix elements $a[m,k]$ with $k < m < M$ are annihilated by means of Givens rotations $G_{k,m}$. This is not unlike LU-decomposition without pivoting, where row k is used as a pivot row for all annihilations in column k. The concurrent LU-decomposition algorithm was obtained by broadcasting the pivot row to all parts of the distributed matrix. Unfortunately, that is not a viable approach here. Each Givens rotation of the form $G_{k,m}$ transforms not only row m, but also the

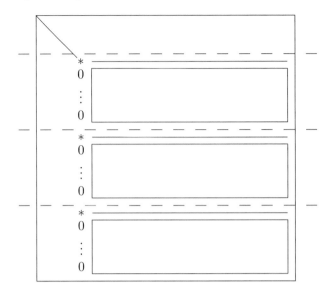

FIGURE 5.1. Illustration of the result of local elimination.

pivot row itself. This precludes using simple broadcasts and sequentializes the Givens rotations of program QR-Givens-1.

To find concurrency, we must return to Section 5.2.2 and the general incremental-update idea, which gives us the freedom to apply any orthogonal transformations in any order. This flexibility will be crucial to derive an efficient concurrent program. One must bear in mind, however, that changing the elementary transformations or their order also changes the numerics: strictly speaking, concurrent QR-decomposition programs do not implement program QR-Givens-1. Somewhat arbitrarily, we restrict our attention to Givens rotations, and we search for an annihilation order in which many Givens rotations may be applied concurrently.

To annihilate elements of the pivot column, begin by introducing one *local pivot* in each process. By means of Givens rotations between the local pivot rows and other rows, all other active elements of the pivot column can be annihilated. This is called the *local elimination*. As can be seen from program QR-Givens-2, this step is virtually identical to the sequential program QR-Givens-1. After the local elimination, each process contains at most one active matrix row with a nonzero leading element. The result of the local elimination is illustrated in Figure 5.1. For the purpose of graphic illustration, the matrix is distributed linearly over the processes. Dashed lines represent process boundaries. In process 0, all rows have been deactivated. In the other processes, only the local pivot row has a nonzero leading element.

The *global elimination* annihilates leading elements of local pivot rows. As a first approximation to the actual global elimination, consider a problem

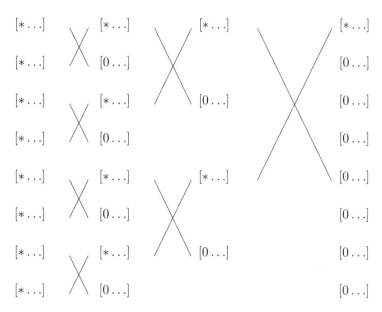

FIGURE 5.2. Concurrent global elimination of eight rows.

with P rows distributed over P processes. Representing rows with a nonzero leading element by $[*\ldots]$ and rows with a zero leading element by $[0\ldots]$, the desired transformation is given by

$$\begin{matrix} [*\ldots] \\ [*\ldots] \\ \vdots \\ [*\ldots] \end{matrix} \longrightarrow \begin{matrix} [*\ldots] \\ [0\ldots] \\ \vdots \\ [0\ldots] \end{matrix}$$

Figure 5.2 shows how, in the case $P = 8$, recursive doubling can be used to compute this transformation by means of Givens rotations. For $P = 2^D$, the procedure consists of D steps. In step d, where $0 \le d < D$, Givens rotations are applied between rows p and $q = p \,\triangledown\, 2^d$. The operations performed on row p depend on whether the leading element of row p or that of row q is to be annihilated. Let α_p and α_q be the leading elements of row vectors $\vec{r}_p$ and $\vec{r}_q$, respectively, and let $t = \sqrt{\alpha_p^2 + \alpha_q^2}$. To annihilate its own leading element, row vector $\vec{r}_p$ is transformed according to

$$\vec{r}_p := (\alpha_q \vec{r}_p - \alpha_p \vec{r}_q)/t.$$

On the other hand, when the leading element of row vector $\vec{r}_q$ is annihilated, row vector $\vec{r}_p$ is modified according to

$$\vec{r}_p := (\alpha_p \vec{r}_p + \alpha_q \vec{r}_q)/t.$$

In program QR-Givens-2, the leading elements α_p and α_q are represented by the variables $a0_{rc}$ and $a1_{rc}$, respectively. The row vectors $\vec{r}_p$ and $\vec{r}_q$ are represented by arrays $a0_r$ and $a1_r$, respectively.

In this *asymmetric recursive-doubling procedure*, process p needs a criterion to decide whether row vector $\vec{r}_p$ is annihilated or modified in step d. If the global pivot is in row 0, step d annihilates the leading element of row p if bit number d of the binary representation of p is set. This criterion corresponds to the following test:

$$\textbf{if } (p \wedge 2^d) = 2^d \textbf{ then} \text{ annihilate } \textbf{else} \text{ modify}$$

Note that process $q = p \bar{\vee} 2^d$ always takes the opposite action of process p. Moreover, only row 0 can have a nonzero leading element after D steps, because $p = 0$ is the only process identifier whose bits are never set.

A nonzero leading element can be produced in any row $\hat{p}$ by replacing p in the above test with a suitable permutation of the process identifiers. This permutation must be such that $\hat{p}$ plays the role of 0. The permutation $p \bar{\vee} \hat{p}$ satisfies this requirement, and it suffices to replace the above test by:

$$\textbf{if } ((p \bar{\vee} \hat{p}) \wedge 2^d) = 2^d \textbf{ then} \text{ annihilate } \textbf{else} \text{ modify}$$

to obtain a nonzero leading element in row $\hat{p}$. A row stops participating in the recursive-doubling procedure as soon as its leading element is annihilated. In program QR-Givens-2, process p sets its boolean variable *active* to **false** as soon as it has annihilated the leading element of its local pivot row.

As the QR-decomposition proceeds, rows are deactivated. Eventually, some processes have no active rows and cannot participate in the global elimination. This is a problem, because the set of participating processes may change from step k to step $k + 1$. To avoid this problem, processes without any active rows use a dummy local pivot row consisting entirely of zero elements. A Givens rotation with a zero row results in at most a permutation and does not change the numerical aspects of the program. While solving one problem, these dummy rows introduce another: because of row permutations during the recursive-doubling procedure, the contents of actual matrix rows may end up in memory reserved for dummy rows! This is illustrated in Figure 5.3. Round brackets represent dummy rows; square brackets represent actual rows. The contents of rows may be any of three types. Empty rows consist of zeroes only and are represented by $\emptyset$. Rows with nonzero leading element are represented by $* \ldots$. Rows with zero leading element are represented by $0 \ldots$. Initially, rows 4 and 6 are empty dummy rows $(\emptyset)$, row 5 is an actual row with zero leading element $[0\ldots]$, and the others are actual rows with nonzero leading element $[*\ldots]$. When the recursive-doubling procedure terminates, the dummy row in process 4 contains actual data: a matrix row with zero leading element. Row 7, an actual row, contains only zeroes. To understand that this is a problem,

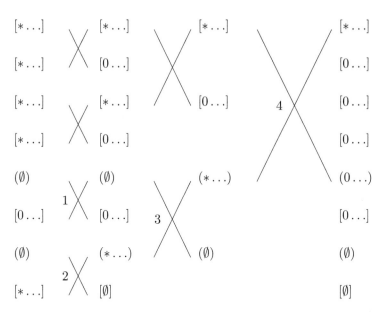

FIGURE 5.3. Concurrent global elimination of eight rows in the presence of dummy rows.

consider the global-elimination segment of program QR-Givens-2. This segment copies the local pivot row to a buffer array $a0_r$, does the recursive doubling, and copies the buffer array back to the local pivot row. The last step is impossible when the buffer array is a dummy, because there is no local pivot row to copy to.

In this example, the remedy is easy: just exchange data between rows 4 and 7. An exchange is an orthogonal transformation and, therefore, allowed by the incremental-update strategy. This case is solved by one exchange, but what happens in general? We need an algorithm to detect, record, and undo spurious data exchanges.

The idea is to record those Givens rotations that put real data into a dummy row. The spurious exchanges of those Givens rotations are then undone in the reverse order of their occurrence. Consider an elementary Givens rotation in any step of the recursive-doubling procedure. Its input consists of two rows, and each row is one of six types: $[*\ldots]$, $[0\ldots]$, $[\emptyset]$,

FIGURE 5.4. Spurious exchanges into dummy rows.

$(*\ldots)$, $(0\ldots)$, or $(\emptyset)$. In all, there are 36 different cases to consider. However, only the 2 cases pictured in Figure 5.4 exchange data into a dummy row. A spurious exchange takes place only if an empty dummy row is used to annihilate a row with nonzero leading element. In program QR-Givens-2, these exchanges are recorded in the bits of integer variable $undo$: bit number d is set if exchange number d needs to be undone. To help identifying spurious exchanges, boolean variables $dummy$ and $empty$ are associated with each buffer array $a0_r$ and $a1_r$. The buffer $a0_r$ contains a copy of the local pivot row of process p, while the buffer $a1_r$ contains the local pivot row of a remote process.

$0..P-1 \times 0..Q-1 \parallel (p,q)$ **program** QR-Givens-2
declare
      a : array $[\mathcal{I}_p \times \mathcal{J}_q]$ of real ;
      $a0_r$, $a1_r$: array $[\mathcal{J}_q]$ of real ;
      a_c : array $[\mathcal{I}_p]$ of real ;
      $a0_{rc}$, $a1_{rc}$, c, s, t : real ;
      k, d, i, j, $\hat{\imath}$, $\hat{\jmath}$, i_r, $\hat{p}$, $\hat{q}$, $undo$: integer ;
      $\mathcal{I}$, $\mathcal{J}$: set of integer ;
      $dummy0$, $dummy1$, $empty0$, $empty1$, $active$: boolean
initially
      $\mathcal{I}, \mathcal{J} = \mathcal{I}_p, \mathcal{J}_q$
assign
      $\langle$; k : $0 \le k < N$ ::
          { Broadcast Pivot Column }
          $\hat{q}, \hat{\jmath} := \nu(k)$;
          if $q = \hat{q}$ **then begin**
              $\mathcal{J} := \mathcal{J} \setminus \{\hat{\jmath}\}$;
              $\langle$; i : $i \in \mathcal{I}$ :: $a_c[i] := a[i, \hat{\jmath}] \rangle$;
              send $\{a_c[i] : i \in \mathcal{I}\}$ **to** $(p, \mathbf{all})$
          end else receive $\{a_c[i] : i \in \mathcal{I}\}$ **from** $(p, \hat{q})$;
          { Choose Local Pivot Row }
          $\hat{p}, \hat{\imath} := \mu(k)$;
          if $p = \hat{p}$ **then** $\mathcal{I}$, $i_r := \mathcal{I} \setminus \{\hat{\imath}\}$, $\hat{\imath}$ **else** $i_r := \mathbf{any}(\mathcal{I})$;
          { LOCAL ELIMINATION }
          $\langle$; i : $i \in \mathcal{I} \setminus \{i_r\}$ ::
              $t := \sqrt{a_c[i_r]^2 + a_c[i]^2}$;
              if $t \ne 0$ **then begin**
                 $a_c[i_r]$, c, $s := t$, $a_c[i_r]/t$, $a_c[i]/t$;
                 $\langle$; j : $j \in \mathcal{J}$ ::

$$\begin{bmatrix} a[i_r, j] \\ a[i, j] \end{bmatrix} := \begin{bmatrix} c & s \\ -s & c \end{bmatrix} \begin{bmatrix} a[i_r, j] \\ a[i, j] \end{bmatrix}$$

                 $\rangle$
             end
        $\rangle$;

```
{ GLOBAL ELIMINATION }
if I = ∅ then begin { Dummy Local Pivot Row }
    ⟨ ; j : j ∈ J :: a0_r[j] := 0 ⟩ ;
    a0_rc, dummy0, empty0 := 0, true, true
end else begin { Actual Local Pivot Row }
    ⟨ ∥ j : j ∈ J :: a0_r[j] := a[i_r, j] ⟩ ;
    a0_rc, dummy0, empty0 := a_c[i_r], false, false
end ;
{ Recursive Doubling }
undo, active := 0, true ;
⟨ ; d : 0 ≤ d < D ::
    if active then begin
        send empty0, a0_rc, {a0_r[j] : j ∈ J} to (p∇2^d, q) ;
        receive empty1, a1_rc, {a1_r[j] : j ∈ J} from (p∇2^d, q) ;
        t := √(a0²_rc + a1²_rc) ;
        if t ≠ 0 then begin
            a0_rc, a1_rc := a0_rc/t, a1_rc/t ;
            if (p∇p̂) ∧ 2^d = 2^d then begin { Annihilate }
                if empty1 and a0_rc ≠ 0 then undo := undo ∨ 2^d ;
                ⟨ ; j : j ∈ J :: a0_r[j] := a1_rc a0_r[j] − a0_rc a1_r[j] ⟩ ;
                a0_rc := 0
            end else begin { Modify }
                if empty0 and a1_rc ≠ 0 then
                    undo, empty0 := undo ∨ 2^d, false ;
                ⟨ ; j : j ∈ J :: a0_r[j] := a0_rc a0_r[j] + a1_rc a1_r[j] ⟩ ;
                a0_rc := t
            end
        end ;
        if (p∇p̂) ∧ 2^d = 2^d then active := false
    end
⟩ ;
{ Undo the Dummy Exchanges }
⟨ ; d : D > d ≥ 0 ::
    if undo ∧ 2^d = 2^d then begin
        send a0_rc, {a0_r[j] : j ∈ J} to (p∇2^d, q) ;
        receive a0_rc, {a0_r[j] : j ∈ J} from (p∇2^d, q)
    end
⟩ ;
{ Map Nonzero Local Pivot Rows back to Matrix }
if not dummy0 then ⟨ ; j : j ∈ J :: a[i_r, j] := a0_r[j] ⟩ ;
if p = p̂ and q = q̂ then a[î, ĵ] := a0_rc
⟩
end
```

Further optimizations are possible. The local elimination can be based on Householder reflections, which require fewer floating-point operations than Givens rotations. This optimization is easily incorporated. In the global elimination, processes are idle after annihilating their leading element. They can do useful work by preparing the next step of the QR-decomposition: it is possible to annihilate some elements of the next pivot column within the recursive-doubling procedure of this step. For coarse-grained computations, this is only a minor gain. Moreover, for matrices distributed over a two-dimensional process mesh, it requires the insertion of broadcasts of the extra pivot columns. That makes it unlikely to pay off, even in fine-grained computations.

5.5   Notes and References

For a concurrent QR-decomposition algorithm based on Householder reflections, see Bowgen and Modi [10]. Chu and George [17, 18] use a combination of Householder reflections and Givens rotations. Bischof [8] studies concurrent QR-decomposition with column pivoting. Sparse QR-decomposition must be based on Givens rotations. All technical difficulties encountered in sparse LU-decomposition are also encountered here, particularly if column pivoting is required. The survey papers mentioned in Section 4.6 are also valuable references for this chapter.

Exercises

Exercise 24 *Develop a performance analysis of program QR-Givens-2.*

Exercise 25 *Develop a multicomputer program for a QR-decomposition that uses Householder reflections.*

Exercise 26 *Combine program QR-Givens-2 and an appropriate triangular solver to obtain a multicomputer least-squares minimization procedure.*

6
Tridiagonal Solvers

We shall develop one direct and one iterative solver for tridiagonal systems. The direct solver is based on a generalization of the recursive-doubling procedure known as *full recursive doubling*. The iterative tridiagonal solver is based on concurrent relaxation and is often a good alternative to any direct solver. Tridiagonal systems of linear equations often occur as part of a larger computation. They occur, for example, in some fast Poisson solvers and in alternating-direction methods. In these applications, one must usually solve many tridiagonal systems. Issues related to solving many tridiagonal systems are postponed until Chapter 8.

6.1   Full Recursive Doubling

The recursive-doubling procedure can be used to compute expressions like

$$\sum_{m=0}^{M-1} \omega_m,$$
$$\max_{0 \leq m < M} \omega_m,$$
$$\min_{0 \leq m < M} \omega_m.$$

These expressions have in common that they use an associative operator to reduce a set of numbers to one number. In generalizing the recursive-doubling procedure, we shall use the addition operator as an example. However, all derivations will be valid for any associative operator.

6.1.1 Specification

Given the array ω_m, where $0 \leq m < M$, define the array σ_m such that

$$\forall m \in 0..M - 1 : \ \sigma_m = \sum_{k=0}^{m} \omega_k.$$

The recursive-doubling procedure can be used to compute σ_{M-1}. Here, we wish to compute all other values σ_m as well. Program Full-Recursive-Doubling-1 specifies the special case $P = M$. After computing σ_m, it assigns this value to the variable $\omega[m]$.

> $0..P - 1 \parallel p$ **program** Full-Recursive-Doubling-1
> **declare**
> q : integer ;
> ω : real
> **initially**
> $\omega[p] = \tilde{\omega}_p$
> **assign**
> $\omega[p] := \langle \ + \ q \ : \ 0 \leq q \leq p \ :: \ \omega[q] \ \rangle$
> **end**

This program uses explicit process identifiers on the ω-variables.

   The *full-recursive-doubling procedure* is obtained by transforming this program into a multicomputer program. The proposed procedure is displayed in Figure 6.1 for the case $P = M = 8$. Each line in the figure represents a message sent from one process to another. Upon receipt of a message, a pairwise summation is performed. Program Full-Recursive-Doubling-2 is obtained by generalizing this example to the case

$$2^D \leq P = M < 2^{D+1}.$$

> $0..P - 1 \parallel p$ **program** Full-Recursive-Doubling-2
> **declare**
> d : integer ;
> ω, t : real
> **initially**
> $\omega = \tilde{\omega}_p$
> **assign**
> $\langle \ ; \ d \ : \ 0 \leq d \leq D \ ::$
>   **if** $p + 2^d < P$ **then send** ω **to** $p + 2^d$;
>   **if** $p - 2^d \geq 0$ **then begin**
>   **receive** t **from** $p - 2^d$;
> $\omega := \omega + t$
>   **end**
> $\rangle$
> **end**

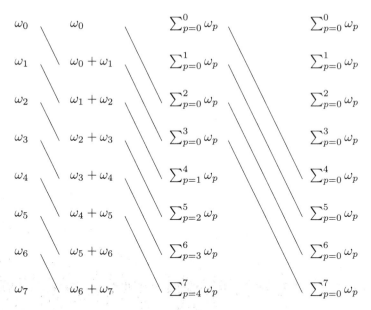

FIGURE 6.1. The full-recursive-doubling procedure for adding eight numbers.

Our first task is to prove the correctness of this program.

Lemma 16 *Program Full-Recursive-Doubling-2 implements program Full-Recursive-Doubling-1.*

We use the same notation as in Lemma 5 of Section 1.7.2 and tag the value of the ω-variables by the value of d, where $0 \leq d \leq D$. The final-state value of variable $\omega[p]$ is given by $\omega_p^{(D+1)} = \omega_p^{(\infty)}$. It follows from the statements of step d of program Full-Recursive-Doubling-2 that

$$
\omega_p^{(d+1)} = \begin{cases} \omega_p^{(d)} + \omega_{p-2^d}^{(d)} & \text{if } p - 2^d \geq 0 \\ \omega_p^{(d)} & \text{if } p - 2^d < 0. \end{cases}
$$

With $\mathcal{D}(p, d) = \{q : \max(0, p + 1 - 2^d) \leq q \leq p\}$, we shall show that

$$
\forall d \in 0..D+1, \ \forall p \in 0..P-1 \ : \ \omega_p^{(d)} = \sum_{q \in \mathcal{D}(p,d)} \omega_q^{(0)}. \tag{6.1}
$$

The proof of the lemma follows by setting $d = D + 1$ and noting that $\mathcal{D}(p, D + 1) = 0..p$ for all $p < 2^{D+1}$. Equation 6.1 is obvious for $d = 0$. To prove the result for arbitrary values, we make the induction hypothesis that Equation 6.1 holds for d and prove the result for $d + 1$.

In the first case, $p - 2^d \geq 0$, the induction hypothesis implies that

$$
\omega_p^{(d+1)} = \sum_{q \in \mathcal{D}(p,d)} \omega_q^{(0)} + \sum_{q \in \mathcal{D}(p-2^d,d)} \omega_q^{(0)}.
$$

This may be rewritten as

$$\omega_p^{(d+1)} = \sum_{q \in \mathcal{D}(p-2^d,d) \cup \mathcal{D}(p,d)} \omega_q^{(0)},$$

because the index sets $\mathcal{D}(p,d)$ and $\mathcal{D}(p-2^d,d)$ are disjoint. It is now easily verified that $\mathcal{D}(p-2^d,d) \cup \mathcal{D}(p,d) = \mathcal{D}(p,d+1)$.

As soon as the condition $p-2^d < 0$ is reached, the value of $\omega[p]$ remains unchanged for the remainder of the program, and

$$\omega_p^{(\infty)} = \omega_p^{(d+1)} = \omega_p^{(d)}.$$

By induction, $\omega_p^{(d)}$ is the sum over the index set:

$$\mathcal{D}(p,d) = \{q : \max(0, p+1-2^d) \leq q \leq p\} = 0..p = \mathcal{D}(p,d+1).$$

This completes the induction.   □

6.1.2 Implementation

To apply the full-recursive-doubling procedure to recursions of arbitrary length M, not just $M = P$, we introduce a P-fold data distribution of the index set $0..M-1$. In process p, store array entries with global indices in the range $M_p..M_{p+1} - 1$, where

$$M_0 = 0, \ M_p \leq M_{p+1}, \text{ and } M_P = M.$$

The local index i corresponding to global index $m \in M_p..M_{p+1} - 1$ is given by $i = m - M_p$. With $I_p = M_{p+1} - M_p$, we have that $i \in 0..I_p - 1$. The sequence of indices M_p defines a linear data distribution, but one that is not necessarily load balanced.

For all $m \in M_p..M_{p+1} - 1$, the sum σ_m can be written as

$$\sigma_m = \sum_{k=M_p}^{m} \omega_k + \sum_{q=0}^{p-1} \sum_{k=M_q}^{M_{q+1}-1} \omega_k = \alpha_m + \sum_{q=0}^{p-1} \beta_q,$$

where

$$\forall m \in M_p..M_{p+1} - 1 : \quad \alpha_m = \sum_{k=M_p}^{m} \omega_k$$

$$\forall p \in 0..P - 1 : \quad \beta_p = \alpha_{M_{p+1}-1}.$$

The values α_m for $m \in M_p..M_{p+1} - 1$ are computed by a linear recursion in process p. Because exactly one β-value is known in each process, the sums $\sum_{q=0}^{p} \beta_q$ can be computed by full recursive doubling over all processes. For $2^D \leq P < 2^{D+1}$, program Full-Recursive-Doubling-3 is obtained. Note that this program has already incorporated the conversion from global to local indices.

$0..P-1 \parallel p$ **program** Full-Recursive-Doubling-3
declare
  i, d : integer ;
  β, t : real ;
  ω : array $[0..I_p - 1]$ of real
initially
  $\langle\ ;\ i\ :\ 0 \leq i < I_p\ ::\ \omega[i] = \tilde{\omega}_{\mu^{-1}(p,i)} \rangle$
assign
  { Sequential Recursion }
  $\langle\ ;\ i\ :\ 0 < i < I_p\ ::\ \omega[i] := \omega[i-1] + \omega[i] \rangle$;

  { Full Recursive Doubling }
  $\beta := \omega[I_p - 1]$;
  $\langle\ ;\ d\ :\ 0 \leq d \leq D\ ::$
    if $p + 2^d < P$ **then send** β **to** $p + 2^d$;
    if $p - 2^d \geq 0$ **then begin**
      receive t **from** $p - 2^d$;
      $\beta := \beta + t$
    end
  $\rangle$;

  { Combine Step }
  send β **to** $(p+1) \bmod P$;
  receive β **from** $(p-1) \bmod P$;
  if $p > 0$ **then**
    $\langle\ ;\ i\ :\ 0 \leq i < I_p\ ::\ \omega[i] := \omega[i] + \beta \rangle$
end

6.1.3 Performance Analysis

The sequential-execution time for a linear recursion of length M is

$$T_1^S = (M - 1)\tau_A.$$

The corresponding multicomputer-execution time for a P-node computation consists of three terms, one for each phase of the program: the sequential linear recursion, the full-recursive-doubling step, and the combine step:

$$T_P = \max_{0 \leq p < P}(I_p - 1)\tau_A + (D + 1)(\tau_C(1) + \tau_A) + \max_{0 \leq p < P}(I_p)\tau_A.$$

The use of $\tau_C()$ is actually an overestimate, because no message exchanges are involved in the computation, only single-direction send–receive pairs. For coarse-grained computations with load-balanced data distributions, we have that $I_p \approx M/P$ and

$$T_P \approx 2\frac{M}{P}\tau_A + (D+1)(\tau_C(1) + \tau_A).$$

This leads to a speed-up

$$S_P = \frac{T_1^S}{T_P} \approx \frac{P}{2} \frac{1}{1 + \frac{P(D+1)}{2M}\left(\frac{\tau_C(1)}{\tau_A} + 1\right)}.$$

The maximum possible speed-up of $P/2$ is due to the combine step. If only some of the variables $w[i]$ are subsequently used, it is better to postpone the last addition until the value of $w[i]$ is actually needed.

6.2 Rational Recursion

Given the arrays α_m, β_m, γ_m, and δ_m, where $0 \le m < M$, and given a value ξ_0, we construct a new array ξ_m via the recursion relation:

$$\forall m \in 1..M : \ \xi_m = \frac{\alpha_m \xi_{m-1} + \beta_m}{\gamma_m \xi_{m-1} + \delta_m}. \tag{6.2}$$

In spite of the nonlinearity of this recursion, full recursive doubling is applicable. The rational function that maps ξ_{m-1} to ξ_m is known as a *Möbius transformation* and is characterized by the matrix:

$$R_m = \left[\begin{array}{cc} \alpha_m & \beta_m \\ \gamma_m & \delta_m \end{array} \right].$$

Consider the value ξ_{m+1} with $m > 0$:

$$\begin{aligned} \xi_{m+1} &= \frac{\alpha_{m+1}\xi_m + \beta_{m+1}}{\gamma_{m+1}\xi_m + \delta_{m+1}} \\ &= \frac{\alpha_{m+1}\frac{\alpha_m \xi_{m-1} + \beta_m}{\gamma_m \xi_{m-1} + \delta_m} + \beta_{m+1}}{\gamma_{m+1}\frac{\alpha_m \xi_{m-1} + \beta_m}{\gamma_m \xi_{m-1} + \delta_m} + \delta_{m+1}} \\ &= \frac{(\alpha_{m+1}\alpha_m + \beta_{m+1}\gamma_m)\xi_{m-1} + (\alpha_{m+1}\beta_m + \beta_{m+1}\delta_m)}{(\gamma_{m+1}\alpha_m + \delta_{m+1}\gamma_m)\xi_{m-1} + (\gamma_{m+1}\beta_m + \delta_{m+1}\delta_m)}. \end{aligned}$$

The mapping from ξ_{m-1} to ξ_{m+1} is also a Möbius transformation, and its corresponding 2×2 matrix is given by the product $R_{m+1}R_m$. By induction, the Möbius transformation that maps ξ_0 into ξ_{m+1} is characterized by the matrix

$$S_{m+1} = R_{m+1}R_m R_{m-1}\ldots R_1.$$

The nonlinear recursion of Equation 6.2 is, therefore, equivalent to the linear recursion of matrices defined by

$$\forall m \in 1..M-1 : \ S_m = R_m S_{m-1},$$

where S_0 is the identity matrix. Full recursive doubling is applicable to this problem, because matrix multiplication is an associative operation. Once the sequence S_m is known, each value ξ_m is computed by applying the Möbius transformation defined by S_m to ξ_0.

6.3   Tridiagonal LU-Decomposition

6.3.1   Specification

The LU-decomposition without pivoting of the tridiagonal matrix

$$A = \begin{bmatrix} \sigma_0 & \tau_1 \\ \rho_1 & \sigma_1 & \tau_2 \\ & \rho_2 & \sigma_2 & \tau_3 \\ & & \ddots & \ddots & \ddots \\ & & & \rho_{M-2} & \sigma_{M-2} & \tau_{M-1} \\ & & & & \rho_{M-1} & \sigma_{M-1} \end{bmatrix}$$

is given by $A = LU$. The factors L and U have the structure

$$L = \begin{bmatrix} 1 & 0 \\ \lambda_1 & 1 & 0 \\ & \lambda_2 & 1 & 0 \\ & & \ddots & \ddots & \ddots \\ & & & \lambda_{M-2} & 1 & 0 \\ & & & & \lambda_{M-1} & 1 \end{bmatrix}$$

$$U = \begin{bmatrix} \mu_0 & \nu_1 \\ 0 & \mu_1 & \nu_2 \\ & 0 & \mu_2 & \nu_3 \\ & & \ddots & \ddots & \ddots \\ & & & \mu_{M-2} & \nu_{M-1} \\ & & & & 0 & \mu_{M-1} \end{bmatrix}.$$

We obtain recursion relations for the coefficients μ_m, λ_m, and ν_m by writing out the product LU and equating the elements of LU to those of A. For $m \in 1..M-1$, we have that

$$\sigma_0 = \mu_0$$
$$\sigma_m = \lambda_m \nu_m + \mu_m \qquad (6.3)$$
$$\tau_m = \nu_m \qquad (6.4)$$
$$\rho_m = \lambda_m \mu_{m-1}. \qquad (6.5)$$

Program Tridiagonal-LU-1 implements the recursion defined by Equations 6.3, 6.4, and 6.5. The recursion can be started at $m = 0$ because of the initializations: $\mu[-1] = 1$ and $\rho[0] = \nu[0] = 0$.

```
program Tridiagonal-LU-1
declare
      m  :  integer ;
      σ, ρ, λ, ν  :  array [0..M − 1] of real ;
      μ  :  array [−1..M − 1] of real
initially
      σ[0], ρ[0], ν[0] = σ̃₀, 0, 0 ;
      ⟨ ; m : 0 < m < M :: σ[m], ρ[m], ν[m] = σ̃ₘ, ρ̃ₘ, τ̃ₘ ⟩
assign
      μ[−1] := 1.0 ;
      ⟨ ; m : 0 ≤ m < M ::
            λ[m] := ρ[m]/μ[m − 1] ;
            μ[m] := σ[m] − λ[m]ν[m]
      ⟩
end
```

The LU-decomposition is usually followed by the solution step, which solves the triangular systems $L\vec{y} = \vec{b}$ and $U\vec{x} = \vec{y}$. With $\vec{y} = [\eta_m]$, $\vec{b} = [\beta_m]$, and $\eta_0 = \beta_0$, the forward solve is defined by the recursion

$$\forall m \in 1..M - 1 :\ \eta_m = \beta_m - \lambda_m \eta_{m-1}. \tag{6.6}$$

The backward solve completes the solution step. It is given by the recursion

$$\forall m \in 0..M - 2 : \xi_m = (\eta_m - \nu_{m+1}\xi_{m+1})/\mu_m, \tag{6.7}$$

where $\vec{x} = [\xi_m]$ and $\xi_{M-1} = \eta_{M-1}/\mu_{M-1}$. The multicomputer implementation of the solution step is analogous to that of the LU-decomposition and is left as an exercise.

6.3.2 Implementation

Substituting Equations 6.4 and 6.5 into Equation 6.3, we obtain a rational recursion for the coefficients μ_m:

$$\begin{aligned}
\mu_m &= \sigma_m - \lambda_m \nu_m \\
&= \sigma_m - \frac{\rho_m}{\mu_{m-1}}\tau_m \\
&= \frac{\sigma_m \mu_{m-1} - \rho_m \tau_m}{\mu_{m-1} + 0}.
\end{aligned}$$

This rational recursion can be solved by a full-recursive-doubling procedure on the sequence of 2×2 matrices

$$R_m = \begin{bmatrix} \sigma_m & -\tau_m \rho_m \\ 1 & 0 \end{bmatrix}.$$

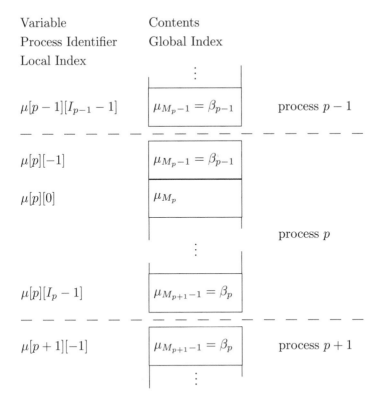

FIGURE 6.2. Data distribution for program Tridiagonal-LU-3.

Setting $\tau_0 = \rho_0 = 0$ and $\mu_{-1} = 1$, we find that the Möbius transformations

$$S_m = R_m R_{m-1} \ldots R_0$$

map $\mu_{-1} = 1$ into μ_m, where $0 \leq m < M$.

In principle, the multicomputer LU-decomposition can now proceed as follows. First, use full recursive doubling to compute all matrices S_m from the sequence of matrices R_m. Subsequently, compute all μ_m by applying the Möbius transformations S_m to $\mu_{-1} = 1$. Finally, compute all λ_m using Equation 6.5. Although this is the basic idea, program Tridiagonal-LU-2 is more efficient in its use of full recursive doubling.

As in program Full-Recursive-Doubling-3, we use a linear data distribution defined by the sequence of global indices $M_0 = 0$, $M_p \leq M_{p+1}$, and $M_P = M$. It is the task of process p to compute the values μ_m and λ_m for $m \in M_p..M_{p+1} - 1$. These values are stored in the variables $\mu[p][i]$ and $\lambda[p][i]$, respectively, where $i = m - M_p$ and $0 \leq i < I_p$. The distributed array μ and the contents of the array variables is displayed in Figure 6.2. For clarity, process identifiers of program variables will be explicit in this section.

$0..P-1 \parallel p$ **program** Tridiagonal-LU-2
declare

    i, d : integer ;

    $\sigma, \rho, \lambda, \nu$: array$[0..I_p - 1]$ of real ;

    μ : array$[-1..I_p - 1]$ of real ;

    S, T : array $[0..1 \times 0..1]$ of real

initially

    $\langle\ ;\ i\ :\ 0 \le i < I_p\ ::\ \sigma[i], \rho[i], \nu[i] = \tilde{\sigma}_{M_p+i}, \tilde{\rho}_{M_p+i}, \tilde{\tau}_{M_p+i} \rangle$

assign

    { Sequential Recursion }

    $S := \begin{bmatrix} 1 & 0 \\ 0 & 1 \end{bmatrix}$;

    $\langle\ ;\ i\ :\ 0 \le i < I_p\ ::\ S := \begin{bmatrix} \sigma[i] & -\nu[i]\rho[i] \\ 1 & 0 \end{bmatrix} S \rangle$;

    { Full Recursive Doubling }

    $\langle\ ;\ d\ :\ 0 \le d \le D\ ::$

        if $p + 2^d < P$ **then send** S **to** $p + 2^d$;

        if $p - 2^d \ge 0$ **then begin**

            receive T **from** $p - 2^d$;

            $S := ST$

        end

    $\rangle$;

    { Combine Step }

    $\mu[I_p - 1] := (S[0,0] + S[0,1])/(S[1,0] + S[1,1])$;

    send $\mu[I_p - 1]$ **to** $(p + 1) \bmod P$;

    receive $\mu[-1]$ **from** $(p - 1) \bmod P$;

    if $p = 0$ **then** $\mu[-1] := 1.0$;

    $\langle\ ;\ i\ :\ 0 \le i < I_p - 1\ ::$

        $\lambda[i] := \rho[i]/\mu[i - 1]$;

        $\mu[i] := \sigma[i] - \lambda[i]\nu[i]$

    $\rangle$

end

As soon as a process has access to the last μ-value of the previous process, the efficient sequential recursion of program Tridiagonal-LU-1 can be used to compute the remainder of the sequences μ_m and λ_m. This explains the last part of the combine step. The rest of program Tridiagonal-LU-2 is concerned with computing the initial condition for the combine step of each process p. As shown in Figure 6.2, the array $\mu[p]$ is extended so that this initial condition may be stored in $\mu[p][-1]$.

It is convenient to introduce the short-hand notation $\beta_p = \mu_{M_{p+1}-1}$ for the last μ-value of process p. The data distribution maps the value β_p to $\mu[p][I_p - 1]$. Before the recursion of the combine step can be started, the value β_{p-1} must be assigned to $\mu[p][-1]$. For initialization purposes, the value β_{-1} is defined, and $\beta_{-1} = \mu_{M_0-1} = \mu_{-1} = 1$. This value is stored in the variable $\mu[0][-1]$.

The first two steps of program Tridiagonal-LU-2 compute the Möbius transformations S_m, which map $\beta_{-1} = \mu_{-1} = 1$ into μ_m. However, only a small subset of these Möbius transformations are needed: the ones that map β_{-1} into $\beta_p = \mu_{M_{p+1}-1}$, where $0 \le p < P$. It suffices, therefore, to compute just one transformation per process.

The sequential-recursion step computes the Möbius transformation

$$R_{M_{p+1}-1} \ldots R_{M_p+1} R_{M_p},$$

which maps β_{p-1} into β_p. This transformation is stored in the 2×2 matrix $S[p]$. Subsequently, the full-recursive-doubling step replaces the contents of $S[p]$ with the Möbius transformation that maps β_{-1} into β_p.

The first assignment of the combine step computes β_p by applying the Möbius transformation to $\beta_{-1} = 1$. The result of this transformation is stored in variable $\mu[p][I_p - 1]$ and, subsequently, handed over to the next process. The value β_{p-1} is stored in variable $\mu[p][-1]$. Process 0 ignores the value β_{P-1}, which it receives from process $P - 1$, and performs the assignment $\mu[0][-1] := 1.0$ instead.

6.3.3 Performance Analysis

The estimated sequential-execution time of program Tridiagonal-LU-1 is

$$T_1^S = 3(M - 1)\tau_A.$$

The multicomputer-execution time of program Tridiagonal-LU-2 using P independent processes is given by:

$$
\begin{aligned}
T_P \quad = \quad & (1+6)I_p\tau_A && \text{(sequential recursion)} \\
+ \quad & (D+1)(\tau_C(4) + 12\tau_A) && \text{(full recursive doubling)} \\
+ \quad & 3\tau_A + \tau_C(1) + 3(I_p - 1)\tau_A. && \text{(combine step)}
\end{aligned}
$$

The product of two 2×2 matrices requires 12 floating-point operations. This is used in the full-recursive-doubling step. In the sequential-recursion step, however, the number of floating-point operations per matrix–matrix product is reduced to 6, because the second row of the matrices R_m is always $[1\ 0]$. Using that $\tau_C(L) \approx \tau_S$ for short messages, the execution-time formula is simplified to:

$$T_P \approx (10I_p + 12(D+1))\tau_A + (D+2)\tau_S.$$

For load-balanced coarse-grained computations with $I_p \approx M/P \gg D+1$, we obtain that

$$T_P \approx 10\frac{M}{P}\tau_A + (D+2)\tau_S. \tag{6.8}$$

In turn, this leads to a speed-up

$$S_P \approx \frac{3}{10} P \frac{1}{1 + \frac{P(D+2)}{10M} \frac{T_S}{T_A}}.$$

Hence, the most one can hope for is a speed-up of about $3P/10$, and this requires the coarse-granularity assumption of $P(D+2) \ll 10M$. This is decidedly poor. If $P < 4$, the multicomputer-execution time is always larger than the sequential-execution time!

This analysis does not include other considerations that have an impact on program performance. The multicomputer program has more index arithmetic and performs more assignments. These operations have real costs that are ignored in the analysis. Another factor to consider in practice: compilers generate more efficient object code for simple programs, like program Tridiagonal-LU-1, than for more complicated programs, like program Tridiagonal-LU-2. Finally, numerical stability requires that one occasionally scale the 2×2 matrices in the full-recursive-doubling procedure. This requires additional floating-point operations and reduces the performance even further.

6.4   An Iterative Tridiagonal Solver

The above data distribution of the coefficients σ_m, τ_m, and ρ_m can also be interpreted as follows. The information mapped to process p consists of a diagonal block of the matrix A given by

$$A_p = \begin{bmatrix} \sigma_{M_p} & \tau_{M_p+1} \\ \rho_{M_p+1} & \sigma_{M_p+1} & \tau_{M_p+2} \\ & \rho_{M_p+2} & \sigma_{M_p+2} & \tau_{M_p+3} \\ & & \ddots & \ddots & \ddots \\ & & & \rho_{M_{p+1}-2} & \sigma_{M_{p+1}-2} & \tau_{M_{p+1}-1} \\ & & & & \rho_{M_{p+1}-1} & \sigma_{M_{p+1}-1} \end{bmatrix}$$

and of the coefficients τ_{M_p} and ρ_{M_p}, which couple the diagonal blocks A_{p-1} and A_p. This data distribution defines a splitting $A = G - H$ with

$$G = \begin{bmatrix} A_0 & 0 \\ 0 & A_1 & 0 \\ & 0 & A_2 & 0 \\ & & \ddots & \ddots & \ddots \\ & & & 0 & A_{P-2} & 0 \\ & & & & 0 & A_{P-1} \end{bmatrix}$$

and the matrix H containing the coupling coefficients τ_{M_p} and ρ_{M_p} only.

As seen in Section 3.2, such a splitting defines the relaxation method

$$G\vec{x}_{k+1} = \vec{b} + H\vec{x}_k$$

for systems of linear equations $A\vec{x} = \vec{b}$. The implementation of this relaxation method on multicomputers consists of a program to multiply a vector with the matrix H and a program to solve systems with the coefficient matrix G. The multiplication of H and a vector requires only minimal communication. Solving systems with coefficient matrix G only requires solving tridiagonal systems local to each process and does not require any communication.

Of course, the real issue is whether and how fast the proposed iteration converges. From Section 3.2, we know that both are determined by the spectral radius of $G^{-1}H$. Because this quantity differs from one application to the next, the convergence needs to be examined in each individual case. Whether or not this relaxation method competes with direct methods depends on the number of iteration steps. In turn, this depends not only on the spectral radius of $G^{-1}H$, but also on the accuracy with which the solution is desired and on the quality of the initial guess. As to the latter, there is a convenient and usually excellent candidate:

$$\vec{x}_0 = G^{-1}\vec{b}.$$

The convergence issue is actually more complicated than it seems. Because the splitting is determined by the data distribution, the numerical properties of the relaxation method change when any aspect of the data distribution changes. In particular, the convergence rate is a function of the number of processes P and of all the indices M_p that define the data distribution. Even for the simplest of problems, it is a challenge to establish convergence for all possible combinations. This makes it difficult to guarantee robustness of the iterative solver. Concurrent direct tridiagonal solvers are very weak competition, however. In practice, the proposed iterative solver often outperforms direct solvers by a substantial margin.

6.5 Notes and References

The direct solver is an algorithm proposed by Stone [76, 77]. Many other concurrent direct solvers for tridiagonal systems have been proposed. Although the implementation details vary widely, most rely on the basic communication structure of recursive doubling. The term *cyclic reduction*, which is frequently used in the literature, refers to an algorithm that differs from recursive doubling in technical implementation details. Cyclic reduction and recursive doubling have the same basic structure, and history seems to be the only reason why any distinction in terminology is made.

From a theoretical perspective, the performance of all direct methods for tridiagonal systems is equivalent. The concurrent-execution time is of the form given by Equation 6.8, but different constants appear for different methods. From a practical perspective, methods have widely varying execution times precisely because of these constants.

Most methods, including the iterative solver, can be generalized to solve block-tridiagonal and narrowly banded systems. The generalization requires replacing scalar arithmetic with matrix arithmetic. Although programs for block-tridiagonal systems are more complicated, they are more efficient, because the increased granularity of these problems leads to an improved ratio of communication over arithmetic time.

Exercises

Exercise 27 *Using Equations 6.6 and 6.7, develop the multicomputer programs that solve $L\vec{y} = \vec{b}$ and $U\vec{x} = \vec{y}$.*

Exercise 28 *Examine the impact on full recursive doubling of the non-associativity of floating-point addition.*

Exercise 29 *Implement the concurrent iterative tridiagonal solver of Section 6.4.*

7

The Fast Fourier Transform

7.1 Fourier Analysis

7.1.1 Fourier Series

We consider the space L_2 of complex-valued functions $f(x)$ on the closed interval $[0, 2\pi]$ that are bounded, continuous, and periodic with period 2π. The inner product of two functions $f(x), g(x) \in L_2$ is defined by

$$(f, g) = \frac{1}{2\pi} \int_0^{2\pi} \overline{f(x)} g(x) dx,$$

where $\overline{f(x)}$ denotes the complex conjugate of $f(x)$. Throughout this chapter, we shall use standard mathematical notation that $i = \sqrt{-1}$.

With respect to this inner product, the set of functions $\{e^{ikx} : k \in \mathbb{Z}\}$ forms an orthonormal basis of L_2. It is, therefore, possible to expand every function $f(x) \in L_2$ into a unique linear combination of the basis functions e^{ikx}, or

$$f(x) = \sum_{k=-\infty}^{+\infty} \hat{f}_k e^{ikx}. \tag{7.1}$$

This linear combination is known as *the Fourier series of $f(x)$*, and the coefficients $\hat{f}_k$ are called the *Fourier-series coefficients*. The orthonormality of the basis functions implies that

$$\hat{f}_k = \frac{(e^{ikx}, f(x))}{(e^{ikx}, e^{ikx})} = \frac{1}{2\pi} \int_0^{2\pi} e^{-ikx} f(x) dx. \tag{7.2}$$

The Fourier-series expansion can be interpreted as a transformation of the function space L_2 into the infinite-dimensional complex vector space ℓ_2, which consists of vectors

$$\hat{\vec{f}} = [\hat{f}_k]_{k=-\infty}^{+\infty}.$$

In ℓ_2, we define the inner product of two vectors $\hat{\vec{f}}$ and $\hat{\vec{g}}$ by:

$$(\hat{\vec{f}}, \hat{\vec{g}}) = \sum_{k=-\infty}^{+\infty} \overline{\hat{f}_k} \hat{g}_k.$$

Using a norm consistent with the inner product defined in each space,

$$\| f(x) \|^2 = (f, f) = \frac{1}{2\pi} \int_0^{2\pi} |f(x)|^2 dx$$

$$\| \hat{\vec{f}} \|^2 = (\hat{\vec{f}}, \hat{\vec{f}}) = \sum_{k=-\infty}^{+\infty} |\hat{f}_k|^2,$$

it is easily verified that

$$\| f(x) \|^2 = \| \hat{\vec{f}} \|^2 . \tag{7.3}$$

In other words, the Fourier-series expansion is a norm-preserving transformation from the space L_2 to the space ℓ_2.

7.1.2   The Discrete Fourier Transform

In numerical computations, functions are approximated by a set of function values on some grid. In certain cases, one can use the *discrete Fourier transform* to compute numerical approximations of Fourier-series coefficients. The precise connection between Fourier series and the discrete Fourier transform will be discussed in Section 7.1.3.

The representation of a 2π-periodic function on an equidistant grid with grid spacing $h = 2\pi/N$ is the set of values $f_j = f(jh)$, where $0 \leq j \leq N$. Because of periodicity, we have that $f_N = f_0$. This leads us to study the complex N-dimensional vector space $\mathbb{C}^N$ of vectors $\vec{f} = [f_j]_{j=0}^{N-1}$.

The inner product of two vectors $\vec{f}, \vec{g} \in \mathbb{C}^N$ is defined by

$$(\vec{f}, \vec{g}) = \sum_{j=0}^{N-1} \overline{f_j} g_j,$$

and the norm implied by this inner product is given by

$$\| \vec{f} \|^2 = (\vec{f}, \vec{f}) = \sum_{j=0}^{N-1} \overline{f_j} f_j = \sum_{j=0}^{N-1} |f_j|^2.$$

Lemma 17 *The set of vectors* $\{\vec{e}_n = [e^{ijnh}]_{j=0}^{N-1} : 0 \leq n < N\}$ *is an orthogonal basis for the space* $\mathbb{C}^N$.

The inner product of any two of the proposed basis vectors is given by:

$$(\vec{e}_n, \vec{e}_m) = \sum_{j=0}^{N-1} \overline{e^{ijnh}} e^{ijmh} = \sum_{j=0}^{N-1} e^{-ijnh} e^{ijmh} = \sum_{j=0}^{N-1} e^{ij(m-n)h}.$$

With $r = e^{i(m-n)h}$ we have that

$$(\vec{e}_n, \vec{e}_m) = \sum_{j=0}^{N-1} r^j = \begin{cases} N & \text{if } r = 1 \\ \frac{r^N-1}{r-1} & \text{if } r \neq 1. \end{cases}$$

The case $r = 1$ arises only if $m = n \bmod N$. Otherwise, $r \neq 1$ and

$$r^N = e^{i(m-n)hN} = e^{i2(m-n)\pi} = 1.$$

It follows that the vectors $\vec{e}_n$ are mutually orthogonal and that

$$\forall m, n \in 0..N-1: \ (\vec{e}_n, \vec{e}_m) = \begin{cases} N & \text{if } n = m \\ 0 & \text{if } n \neq m. \end{cases}$$

$\square$

Corollary 1 *Every vector* $\vec{f} \in \mathbb{C}^N$ *has a unique representation of the form*

$$\vec{f} = [f_j]_{j=0}^{N-1} = \sum_{n=0}^{N-1} \hat{f}_n \vec{e}_n,$$

where the discrete Fourier coefficients are given by

$$\hat{\vec{f}} = [\hat{f}_n]_{n=0}^{N-1} = \left[\frac{(\vec{e}_n, \vec{f})}{(\vec{e}_n, \vec{e}_n)} \right]_{n=0}^{N-1}.$$

(Without proof.)

$\square$

The vector $\hat{\vec{f}}$ is called the discrete Fourier transform of $\vec{f}$. The inverse discrete Fourier transform maps $\hat{\vec{f}}$ into $\vec{f}$. The equations of Corollary 1 are the discrete analogs of Equations 7.1 and 7.2. When written out component by component, they become:

$$\forall j \in 0..N-1: \quad f_j = \sum_{n=0}^{N-1} \hat{f}_n e^{ijnh} \tag{7.4}$$

$$\forall n \in 0..N-1: \quad \hat{f}_n = \frac{1}{N} \sum_{j=0}^{N-1} f_j e^{-ijnh}. \tag{7.5}$$

The analog of Equation 7.3, the norm-preservation property, is easily established. We have that:

$$\| \hat{\vec{f}} \| = \frac{1}{\sqrt{N}} \| \vec{f} \| . \tag{7.6}$$

The norms of a vector and its discrete Fourier transform differ by a multiplicative factor, which could be eliminated by using an orthonormal instead of an orthogonal basis for $\mathbb{C}^N$. For all practical purposes, the discrete Fourier transform is a norm-preserving transformation of $\mathbb{C}^N$ onto itself.

7.1.3 Aliasing

Lemma 18 establishes the relation between the discrete Fourier transform and Fourier series.

Lemma 18 *Let $\hat{\vec{f}} \in \mathbb{C}^N$ be the discrete Fourier transform of $\vec{f} \in \mathbb{C}^N$. The trigonometric polynomial*

$$p_N(x) = \sum_{n=0}^{N-1} \hat{f}_n e^{inx}$$

interpolates the N data points (jh, f_j), where $0 \le j < N$.

This is an immediate consequence of Equation 7.4. $\square$

Let $f(x) \in L_2$ and $\vec{f} = [f_j]_{j=0}^{N-1} \in \mathbb{C}^N$ such that $f_j = f(jh)$ with $h = 2\pi/N$. The Fourier series of $f(x)$ and the trigonometric polynomial $p_N(x)$ that interpolates $\vec{f}$ are, respectively, given by

$$f(x) = \sum_{k=-\infty}^{+\infty} \hat{f}_k^c e^{ikx}$$

$$p_N(x) = \sum_{n=0}^{N-1} \hat{f}_n^d e^{inx}.$$

We use superscripts c and d to distinguish between continuous and discrete Fourier coefficients. For sufficiently smooth functions $f(x)$, the sequence $p_N(x)$ converges uniformly to $f(x)$ as the number of interpolation points N increases. However, the precise relationship between the Fourier-series coefficients $\hat{f}_k^c$ and the discrete Fourier coefficients $\hat{f}_n^d$ is more complicated.

The trigonometric polynomial $p_N(x)$ contains only terms e^{inx} with positive n. This asymmetry with respect to the Fourier series of $f(x)$ is easily resolved as follows. For all integer ℓ, the function $e^{i\ell Nx}$ equals one at all interpolation points jh. As a result, a trigonometric polynomial obtained by replacing any basis function e^{inx} by $e^{i(\ell N+n)x}$ still interpolates all points.

This property is known as *aliasing*. With $N = 2M$, we perform the substitution

$$\forall n \in M..N - 1 : \quad e^{inx} \rightarrow e^{i(-N+n)x}.$$

The interpolating trigonometric polynomial so obtained is given by

$$
\begin{aligned}
q_N(x) &= \sum_{n=0}^{M-1} \hat{f}_n^d e^{inx} + \sum_{n=M}^{N-1} \hat{f}_n^d e^{i(-N+n)x} \\
&= \sum_{n=0}^{M-1} \hat{f}_n^d e^{inx} + \sum_{n=-M}^{-1} \hat{f}_{N+n}^d e^{inx}.
\end{aligned}
$$

It can be shown that the discrete Fourier coefficients $\hat{f}_n^d$ with $n \in 0..M - 1$ converge to the Fourier-series coefficients $\hat{f}_n^c$ as N increases. Similarly, the coefficients $\hat{f}_n^d$ with $n \in M..N - 1$ converge to $\hat{f}_{n-N}^c$.

Because of aliasing, one might be led to the mistaken conclusion that trigonometric interpolation is not unique. Finding the interpolating polynomial through N points is uniquely determined, once the set of basis vectors is defined. Among all trigonometric polynomials that are a linear combination of these basis functions, there is exactly one polynomial that interpolates all function values. Aliasing is the effect of changing from one basis to another, for example, from $\{e^{inx} : 0 \leq n < N\}$ to $\{e^{inx} : -M \leq n < M\}$.

Aliasing is particularly important when one wants to distinguish between high and low frequencies. In Fourier series, the absolute value of n determines whether the term $\hat{f}_n^c e^{inx}$ is of high or low frequency. To make this distinction among the discrete Fourier modes, one must take the corresponding terms in $q_N(x)$, not in $p_N(x)$. Thus, the highest frequency on a regular grid of size N is $M = N/2$. To divide up the discrete spectrum equally between low and high, the following criterion can be used: n is a high frequency if $|\theta_n| \geq \frac{\pi}{2}$, where $\theta_n = nh$ is interpreted as an angle and is reduced to the interval $[-\pi, \pi]$.

7.2   Divide and Conquer

Comparison of Equations 7.4 (the inverse transform) and 7.5 (the forward transform) shows that the inverse transform differs from the forward transform only by a multiplicative factor and a sign in the exponents. Both transforms map vectors of $\mathbb{C}^N$ into vectors of $\mathbb{C}^N$. For the purpose of algorithm development, the discrete Fourier transform is identical to its inverse. Without loss of generality, we shall concentrate on developing an algorithm for the inverse transform. Slightly changing notation, we must compute

$$\forall k \in 0..N - 1 : \quad \hat{f}_k = \sum_{n=0}^{N-1} f_n e^{iknh}. \tag{7.7}$$

Each of the N discrete Fourier coefficients $\hat{f}_k$ could be computed by summing N terms. This would require $2N^2$ complex floating-point operations. The cost of evaluating the expressions e^{iknh} is not counted, because it can be amortized over many transforms. In practice, the exponential constants are saved in a look-up table.

Given the trigonometric polynomial

$$p(x) = \sum_{n=0}^{N-1} f_n e^{inx} \tag{7.8}$$

and the N equidistant points $x_k = kh$, where $h = \frac{2\pi}{N}$ and $k \in 0..N-1$, it is clear that $\hat{f}_k = p(x_k)$. The problem of computing the N discrete Fourier coefficients $\hat{f}_k$ is thus equivalent to evaluating $p(x)$ at the N points x_k. For the purpose of starting an induction argument, we assume that for all $M < N$ a procedure exists to evaluate trigonometric polynomials of degree $M - 1$ at M equidistant points $y_\ell = \ell\frac{2\pi}{M}$, where $\ell \in 0..M-1$. Our goal is to reduce the problem of size N into several smaller problems.

With $N = 2M$, split $p(x)$ into an even and an odd part:

$$
\begin{aligned}
p(x) &= \sum_{m=0}^{M-1} f_{2m} e^{i2mx} + \sum_{m=0}^{M-1} f_{2m+1} e^{i(2m+1)x} \\
&= \sum_{m=0}^{M-1} f_{2m} e^{i2mx} + e^{ix} \sum_{m=0}^{M-1} f_{2m+1} e^{i2mx} \\
&= q(2x) + e^{ix} r(2x).
\end{aligned}
$$

By induction, $q(y_k)$ and $r(y_k)$ are known quantities for all $k \in 0..M-1$. Moreover, it follows from $N = 2M$ that

$$\forall k \in 0..M-1 : \begin{cases} 2x_k &= y_k \\ 2x_{k+M} &= y_k + 2\pi. \end{cases}$$

This and the 2π-periodicity of trigonometric polynomials imply that

$$\forall k \in 0..M-1 : \begin{cases} p(x_k) &= q(y_k) + e^{ik\frac{\pi}{M}} r(y_k) \\ p(x_{k+M}) &= q(y_k) - e^{ik\frac{\pi}{M}} r(y_k). \end{cases} \tag{7.9}$$

The evaluation of a trigonometric polynomial of degree $N - 1$ at N equidistant points is thus reduced to the evaluation of two trigonometric polynomials of degree $N/2 - 1$ at $N/2$ equidistant points. If $N = 2^n$, this dividing process can be continued until the trigonometric polynomials are just constants.

Before developing this method further, let us examine the operation count of the divide-and-conquer strategy. Let $W(N)$ be the number of complex floating-point operations to evaluate $p(x)$ at N equidistant points by this procedure. We must evaluate $q(x)$ and $r(x)$ at $M = N/2$ equidistant points and perform the work specified by Equation 7.9. This implies that

$$W(2M) = 2W(M) + 4M.$$

Again, the cost of evaluating exponential constants like $e^{ik\frac{\pi}{M}}$ is ignored, because it can be amortized over many transforms. Introducing $w_j = W(2^j)$, we have that $w_0 = W(1) = 0$ and that

$$\forall j \in 1..n : \ w_j = 2w_{j-1} + 2 \times 2^j.$$

Multiply the equation for w_j by 2^{n-j}, and sum for $j \in 1..n$:

$$\sum_{j=1}^{n} 2^{n-j} w_j = 2 \sum_{j=1}^{n} 2^{n-j} (w_{j-1} + 2^j)$$

$$= 2n2^n + \sum_{j=0}^{n-1} 2^{n-j} w_j.$$

This implies that $w_n = 2n2^n$ and $W(N) = 2N \log_2 N$. This is a substantial reduction over the primitive procedure, which required $2N^2$ complex floating-point operations.

7.3   The Fast-Fourier-Transform Algorithm

Methods that compute the discrete Fourier transform in $O(N \log N)$ complex floating-point operations are referred to as *fast Fourier transforms*, FFT for short.

Based on the odd-even decomposition of a trigonometric polynomial, a problem of size 2^n is reduced to two problems of size 2^{n-1}. Subsequently, two problems of size 2^{n-1} are reduced to four problems of size 2^{n-2}. Ultimately, 2^n problems of size 1 are obtained, each of which is solved trivially. The implementation hinges on a precise, albeit cumbersome, notation to identify each subproblem. The term subproblem in this context refers to the evaluation of a trigonometric polynomial on a set of equidistant points. The size of a subproblem is the number of points at which the trigonometric polynomial is evaluated.

If the main problem has size 2^n, there are 2^{n-m} subproblems of size 2^m. The trigonometric polynomials corresponding to these subproblems are denoted by:

$$\forall m \in 0..n, \forall r \in 0..2^{n-m} - 1 : \ p_r^{(m)}(x). \tag{7.10}$$

There are twice as many subproblems of size 2^m as there are subproblems of size 2^{m+1}. Each problem of size 2^m is either the even or the odd part of a problem of size 2^{m+1}. The indices of the polynomials are chosen such that $p_r^{(m)}(x)$ is the even part of $p_r^{(m+1)}(x)$ and $p_{r+2^{n-m-1}}^{(m)}(x)$ its odd part. With this choice of indices, we have that

$$\forall m \in 0..n-1, \forall r \in 0..2^{n-m-1}-1 :$$
$$p_r^{(m+1)}(x) = p_r^{(m)}(2x) + e^{ix}p_{r+2^{n-m-1}}^{(m)}(2x). \qquad (7.11)$$

Consider the first three odd-even decompositions. For $m = n-1$, we obtain the odd-even decomposition of the main problem:

$$p_0^{(n)}(x) = p_0^{(n-1)}(2x) + e^{ix}p_1^{(n-1)}(2x).$$

The decomposition of the problems of size 2^{n-1} is given by:

$$p_0^{(n-1)}(x) = p_0^{(n-2)}(2x) + e^{ix}p_{0+2}^{(n-2)}(2x)$$
$$p_1^{(n-1)}(x) = p_1^{(n-2)}(2x) + e^{ix}p_{1+2}^{(n-2)}(2x),$$

and for the problems of size 2^{n-2} we obtain:

$$p_0^{(n-2)}(x) = p_0^{(n-3)}(2x) + e^{ix}p_{0+4}^{(n-3)}(2x)$$
$$p_1^{(n-2)}(x) = p_1^{(n-3)}(2x) + e^{ix}p_{1+4}^{(n-3)}(2x)$$
$$p_2^{(n-2)}(x) = p_2^{(n-3)}(2x) + e^{ix}p_{2+4}^{(n-3)}(2x)$$
$$p_3^{(n-2)}(x) = p_3^{(n-3)}(2x) + e^{ix}p_{3+4}^{(n-3)}(2x).$$

These relations define a hierarchy between these polynomials, which is represented by a binary tree:

$$p_0^{(n)}(x) \begin{cases} p_0^{(n-1)}(x) \begin{cases} p_0^{(n-2)}(x) \begin{cases} p_0^{(n-3)}(x) \\ p_4^{(n-3)}(x) \end{cases} \\ p_2^{(n-2)}(x) \begin{cases} p_2^{(n-3)}(x) \\ p_6^{(n-3)}(x) \end{cases} \end{cases} \\ p_1^{(n-1)}(x) \begin{cases} p_1^{(n-2)}(x) \begin{cases} p_1^{(n-3)}(x) \\ p_5^{(n-3)}(x) \end{cases} \\ p_3^{(n-2)}(x) \begin{cases} p_3^{(n-3)}(x) \\ p_7^{(n-3)}(x) \end{cases} \end{cases} \end{cases}$$

The top and bottom descendants in this binary tree correspond to the even and odd part of the parent, respectively.

The trigonometric polynomials $p_r^{(m)}(x)$, expressed in terms of the original data f_j, are given by

$$\forall m \in 0..n, \forall r \in 0..2^{n-m}-1 : \quad p_r^{(m)}(x) = \sum_{j=0}^{2^m-1} f_{2^{n-m}j+r}e^{ijx}. \qquad (7.12)$$

Two important special cases of this formula occur for $m = 0$ and $m = n$, for which we obtain, respectively,

$$\forall r \in 0..2^n - 1 : \quad p_r^{(0)}(x) = f_r \tag{7.13}$$

$$\forall k \in 0..2^n - 1 : \quad p_0^{(n)}(x_k^{(n)}) = \hat{f}_k. \tag{7.14}$$

The points $x_k^{(n)}$ define the equidistant grid on which $p_0^{(n)}(x)$ is evaluated.

We must evaluate $p_r^{(m+1)}(x)$ at 2^{m+1} equidistant points in the interval $[0, 2\pi]$. These points are given by the familiar formula:

$$\forall k \in 0..2^{m+1} - 1 : \quad x_k^{(m+1)} = k\frac{2\pi}{2^{m+1}}. \tag{7.15}$$

It is clear that

$$\forall k \in 0..2^m - 1 : \quad \begin{cases} 2x_k^{(m+1)} & = & x_k^{(m)} \\ 2x_{k+2^m}^{(m+1)} & = & 2\pi + x_k^{(m)}. \end{cases} \tag{7.16}$$

We can now rewrite the fundamental divide-and-conquer formulas of Equation 7.9, using the notation of Equations 7.10, 7.11, 7.15, and 7.16. This results in the following identities:

$$\forall m \in 0..n - 1, \forall r \in 0..2^{n-m-1} - 1, \forall k \in 0..2^m - 1 : \tag{7.17}$$

$$\begin{bmatrix} p_r^{(m+1)}(x_k^{(m+1)}) \\ p_r^{(m+1)}(x_{k+2^m}^{(m+1)}) \end{bmatrix} = \begin{bmatrix} 1 & e^{ik\frac{2\pi}{2^{m+1}}} \\ 1 & -e^{ik\frac{2\pi}{2^{m+1}}} \end{bmatrix} \begin{bmatrix} p_r^{(m)}(x_k^{(m)}) \\ p_{r+2^{n-m-1}}^{(m)}(x_k^{(m)}) \end{bmatrix}.$$

A program is derived from these formulas by assigning each value to a program variable. As a first step, store

$$p_r^{(m)}(x_k^{(m)}) \quad \text{in variable} \quad \beta[m, r, k]$$

$$e^{ik\frac{2\pi}{2^{m+1}}} = e^{i2^{n-m-1}k\frac{2\pi}{2^n}} \quad \text{in variable} \quad \epsilon[2^{n-m-1}k].$$

Equation 7.17 immediately implies the main part of the program: an assignment to a vector of two variables within a concurrent quantification over r and k. The quantification over the index m is sequential, however, because subproblems of size 2^{m+1} depend on subproblems of size 2^m. Equation 7.13 specifies the initialization of the variables $\beta[0, r, 0]$. Equation 7.14 tells us which variables hold the values $\hat{f}_k$ upon termination.

$\langle \parallel r : 0 \leq r < 2^n :: \beta[0, r, 0] := f[r] \rangle$;
$\langle ; m : 0 \leq m < n ::$
$\quad \langle \parallel r, k : 0 \leq r < 2^{n-m-1} \text{ and } 0 \leq k < 2^m ::$
$\quad \begin{bmatrix} \beta[m+1, r, k] \\ \beta[m+1, r, k+2^m] \end{bmatrix} := \begin{bmatrix} 1 & \epsilon[2^{n-m-1}k] \\ 1 & -\epsilon[2^{n-m-1}k] \end{bmatrix} \begin{bmatrix} \beta[m, r, k] \\ \beta[m, r+2^{n-m-1}, k] \end{bmatrix}$
$\quad \rangle$
$\rangle$;
$\langle \parallel k : 0 \leq k < 2^n :: \hat{f}[k] := \beta[n, 0, k] \rangle$

The remainder of the development of the fast-Fourier-transform algorithm is concerned with efficient use of memory. Our goal is to reorganize the computations such that the transformed data and all intermediate results share identical memory locations. This program transformation must preserve the concurrency of the inner quantification.

The key observation is that the innermost assignment in the above program uses two entries of the array $\beta[m]$ to evaluate two new entries of the array $\beta[m+1]$. Once the assignment has been performed, the values of the right-hand-side variables may be discarded. To replace the array β by a one-dimensional array b, we need a map τ from the β-indices $[m, r, k]$ to a b-index $[t]$ such that each assignment overwrites its own right-hand-side variables. Such a map $t = \tau(m, r, k)$ must satisfy:

$$\forall m \in 0..n-1, \forall r \in 0..2^{n-m-1}-1, \forall k \in 0..2^m-1 : \qquad (7.18)$$

$$\text{either} \quad \left\{ \begin{array}{lll} \tau(m+1, r, k) & = & \tau(m, r, k) \\ \tau(m+1, r, k+2^m) & = & \tau(m, r+2^{n-m-1}, k) \end{array} \right.$$

$$\text{or} \quad \left\{ \begin{array}{lll} \tau(m+1, r, k) & = & \tau(m, r+2^{n-m-1}, k) \\ \tau(m+1, r, k+2^m) & = & \tau(m, r, k). \end{array} \right.$$

If such a map τ is available, the program can be transformed as follows.

$$\langle \ \| \ r \ : \ 0 \le r < 2^n \ :: \ b[\tau(0, r, 0)] := f_r \ \rangle \ ;$$
$$\langle \ ; \ m \ : \ 0 \le m < n \ ::$$
$$\quad \langle \ \| \ r, k \ : \ 0 \le r < 2^{n-m-1} \ \textbf{and} \ 0 \le k < 2^m \ ::$$
$$\quad \begin{bmatrix} b[\tau(m+1, r, k)] \\ b[\tau(m+1, r, k+2^m)] \end{bmatrix} := \begin{bmatrix} 1 & \epsilon[2^{n-m-1}k] \\ 1 & -\epsilon[2^{n-m-1}k] \end{bmatrix} \begin{bmatrix} b[\tau(m, r, k)] \\ b[\tau(m, r+2^{n-m-1}, k)] \end{bmatrix}$$
$$\quad \rangle$$
$$\rangle \ ;$$
$$\langle \ \| \ k \ : \ 0 \le k < 2^n \ :: \ \hat{f}[k] := b[\tau(n, 0, k)] \ \rangle$$

One particular map $t = \tau(m, r, k)$ that satisfies Equation 7.18 relies on the following definition.

Definition 6 *The* bit-reversal map ρ_n *is a bijection of* $0..2^n-1$ *to itself, such that:*

$$\rho_n\left(\sum_{j=0}^{n-1} \alpha_j 2^j\right) = \sum_{j=0}^{n-1} \alpha_{n-1-j} 2^j,$$

where $\forall j \in 0..n-1 : \ \alpha_j \in \{0, 1\}$.

As implied by its name, the bit-reversal map of an integer with binary representation '$\alpha_{n-1}\alpha_{n-2}\ldots\alpha_1\alpha_0$' is the integer with binary representation '$\alpha_0\alpha_1\ldots\alpha_{n-2}\alpha_{n-1}$'.

The following lemma constructs a map $t = \tau(m, r, k)$ that satisfies Equation 7.18.

Lemma 19 *The map*

$$t = \tau(m, r, k) = \rho_n(r) + k \qquad (7.19)$$

satisfies $\forall m \in 0..n-1, \forall r \in 0..2^{n-m-1} - 1, \forall k \in 0..2^m - 1:$

$$\tau(m+1, r, k) = \tau(m, r, k) \qquad (7.20)$$
$$\tau(m+1, r, k+2^m) = \tau(m, r+2^{n-m-1}, k). \qquad (7.21)$$

The proof of Equation 7.20 is trivial. To prove Equation 7.21, let the binary representation of r be given by '$\alpha_{n-m-2}\alpha_{n-m-3}\ldots\alpha_0$'. In this case, we have that

$$
\begin{aligned}
\tau(m, r+2^{n-m-1}, k) &= \rho_n(r+2^{n-m-1}) + k \\
&= \rho_n\Big(2^{n-m-1} + \sum_{j=0}^{n-m-2} \alpha_j 2^j\Big) + k \\
&= 2^m + \sum_{\ell=m+1}^{n-1} \alpha_{n-1-\ell} 2^\ell + k \\
&= \rho_n(r) + 2^m + k.
\end{aligned}
$$

This and Equation 7.19 imply Equation 7.21.   □
   Equation 7.19 also implies that

$$
\begin{aligned}
\tau(n, 0, k) &= k \\
\tau(0, r, 0) &= \rho_n(r).
\end{aligned}
$$

Together with Equations 7.20 and 7.21, this leads to the following program.

$\langle \parallel r : 0 \le r < 2^n :: b[\rho_n(r)] := f_r \rangle$;
$\langle ; m : 0 \le m < n ::$
  $\langle \parallel r, k : 0 \le r < 2^{n-m-1} \text{ and } 0 \le k < 2^m ::$
    $\begin{bmatrix} b[\rho_n(r) + k] \\ b[\rho_n(r) + k + 2^m] \end{bmatrix} := \begin{bmatrix} 1 & \epsilon[2^{n-m-1}k] \\ 1 & -\epsilon[2^{n-m-1}k] \end{bmatrix} \begin{bmatrix} b[\rho_n(r) + k] \\ b[\rho_n(r) + k + 2^m] \end{bmatrix}$
  $\rangle$
$\rangle$;
$\langle \parallel k : 0 \le k < 2^n :: \hat{f}[k] := b[k] \rangle$

We conclude the development of the fast-Fourier-transform algorithm by simplifying the index arithmetic of the program. From the definition of the bit-reversal map, it follows that

$$\{\rho_n(r) : 0 \le r < 2^{n-m-1}\} = \{2^{m+1}j : 0 \le j < 2^{n-m-1}\}.$$

Using the variable j instead of r, all occurrences of $\rho_n(r)$ may be replaced by $2^{m+1}j$, while j ranges over the index set $0..2^{n-m-1} - 1$. The expression

$\rho_n(r) + k = 2^{m+1}j + k$ takes on all values in the range $0..2^n - 1$ that have a binary representation in which bit number m is zero. This observation allows us to replace the two indices k and r by one index $t = \rho_n(r) + k$.

The assignments to $b[t]$ and $b[t + 2^m]$ for $t = \rho_n(r) + k$ require the exponential constant $\epsilon[2^{n-m-1}k]$. To replace k and r by the single index t, one must express $2^{n-m-1}k$ in terms of t only. It follows from $t = 2^{m+1}j + k$ that

$$2^{n-m-1}k = 2^{n-m-1}t - 2^n j.$$

However, we already know that $0 \leq k < 2^m$ and $0 \leq 2^{n-m-1}k < 2^{n-1}$. The term $2^n j$ merely shifts $2^{n-m-1}t$ into the right range. This can also be accomplished by modulo arithmetic, and we obtain that

$$2^{n-m-1}k = 2^{n-m-1}t \bmod 2^{n-1}.$$

The binary representation of $2^{n-m-1}t \bmod 2^{n-1}$ is obtained from that of t by zeroing the $n-m$ most significant bits of t and, subsequently, performing $n - m - 1$ shifts to the left.

After incorporating the above changes in index arithmetic, we obtain program Fast-Fourier-1.

program Fast-Fourier-1
declare
  m, t : integer ;
  ϵ : array$[0..2^{n-1} - 1]$ of complex ;
  b : array$[0..2^n - 1]$ of complex
initially
  $\langle\ ;\ t\ :\ 0 \leq t < 2^{n-1}\ ::\ \epsilon[t] = e^{it\frac{2\pi}{2^n}}\ \rangle$;
  $\langle\ ;\ t\ :\ 0 \leq t < 2^n\ ::\ b[\rho_n(t)] = f_t\ \rangle$
assign
  $\langle\ ;\ m\ :\ 0 \leq m < n\ ::$
    $\langle\ \|\ t\ :\ t \in 0..2^n - 1 \text{ and } t \wedge 2^m = 0\ ::$
      $\begin{bmatrix} b[t] \\ b[t + 2^m] \end{bmatrix} := \begin{bmatrix} 1 & \epsilon[2^{n-m-1}t \bmod 2^{n-1}] \\ 1 & -\epsilon[2^{n-m-1}t \bmod 2^{n-1}] \end{bmatrix} \begin{bmatrix} b[t] \\ b[t + 2^m] \end{bmatrix}$
    $\rangle$
  $\rangle$
end

The information flow of program Fast-Fourier-1 is identical to that of the recursive-doubling procedure: in step m, information is exchanged between array entries $b[t]$ and $b[t \bar\vee 2^m]$. (Note that $t \bar\vee 2^m = t + 2^m$ when $t \wedge 2^m = 0$.) However, the fast Fourier transform differs from the standard recursive-doubling procedure, because the assignment to $b[t]$ differs from the assignment to $b[t \bar\vee 2^m]$. Like the recursive-doubling procedure of the concurrent QR-decomposition (see Section 5.4), the fast-Fourier-transform algorithm is an asymmetric recursive-doubling procedure.

The asymmetry can be made more explicit by rewriting the concurrent quantification over t as follows.

$$\langle \parallel t \; : \; t \in 0..2^n - 1 \; ::$$
$$\quad \textbf{if } t \wedge 2^m = 0 \textbf{ then}$$
$$\qquad b[t] := b[t] + \epsilon[2^{n-m-1}t \bmod 2^{n-1}]b[t\bar{\vee}2^m]$$
$$\quad \textbf{else}$$
$$\qquad b[t] := b[t\bar{\vee}2^m] - \epsilon[2^{n-m-1}t \bmod 2^{n-1}]b[t]$$
$$\rangle$$

The case $t \wedge 2^m \neq 0$ exploits the fact that the expression $2^{n-m-1}t \bmod 2^{n-1}$ does not depend on bit number m of t.

7.4 Multicomputer Implementation

It is our intention to compute the discrete Fourier transform of vectors of dimension $N = 2^n$ using $P = 2^D$ processes, where $P \leq N$ and $D \leq n$. The transformation begins, as usual, with a duplication step. Subsequently, a data distribution is imposed in the selection step. Because P divides N, we shall use perfectly load-balanced data distributions. We shall not impose any other requirements on the data distribution.

The binary representation of global index $t \in 0..2^n - 1$ is a bit string of length n, say

$$t = \text{`}\alpha_{n-1}\alpha_{n-2}\ldots\alpha_1\alpha_0\text{'}.$$

Any P-fold perfectly load-balanced data distribution is defined by a permutation of the bit positions. Given the permutation

$$k_0, k_1, \ldots, k_{D-1}, k_D, \ldots, k_{n-1},$$

the global index t is mapped to process identifier p and local index j, with

$$p = \text{`}\alpha_{k_0}\alpha_{k_1}\ldots\alpha_{k_{D-1}}\text{'}$$
$$j = \text{`}\alpha_{k_D}\alpha_{k_{D+1}}\ldots\alpha_{k_{n-1}}\text{'}.$$

For example, the perfectly load-balanced linear data distribution is based on the permutation

$$n-1, n-2, \ldots, n-D, n-D-1, \ldots, 0,$$

while the perfectly load-balanced scatter data distribution is based on

$$D-1, D-2, \ldots, 0, n-1, \ldots, D.$$

The set $\mathcal{T}_D = \{k_0, k_1, \ldots, k_{D-1}\}$ is the set of D leading bit positions, the set of bit positions that forms the process identifier.

A perfectly load-balanced data distribution splits the binary representation of an integer into two parts. To develop the multicomputer version of program Fast-Fourier-1, we shall use the short-hand notation

$$t = t_D, t_l$$

to denote such a splitting. For the global index t, the "leading part" t_D will become the process identifier, and the "trailing part" t_l will become the local index. To split the integer 2^m in this fashion, we write

$$2^m = (2^m)_D, (2^m)_l.$$

The values of $(2^m)_D$ and $(2^m)_l$ are determined by the rank of m in the permutation of the bit positions. Either $(2^m)_D$ or $(2^m)_l$ must vanish; the other must be a power of two.

Before performing a P-fold duplication of program Fast-Fourier-1, replace its quantification over t by the equivalent quantification listed at the end of Section 7.3. This makes explicit the asymmetric nature of the recursive-doubling procedure. We shall focus on the transformation of this quantification. When studying the program segments that follow, bear in mind that those segments are surrounded by a sequential quantification over m and, because of the P-fold duplication, by a concurrent quantification over p. Moreover, all variables have an implicit process identifier p. The inner quantification of the duplicated program is given by:

$$\langle \parallel t : t \in 0..2^n - 1 ::$$
$$\quad \text{if } t \wedge 2^m = 0 \text{ then}$$
$$\quad \quad b[t] := b[t] + \epsilon[t_m]b[t\bar\vee 2^m]$$
$$\quad \text{else}$$
$$\quad \quad b[t] := b[t\bar\vee 2^m] - \epsilon[t_m]b[t]$$
$$\rangle$$

This program segment uses the short-hand notation t_m for the expression $2^{n-m-1}t \bmod 2^{n-1}$.

To prepare for the selection step, which will distribute the array b over the processes, the global index t is split into its leading and trailing parts. To do this, we must transform three expressions:

- $t \in 0..2^n - 1$, which is used in the quantification,

- $t \wedge 2^m = 0$, which is used in the **if** test, and

- $t\bar\vee 2^m$, which is used as an index for b.

Because the constant array ϵ remains duplicated, we need not consider the transformation of the expression $t_m = 2^{n-m-1}t \bmod 2^{n-1}$ in detail: given t_D and t_l, it is always possible to compute t and, subsequently, t_m.

As t ranges over all values of the index set $0..2^n - 1$, its binary representation ranges over all possible bit strings of length n. It follows that $t \in 0..2^n - 1$ is equivalent to

$$t_D \in 0..2^D - 1 \text{ and } t_l \in 0..2^{n-D} - 1.$$

The expression $t \wedge 2^m = 0$ tests whether or not bit number m of t is set. The position of bit number m in the splitting t_D, t_l depends on the rank of m in the permutation of the bit positions. If m is one of the leading positions, say $m \in \mathcal{T}_D$, then bit number m is mapped to the leading part, and the test **if** $t \wedge 2^m = 0$ **then** ... is transformed into

$$\textbf{if } t_D \wedge (2^m)_D = 0 \textbf{ then} \dots$$

If m is one the trailing positions, say $m \notin \mathcal{T}_D$, then bit number m is mapped to the trailing part, and the test is transformed into

$$\textbf{if } t_l \wedge (2^m)_l = 0 \textbf{ then} \dots$$

The expression $t \bar{\vee} 2^m$ reverses bit number m of t. After locating bit number m in the splitting of t, that bit is reversed. We find that

$$m \in \mathcal{T}_D \;\Rightarrow\; \left\{ \begin{array}{rcl} (t\bar{\vee}2^m)_D & = & t_D \bar{\vee}(2^m)_D \\ (t\bar{\vee}2^m)_l & = & t_l \end{array} \right.$$

$$m \notin \mathcal{T}_D \;\Rightarrow\; \left\{ \begin{array}{rcl} (t\bar{\vee}2^m)_D & = & t_D \\ (t\bar{\vee}2^m)_l & = & t_l \bar{\vee}(2^m)_l. \end{array} \right.$$

To incorporate the index splitting into the previous program segment, one must consider separately two important cases: $m \in \mathcal{T}_D$ and $m \notin \mathcal{T}_D$. The remainder of the transformation follows by mere substitution.

> **if** $m \notin \mathcal{T}_D$ **then**
> $\langle \parallel t_D, t_l \; : \; t_D \in 0..2^D - 1 \text{ and } t_l \in 0..2^{n-D} - 1 \; ::$
> **if** $t_l \wedge (2^m)_l = 0$ **then**
> $b[t_D, t_l] := b[t_D, t_l] + \epsilon[t_m]b[t_D, t_l\bar{\vee}(2^m)_l]$
> **else**
> $b[t_D, t_l] := b[t_D, t_l\bar{\vee}(2^m)_l] - \epsilon[t_m]b[t_D, t_l]$
> $\rangle$
> **else**
> $\langle \parallel t_D, t_l \; : \; t_D \in 0..2^D - 1 \text{ and } t_l \in 0..2^{n-D} - 1 \; ::$
> **if** $t_D \wedge (2^m)_D = 0$ **then**
> $b[t_D, t_l] := b[t_D, t_l] + \epsilon[t_m]b[t_D\bar{\vee}(2^m)_D, t_l]$
> **else**
> $b[t_D, t_l] := b[t_D\bar{\vee}(2^m)_D, t_l] - \epsilon[t_m]b[t_D, t_l]$
> $\rangle$

Remember that the above program segment is surrounded by a sequential quantification over m. It is also implicitly duplicated.

The selection step imposes the data distribution on array b but leaves the entire constant array ϵ duplicated in every process. After data distribution, process p may access only the variables $b[p][t_D, t_l]$ with $t_D = p$. As long as $m \notin \mathcal{T}_D$, it is easy to incorporate this restriction, because assignments involve only entries of b whose indices have the same leading part. When $m \in \mathcal{T}_D$, however, assignments are between entries of b whose indices have different leading parts. Again, process p may assign new values to $b[p][t_D, t_l]$ only if $t_D = p$. However, assignments to $b[p][t_D, t_l]$ require access to $b[p\bar{\vee}(2^m)_D][t_D\bar{\vee}(2^m)_D, t_l]$. The latter must be obtained by communication with process $p\bar{\vee}(2^m)_D$. The following program segment results.

> **if** $m \notin \mathcal{T}_D$ **then**
> $\langle\; \|\; t_D, t_l\; :\; t_D = p$ **and** $t_l \in 0..2^{n-D} - 1\; ::$
>   **if** $t_l \wedge (2^m)_l = 0$ **then**
> $b[t_D, t_l] := b[t_D, t_l] + \epsilon[t_m]b[t_D, t_l\bar{\vee}(2^m)_l]$
>   **else**
> $b[t_D, t_l] := b[t_D, t_l\bar{\vee}(2^m)_l] - \epsilon[t_m]b[t_D, t_l]$
> $\rangle$
>
> **else begin**
>   **send** $\{b[t_D, t_l] : t_D = p$ **and** $t_l \in 0..2^{n-D} - 1\}$ **to** $p\bar{\vee}(2^m)_D$;
>   **receive** $\{x[t_D, t_l] : t_D = p$ **and** $t_l \in 0..2^{n-D} - 1\}$ **from** $p\bar{\vee}(2^m)_D$;
> $\langle\; \|\; t_D, t_l\; :\; t_D = p$ **and** $t_l \in 0..2^{n-D} - 1\; ::$
>   **if** $t_D \wedge (2^m)_D = 0$ **then**
> $b[t_D, t_l] := b[t_D, t_l] + \epsilon[t_m]x[t_D, t_l]$
>   **else**
> $b[t_D, t_l] := x[t_D, t_l] - \epsilon[t_m]b[t_D, t_l]$
> $\rangle$
> **end**

A few additional transformations, mostly cosmetic, lead to the final version. The leading part of the global index has become superfluous, since it is always equal to p in process p. The trailing part t_l becomes the local index j. Given $j(= t_l)$ and $p(= t_D)$, it is always possible to compute the corresponding global index t. Subsequently, the index t_m used to retrieve exponential constants from the array ϵ can be computed. The details of the calculation of t_m depend, of course, on the permutation of the bit positions that defines the data distribution. In the program, we denote the calculation of t_m as a function of p, j, and m by $t_m(p, j)$. The test

$$\text{\textbf{if} } t_D \wedge (2^m)_D = 0 \text{ \textbf{then} } \ldots$$

is independent of t_l and can be taken outside of the last quantification.

Because the individual processes of a multicomputer computation are sequential, we convert all concurrent into sequential quantifications. This is easy for the case $m \in \mathcal{T}_D$. In the other case, the quantification is first reformulated as in program Fast-Fourier-1. Subsequently, the concurrent separator is replaced by the sequential separator.

$0..P-1 \parallel p$ **program** Fast-Fourier-2
declare
 $m,\ j\ :$ integer ;
 $\epsilon\ :$ array $[0..2^{n-1}-1]$ of complex ;
 $b,\ x\ :$ array $[0..2^{n-D}]$ of complex
assign
 $\langle\ ;\ m\ :\ 0 \le m < n\ ::$
  if $m \notin \mathcal{T}_D$ **then**
   $\langle\ ;\ j\ :\ j \in 0..2^{n-D}-1 \textbf{ and } j \wedge (2^m)_l = 0\ ::$
$$\begin{bmatrix} b[j] \\ b[j\bar{\vee}(2^m)_l] \end{bmatrix} := \begin{bmatrix} 1 & \epsilon[t_m(p,j)] \\ 1 & -\epsilon[t_m(p,j)] \end{bmatrix} \begin{bmatrix} b[j] \\ b[j\bar{\vee}(2^m)_l] \end{bmatrix}$$
   $\rangle$
  else begin
   send $\{b[j] : j \in 0..2^{n-D}-1\}$ **to** $p\bar{\vee}(2^m)_D$;
   receive $\{x[j] : j \in 0..2^{n-D}-1\}$ **from** $p\bar{\vee}(2^m)_D$;
   if $p \wedge (2^m)_D = 0$ **then**
    $\langle\ ;\ j\ :\ j \in 0..2^{n-D}-1\ ::\ b[j] := b[j] + \epsilon[t_m(p,j)]x[j]\ \rangle$
   else
    $\langle\ ;\ j\ :\ j \in 0..2^{n-D}-1\ ::\ b[j] := x[j] - \epsilon[t_m(p,j)]b[j]\ \rangle$
  end
 $\rangle$
end

The performance analysis can be brief. Because the set $\mathcal{T}_D$ of leading bit positions contains exactly D values in the range of m, there are D steps with and $n-D$ steps without communication, and

$$\begin{aligned} T_P &= (n-D)2^{n-D}2\tau_A + D(\tau_C(2^{n-D}) + 2^{n-D}2\tau_A) \\ &= \frac{2nN}{P}\tau_A + D\tau_C(\frac{N}{P}). \end{aligned} \tag{7.22}$$

The corresponding speed-up is given by:

$$S_P = \frac{T_1^S}{T_P} = P\frac{1}{1 + \frac{D}{2n}(\frac{P}{N}\frac{\tau_S}{\tau_A} + \frac{\beta}{\tau_A})}.$$

Coarse-grained computations with $P \ll N$ can achieve excellent speed-up. For fine-grained computations with $P = N$ and $D = n$, we obtain that

$$S_P = P\frac{1}{1 + \frac{\tau_S + \beta}{2\tau_A}}.$$

The ratio $\frac{\tau_S}{\tau_A}$ can be quite large and severely limits the efficiency of fine-grained computations (under our standard assumption of one process per node).

7.5   Generalizations

7.5.1   Bit-Reversal Maps

Programs Fast-Fourier-1 and Fast-Fourier-2 compute the values $\hat{f}_k$ in natural order, provided the values f_r are stored in bit-reversed order. By using a different map τ (recall Equation 7.18), a version can be obtained that computes the values $\hat{f}_k$ in a bit-reversed order, provided the values f_r are in natural order. Both versions are useful. If the forward transform computes the frequencies in bit-reversed order from the values stored in natural order, it is convenient to use an inverse transform that takes the bit-reversed frequencies as input to transform them into natural-order values. When using this combination of forward and inverse fast-Fourier-transform programs, one merely has to adhere to the convention that vectors in physical space are stored in natural order and those in Fourier space in bit-reversed order. With this convention, the inverse-transform program uses the map τ of Equation 7.19. For the map used in the forward-transform program, see Exercise 30.

If it is acceptable to use an extra array for intermediate and final results, no bit-reversal map is necessary at all. With an auxiliary array, the order in which quantities are stored becomes irrelevant for correctness. However, the storage order remains a performance issue, because it determines memory-access and communication patterns. With the bit-reversal maps, we obtained an efficient recursive-doubling structure. This is probably a far more important reason for using bit-reversal maps than saving memory.

7.5.2   General Dimensions

The divide-and-conquer strategy behind the fast-Fourier-transform algorithm can be generalized to transform vectors of any dimension N, not just dimensions that are a power of 2. For $N = rM$ with r a prime number, a trigonometric polynomial $p(x)$ of degree $N - 1$ is decomposed as follows:

$$p(x) = \sum_{n=0}^{N-1} f_n e^{inx} = \sum_{m=0}^{M-1}\sum_{\ell=0}^{r-1} f_{rm+\ell} e^{i(rm+\ell)x}$$

$$= \sum_{\ell=0}^{r-1} e^{i\ell x} \sum_{m=0}^{M-1} f_{rm+\ell} e^{irmx}$$

$$= \sum_{\ell=0}^{r-1} e^{i\ell x} q_\ell(rx).$$

As in Section 7.2, this decomposition can be used as the basis for a divide-and-conquer strategy. This generalization is effective as long as N can be factored into a large number of small primes.

7.5.3   The Discrete Real Fourier Transform

Equation 7.7 implies that the discrete Fourier transform $\tilde{\vec{f}} = [\tilde{f}_k]_{k=0}^{M-1}$ of a real M-dimensional vector $\vec{f} = [f_m]_{m=0}^{M-1}$ satisfies the symmetry:

$$\forall k \in 0..M - 1: \tilde{f}_k = \overline{\tilde{f}_{M-k}}. \tag{7.23}$$

Using this symmetry, the transform of a real M-dimensional vector will be reduced to the transform of a complex N-dimensional vector with $N = M/2$. Throughout this section, the tilde notation will be used for transforms of M-dimensional vectors and the hat notation for transforms of N-dimensional vectors.

Writing $\tilde{f}_k = \tilde{a}_k + i\tilde{b}_k$ with $\tilde{a}_k$ and $\tilde{b}_k$ real, Equation 7.23 implies that the discrete Fourier transform of a real vector of dimension $M = 2N$ is defined by the M real coefficients:

$$\tilde{a}_0, \tilde{a}_1, \tilde{b}_1, \ldots, \tilde{a}_{N-1}, \tilde{b}_{N-1}, \tilde{a}_N.$$

Noting that $\tilde{b}_0 = \tilde{b}_N = 0$ or, equivalently, that $\tilde{f}_0$ and $\tilde{f}_N$ are real, it is thus sufficient to compute the N complex coefficients:

$$\tilde{f}_0 + i\tilde{f}_N, \tilde{f}_1, \ldots, \tilde{f}_{N-1}.$$

Consider the discrete Fourier transform of the complex N-dimensional vector $\vec{u} = \vec{v} + i\vec{w} = [f_{2n}] + i[f_{2n+1}]$. Because the discrete Fourier transform is a linear operator, we have that

$$\forall k \in 0..N - 1: \hat{u}_k = \hat{v}_k + i\hat{w}_k. \tag{7.24}$$

On the other hand, the symmetry expressed by Equation 7.23 and applied to the real N-dimensional vectors $\vec{v}$ and $\vec{w}$ implies that

$$\forall k \in 0..N - 1: \overline{\hat{u}_{N-k}} = \hat{v}_k - i\hat{w}_k. \tag{7.25}$$

From Equations 7.24 and 7.25, it follows that

$$\forall k \in 0..N - 1: \begin{cases} \hat{v}_k &= (\hat{u}_k + \overline{\hat{u}_{N-k}})/2 \\ \hat{w}_k &= (\hat{u}_k - \overline{\hat{u}_{N-k}})/2i. \end{cases} \tag{7.26}$$

The relation between $\tilde{f}_k$, $\hat{v}_k$, and $\hat{w}_k$ follows from the definition of discrete Fourier transforms of the M-dimensional vector $\vec{f}$ and the N-dimensional vectors $\vec{v}$ and $\vec{w}$. With

$$H = \frac{2\pi}{M} = \frac{2\pi}{2N} = \frac{h}{2},$$

we find that

$$\begin{cases} & \tilde{f}_0 &= \hat{v}_0 + \hat{w}_0 \\ \forall k \in 1..N - 1: & \tilde{f}_k &= \hat{v}_k + e^{ikH}\hat{w}_k \\ & \tilde{f}_N &= \hat{v}_0 - \hat{w}_0. \end{cases}$$

Combined with Equation 7.26, this leads to the relation between the discrete Fourier transforms of the real vector $\vec{f}$ and the complex vector $\vec{u}$:

$$\begin{cases} & \tilde{f}_0 + i\tilde{f}_N = (1+i)\overline{\tilde{u}_0} \\ \forall k \in 1..N-1: & \tilde{f}_k = \tfrac{1}{2}\left(\hat{u}_k + \overline{\hat{u}_{N-k}} - ie^{ikH}(\hat{u}_k - \overline{\hat{u}_{N-k}})\right). \end{cases}$$

These equations imply program Fast-Real-Fourier.

> **program** Fast-Real-Fourier
> **declare**
> m, t : integer ;
> ϵ : array$[0..2^{n-1}-1]$ of complex ;
> b : array$[0..2^n-1]$ of complex
> **initially**
> $\langle\, ; t : 0 \le t < 2^{n-1} :: \epsilon[t] = e^{it\frac{2\pi}{2^n}} \,\rangle$;
> $\langle\, ; t : 0 \le t < 2^n :: b[\rho_n(t)] = f_{2t} + if_{2t+1} \,\rangle$
> **assign**
> $\langle\, ; m : 0 \le m < n ::$
> $\langle\, \| \ t : t \in 0..2^n-1 \text{ and } t \wedge 2^m = 0 ::$
> $$\begin{bmatrix} b[t] \\ b[t+2^m] \end{bmatrix} := \begin{bmatrix} 1 & \epsilon[2^{n-m-1}t \bmod 2^{n-1}] \\ 1 & -\epsilon[2^{n-m-1}t \bmod 2^{n-1}] \end{bmatrix} \begin{bmatrix} b[t] \\ b[t+2^m] \end{bmatrix}$$
> $\rangle$
> $\rangle$;
> $\langle\, \| \ t : t \in 1..2^{n-1}-1 ::$
> $b[t] := (b[t] + \overline{b[N-t]} - ie^{itH}(b[t] - \overline{b[N-t]}))/2 \ \|$
> $b[N-t] := (b[N-t] + \overline{b[t]} + ie^{-itH}(b[N-t] - \overline{b[t]}))/2$
> $\rangle$;
> $b[0] := (1+i)\overline{b[0]}$
> **end**

As indicated by the **initially** section, the values $u_t = f_{2t} + if_{2t+1}$, where $0 \le t < N$, are stored in the array b according to the bit-reversal map $\rho_n(t)$. After the complex transform, the array b contains the values of the discrete Fourier coefficients in natural order. Subsequently, the real-transform coefficients $\tilde{f}_k$ are computed in the same array b. To do this in place, the assignments to $b[t]$ and $b[N-t]$ must be done simultaneously.

7.5.4   The Discrete Fourier-Sine Transform

The Fourier series of an odd function $f(x)$ reduces to a Fourier-sine series. More precisely, if $f(x)$ is real, periodic with period 2π, and odd, Equation 7.1 becomes

$$f(x) = \sum_{k=1}^{+\infty} \tilde{f}_k \sin(kx).$$

The Fourier-sine coefficients $\check{f}_k$ in this expansion are given by:

$$\check{f}_k = \frac{2}{\pi} \int_0^\pi f(x) \sin(kx) dx.$$

Discretize the function $f(x)$ on an equidistant grid of the interval $[0, \pi]$. Let $H = \pi/M$ and $f_j = f(jH)$ for all $j \in 0..M$. Because $f(x)$ is periodic and odd, we have that $f(0) = f(\pi) = 0$ and, consequently, that $f_0 = f_M = 0$. The discretized function is, therefore, represented by the real $(M-1)$-dimensional vector $\vec{f} = [f_j]_{j=1}^{M-1}$. The following lemma, which is analogous to Lemma 17, establishes the discrete Fourier-sine transform.

Lemma 20 *The set of vectors $\{\vec{e}_m = [\sin(jmH)]_{j=1}^{M-1} : 0 < m < M\}$, with $H = \pi/M$, is an orthogonal basis for the space $\mathbb{R}^{M-1}$.*

(Without proof.) $\square$

Every vector of $\mathbb{R}^{M-1}$ has a unique representation as a linear combination of the proposed basis vectors. Using that

$$\forall m \in 1..M - 1: \ (\vec{e}_m, \vec{e}_m) = M/2,$$

the following identities are easily obtained:

$$\forall m \in 1..M - 1: \quad f_m = \sum_{k=1}^{M-1} \check{f}_k \sin(mkH) \tag{7.27}$$

$$\forall k \in 1..M - 1: \quad \check{f}_k = \frac{2}{M} \sum_{m=1}^{M-1} f_m \sin(kmH). \tag{7.28}$$

A scalar factor aside, the discrete Fourier-sine transform is its own inverse.

An elementary calculation shows that the discrete Fourier-sine transform of $\vec{f} = [f_m]_{m=1}^{M-1}$ is related to the discrete Fourier transform of the real M-dimensional vector $\vec{y} = [y_m]_{m=0}^{M-1}$ with

$$\forall m \in 0..M - 1: \ y_m = \frac{2}{M} \sin(mH)(f_m + f_{M-m}) + \frac{1}{M}(f_m - f_{M-m}).$$

Note that $y_0 = 0$. Because the vector $\vec{y}$ is real, program Fast-Real-Fourier can be used to compute the $N = M/2$ complex coefficients:

$$\hat{y}_0 + i\hat{y}_N, \hat{y}_1, \ldots, \hat{y}_{N-1},$$

where

$$\forall k \in 0..N: \ \hat{y}_k = \sum_{m=0}^{M-1} y_m e^{ikmh},$$

with $h = 2\pi/M = 2H$. Note that we use the tilde notation for the discrete Fourier-sine transform of real $(M-1)$-dimensional vectors and the hat notation for the discrete Fourier transform of real M-dimensional vectors. The discrete Fourier-sine coefficients $\tilde{f}_k$ are related to the discrete Fourier coefficients $\hat{y}_k$ according to

$$\forall k \in 1..N-1: \; \hat{y}_k = \tilde{f}_{2k+1} - \tilde{f}_{2k-1} + i\tilde{f}_{2k} \tag{7.29}$$

and

$$\hat{y}_0 + i\hat{y}_N = 2(\tilde{f}_1 - i\tilde{f}_{M-1}) \tag{7.30}$$

The imaginary parts of the coefficients $\hat{y}_k$ determine all discrete Fourier-sine coefficients with even indices. Those with odd indices are computed by means of a simple recursion relation, which is derived from the real parts of Equation 7.29 and is initialized by either the real or the imaginary part of Equation 7.30. In a multicomputer program, this recursion could be computed by full recursive doubling.

7.5.5   The Discrete Fourier-Cosine Transform

If the function $f(x)$ is real, periodic with period 2π, and even, Equation 7.1 reduces to the Fourier-cosine series expansion

$$f(x) = \frac{1}{2}\check{f}_0 + \sum_{k=1}^{+\infty} \check{f}_k \cos(kx),$$

with coefficients given by

$$\forall k \geq 0: \; \check{f}_k = \frac{2}{\pi} \int_0^\pi f(x) \cos(kx) dx.$$

To derive the discrete Fourier-cosine transform, discretize the function $f(x)$ on an equidistant grid of the interval $[0, 2\pi]$. Let $H = 2\pi/(2M)$ and $f_j = f(jH)$ for all $j \in 0..2M - 1$. Because $f(x)$ is even, $f(x) = f(2\pi - x)$ and, consequently, $f_j = f_{2M-j}$ for all $j \in 0..2M - 1$. The following lemma, which is analogous to Lemmas 17 and 20, gives us an orthogonal basis for the subspace of even vectors of $\mathbb{R}^{2M}$.

Lemma 21 *The set of vectors* $\{\vec{e}_m = [\cos(jmH)]_{j=0}^{2M-1} : 0 \leq m \leq M\}$, *with* $H = 2\pi/(2M)$, *is an orthogonal basis for the* $(M+1)$-*dimensional subspace of even vectors of* $\mathbb{R}^{2M}$.

(Without proof.)   □

Every even vector $\vec{f}$ in $\mathbb{R}^{2M}$ has a unique representation as a linear combination of the proposed basis vectors. As before, the coefficient of $\vec{e}_k$ in this linear combination is given by

$$\frac{(\vec{e}_k, \vec{f})}{(\vec{e}_k, \vec{e}_k)}.$$

An elementary calculation shows that $(\vec{e}_k, \vec{e}_k) = M$ for $k \neq 0$ and that $(\vec{e}_0, \vec{e}_0) = 2M$. We focus, therefore, on the numerators and compute

$$\forall k \in 0..M : \quad \tilde{f}_k = (\vec{e}_k, \vec{f}) = \sum_{m=0}^{2M-1} f_m \cos(kmH).$$

Because the vector $\vec{f}$ is even, it is easily derived that

$$\forall k \in 0..M : \quad \tilde{f}_k = f_0 + 2 \sum_{j=0}^{M-1} f_j \cos(jkH) + (-1)^k f_M. \qquad (7.31)$$

The $2M$-dimensional vector $\vec{f}$ has only $M + 1$ degrees of freedom, say the components $f_0, f_1, \ldots, f_M$. In this sense, we may refer to the discrete Fourier-cosine transformation of $\vec{f}$ as a transformation of a real $(M + 1)$-dimensional vector.

The discrete Fourier-cosine transform of a real $(M + 1)$-dimensional vector can be computed via the discrete Fourier transform of an auxiliary real M-dimensional vector $\vec{y} \in \mathbb{R}^M$ with components given by:

$$\forall m \in 0..M - 1 : \quad y_m = \frac{1}{2}(f_m + f_{M-m}) - \sin(mh)(f_m - f_{M-m}).$$

An elementary calculation shows that

$$\forall k \in 0..N : \quad 2\hat{y}_k = \tilde{f}_{2k} + i(\tilde{f}_{2k+1} - \tilde{f}_{2k-1}), \qquad (7.32)$$

where $N = M/2$. The tilde notation is used for the discrete Fourier-cosine transform of real $(M + 1)$-dimensional vectors and the hat notation for the discrete Fourier transform of real M-dimensional vectors.

As seen in Section 7.5.3, Program Fast-Real-Fourier computes the N complex coefficients

$$\hat{y}_0 + i\hat{y}_N, \hat{y}_1, \ldots, \hat{y}_{N-1}.$$

It follows from Equation 7.32 that their real parts trivially determine the Fourier-cosine coefficients with even indices. (This holds for all but $\tilde{f}_M$, which is determined by the imaginary part of $\hat{y}_0 + i\hat{y}_N$.) To compute the discrete Fourier-cosine coefficients with odd indices, a recursion involving the imaginary part of the discrete Fourier coefficients is required. This recursion must be initialized with $\tilde{f}_1$, which is computed directly from Equation 7.31.

7.5.6 Multivariate Discrete Fourier Transforms

Multivariate Fourier transforms of functions $f(x, y)$ over a domain $\Omega \subset \mathbb{R}^2$ are defined by successively applying univariate Fourier transforms. For example, let $\Omega = [0, 2\pi) \times [0, \pi]$, $f(x, y)$ periodic in x with period 2π, and $f(x, 0) = f(x, \pi) = 0$.

For every fixed y, the function $f(x, y)$ is a periodic function of x with period 2π, which can be expanded into a Fourier series. It follows that

$$f(x, y) = \sum_{k=-\infty}^{+\infty} \tilde{f}_k(y) e^{ikx}.$$

Because $f(x, y)$ vanishes for $y = 0$ and $y = \pi$, we obtain that

$$\forall k \in \mathbb{Z}: \tilde{f}_k(0) = \tilde{f}_k(\pi) = 0.$$

Every $\tilde{f}_k(y)$ can be extended into an odd function with period 2π and can, therefore, be expanded into a Fourier-sine series. The multivariate Fourier expansion

$$f(x, y) = \sum_{k=-\infty}^{+\infty} \sum_{\ell=1}^{\infty} \hat{f}_{k,\ell} e^{ikx} \sin(\ell y) \tag{7.33}$$

is obtained.

Consider the discrete analog of the above continuous example. The function $f(x, y)$ is represented by function values on a regular rectangular grid defined by the grid spacings $h_x = 2\pi/M$ and $h_y = \pi/N$:

$$\forall (m, n) \in 0..M - 1 \times 1..N - 1: f_{m,n} = f(mh_x, nh_y).$$

To obtain the discrete transform corresponding to Equation 7.33, one must apply transforms in the x- and in the y-direction. In the x-direction, the discrete Fourier transform is applied to $N - 1$ real vectors of dimension M. In the y-direction, the discrete Fourier-sine transform is applied to M vectors of dimension $N - 1$. The bit-reversal map and the packing of real values into complex arrays, one or both of which may have occurred in the first step, complicate the index arithmetic of the second transformation.

Another complication is that the univariate fast-Fourier-transform programs need some reorganization to perform optimally in a multivariate context. Consider program Multivariate-Fourier, which applies $N - 1$ discrete Fourier transforms in the x-direction and $M - 1$ discrete Fourier-sine transforms in the y-direction.

program Multivariate-Fourier
declare
    m, n, t : integer ;
    f : array$[0..M - 1 \times 1..N - 1]$ of real
assign
    $\langle \,\|\, n : 0 < n < N :: \text{Real-Fourier}([f[m, n]]_{m=0}^{M-1}) \,\rangle$;
    $\langle \,\|\, m : 0 \leq m < M :: \text{Fourier-Sine}([f[m, n]]_{n=1}^{N-1}) \,\rangle$
end

Program Real-Fourier performs discrete real Fourier transforms in the x-direction and program Fourier-Sine performs discrete Fourier-sine transforms in the y-direction. Both programs are left unspecified. The array passed to program Real-Fourier consists of array entries $f[m, n]$ with constant second index, while the array passed to program Fourier-Sine consists of array entries $f[m, n]$ with constant first index. For programs Real-Fourier and Fourier-Sine, successive array entries are, therefore, not necessarily stored contiguously in memory. In fact, the stride between successive array entries should be an argument to these programs, because it is unpredictable in which direction one might want to apply which transform. Avoiding the stride problem by copying the entries to a temporary array is not feasible, particularly if the following optimization is also incorporated.

Inside program Real-Fourier, there are several quantifications that involve the first index of array f. For every value of the second index, identical index arithmetic for the first index occurs. Substantial savings can be accomplished by moving the quantification over n inside the fast-Fourier-transform quantifications used by program Real-Fourier. (From a program-correctness point of view, the position of the concurrent quantification over n is arbitrary.) This program transformation also increases performance for two other reasons besides reducing index arithmetic. First, communication of individual transforms can be grouped into combined messages to amortize latency over many transforms. Second, this program transformation allows for pipelining. Consider two successive instances of the quantification over n, say $n = n_0$ and $n = n_0 + 1$. The first instance uses the variables $\{f[m, n_0] \; : \; 0 \leq m < M\}$, while the second uses $\{f[m, n_0 + 1] \; : \; 0 \leq m < M\}$. The variables $f[m, n_0 + 1]$ and $f[m, n_0]$ are separated in memory by a certain stride that is independent of the value of n_0. Quantifications involving data separated by constant strides can be implemented efficiently on many pipelined processors; see Section 12.1.

For the same reasons, it makes sense to transform program Fourier-Sine such that its innermost quantification is the concurrent quantification over m.

The multicomputer implementation requires a data distribution over a $P \times Q$ process mesh. The process-mesh dimensions, P and Q, and the dimensions of the computational grid, M and N, must be powers of two. It is then possible to introduce perfectly load-balanced data distributions for both indices m and n and to obtain a multicomputer multivariate transform based on the above multicomputer univariate transforms.

Another approach obtains a multivariate transform for multicomputers by using only sequential univariate transforms. If either P or Q is one, all univariate transforms in one direction can be computed sequentially. Before computing the univariate transforms in the other direction, a grid transpose is performed, after which the univariate transforms in the other direction can be computed sequentially.

It is possible, even easy, to write down a naive program for the multi-computer grid-transpose operation and to analyze it under our standard performance assumptions. The resulting execution time is overly optimistic, however, because the naive implementation almost always overloads the capacity of the communication network if applied to grids of a size occurring in typical supercomputing applications. Under conditions of network contention, our simple communication-performance assumptions do not hold, and they underestimate the true cost of the operation. A usable implementation of the grid-transpose operation requires some computer-dependent programming to avoid network contention. A more detailed discussion is found in Section 12.3.3, where a realistic estimate for the communication time required by the grid-transpose operation is obtained.

Both approaches to multicomputer multivariate Fourier transform require the same number of floating-point operations. It is, therefore, sufficient to compare the communication times of both approaches. To this end, assume R nodes are available to compute a problem on an $M \times M$ grid with R processes.

When using the one-dimensional data distribution, all the communication occurs during the grid transpose. The result obtained in Section 12.3.3 allows us to estimate the communication time by

$$\tau_C (M^2/R) \log_2 R.$$

The two-dimensional data distribution uses a $P \times Q$ process mesh with $PQ = R$. From Equation 7.22, it follows that the concurrent fast Fourier transform of a vector of length M over P processes requires a communication time of $\tau_C (M/P) \log_2 P$. However, M such transforms are done simultaneously in Q process columns, and one message can carry communicated values of N/Q transforms, thereby avoiding latency. The fast Fourier transform in the x-direction requires a communication time of

$$\tau_C ((M/Q)(M/P)) \log_2 P = \tau_C (M^2/R) \log_2 P.$$

Adding to this the communication time for the transform in the y-direction, we obtain that the communication time for the two computational approaches is identical.

The main disadvantage of one-dimensional data distributions is that fine-granularity effects become important as soon as $R \approx M$, instead of $R \approx M^2$ for two-dimensional data distributions. It also must be stressed that the data distribution is often a given and cannot be chosen: the multivariate Fourier transform is just one part of a larger computation, which may impose a two-dimensional data distribution.

7.6   Notes and References

The fast Fourier transform was introduced by Cooley and Tukey [19]. Subsequent developments addressed issues of packing and unpacking of the vector components; the algorithm of Temperton [80], for example, is both self-sorting and in-place. The first concurrent algorithm was proposed by Pease [72]. There exist many other fast-Fourier-transform algorithms, which differ in the details of data organization and order of computation. Van Loan [84] gives a detailed overview of fast-Fourier-transform algorithms.

Writing a fast-Fourier-transform package remains a challenge for any writer of software libraries, because the context in which the package is used has such an important impact on the details of the computation. Consider, for example, how the multivariate transforms of Section 7.5.6 lead to considerable changes of the univariate transforms. These difficulties are amplified by the large number of different, but related, transforms. Any package that claims to be complete must incorporate these related transforms. Moreover, forward and backward transformations use different bit-reversal maps, and all transformations should work for a variety of data distributions. A considerable task indeed!

Exercises

Exercise 30 *Prove that the map*

$$\tau(m, r, k) = r + \rho_n(k) \qquad (7.34)$$

also satisfies Equation 7.18. Moreover, this map results in a program that computes the Fourier coefficients in bit-reversed order, starting from a vector stored in natural order.

Exercise 31 *Develop a forward-transform program using the bit-reversal map τ of Equation 7.34.*

Exercise 32 *Develop a multicomputer fast Fourier transform that leaves the vector components of both transformed and nontransformed vectors in natural order. Use extra memory if necessary.*

Exercise 33 *Develop forward and backward transforms for real vectors. Particularly, take care to implement the packing and unpacking of the real data into the complex arrays such that no extra memory is required. Also develop the multicomputer version.*

Exercise 34 *Why is the discrete Fourier-cosine transform different from the real part of the discrete Fourier transform? Why is the discrete Fourier-sine transform different from the imaginary part of a discrete Fourier transform?*

Exercise 35 *Apply the divide-and-conquer technique directly to multivariate Fourier transforms by splitting multivariate trigonometric polynomials into its odd-even parts. Consider, for example, the case of a transformation over the two variables x and y. With $M = 2K$, $N = 2L$, and*

$$p(x,y) = \sum_{m=0}^{M-1} \sum_{n=0}^{N-1} f_{m,n} e^{i(mx+ny)},$$

we have that

$$
\begin{aligned}
p(x,y) \; = \; & \sum_{k=0}^{K-1} \sum_{\ell=0}^{L-1} f_{2k,2\ell}\, e^{i2(kx+\ell y)} \\
+ \; & e^{ix} \sum_{k=0}^{K-1} \sum_{\ell=0}^{L-1} f_{2k+1,2\ell}\, e^{i2(kx+\ell y)} \\
+ \; & e^{iy} \sum_{k=0}^{K-1} \sum_{\ell=0}^{L-1} f_{2k,2\ell+1}\, e^{i2(kx+\ell y)} \\
+ \; & e^{i(x+y)} \sum_{k=0}^{K-1} \sum_{\ell=0}^{L-1} f_{2k+1,2\ell+1}\, e^{i2(kx+\ell y)}
\end{aligned}
$$

or

$$p(x,y) = q_{0,0}(2x,2y) + e^{ix} q_{1,0}(2x,2y) + e^{iy} q_{0,1}(2x,2y) + e^{i(x+y)} q_{1,1}(2x,2y).$$

Use this identity as the basis for a divide-and-conquer strategy for multivariate Fourier transforms. Develop the corresponding specification and multicomputer programs.

8
Poisson Solvers

In this chapter, some elementary solution methods for the Poisson equation on a rectangular domain will be introduced. The algorithms used in the implementation of these elementary methods are fundamental building blocks in the construction of programs that solve other, more general, partial-differential equations.

8.1 The Poisson Problem

The *Poisson partial-differential equation* on a domain Ω in $\mathbb{R}^2$ is given by

$$\forall (x, y) \in \Omega : \ -\Delta u = f(x, y). \tag{8.1}$$

The partial-differential operator Δ is called the *Laplace operator* and is defined on twice-differentiable functions $u(x, y)$ by

$$\Delta u \equiv \frac{\partial^2 u}{\partial x^2} + \frac{\partial^2 u}{\partial y^2}.$$

We shall consider only the case of a two-dimensional rectangular domain $\Omega = (a, b) \times (c, d)$.

Equation 8.1 must be supplemented with boundary conditions. *Dirichlet-boundary conditions* fix the value of $u(x, y)$ at boundary points, for example,

$$\forall y \in (c, d) : \quad u(a, y) = g(y).$$

Neumann-boundary conditions specify the value of the derivative of $u(x, y)$ in a direction normal to the boundary. The following are typical Neumann-boundary conditions:

$$\forall x \in (a, b) : \quad \frac{\partial u}{\partial y}(x, d) = g(x)$$

$$\forall y \in (c, d) : \quad \frac{\partial u}{\partial x}(b, y) = g(y).$$

Periodic-boundary conditions require that the function $u(x, y)$ be periodic in the x- and/or the y-direction. For example, the statement

$$\forall y \in (c, d) : \quad u(a, y) = u(b, y)$$

expresses periodicity in x.

There is an infinity of possibilities to combine these three fundamental types (and other types) of boundary conditions. A particular application might require periodicity in one direction, Neumann-boundary conditions on part of the remaining boundary, and Dirichlet-boundary conditions on the rest. The endless array of possibilities complicates the development of efficient black-box solvers. We shall consider only the case of Dirichlet-boundary conditions. This restriction should not be interpreted as an indication that other boundary conditions are not important or are similar. In the practice of developing partial-differential-equation solvers, many implementation and numerical-analysis problems arise due to complicated boundary conditions. A serious treatment of this subject is outside the scope of this text.

The Poisson problem with Dirichlet-boundary conditions is defined by

$$\begin{cases} \forall (x, y) \in \Omega : & -\Delta u = f(x, y) \\ \forall (x, y) \in \partial\Omega : & u(x, y) = g(x, y). \end{cases} \tag{8.2}$$

The function $f(x, y)$ is defined for all $(x, y) \in \Omega$, while the function $g(x, y)$ is defined only on the boundary $\partial\Omega$.

The Poisson problem on the rectangular domain $\Omega = (0, \pi) \times (0, \pi)$ is discretized on an equidistant rectangular $M \times N$ grid. Although it is natural to choose $M = N$ in this case, we keep separate symbols M and N for the number of grid cells in the x- and y-direction, respectively. With $Mh = Nh = \pi$, the grid points are given by

$$\forall (m, n) \in 0..M \times 0..N : \quad (x_m, y_n) = (mh, nh).$$

The boundary points satisfy $m = 0$, $m = M$, $n = 0$, or $n = N$.

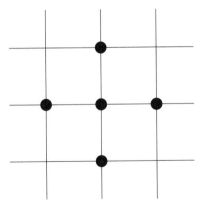

FIGURE 8.1. Five-point stencil corresponding to the second-order discretization of the Laplace operator.

For sufficiently smooth functions $u(x, y)$, the Taylor-series expansions of $u(x+h, y)$, $u(x-h, y)$, $u(x, y+h)$, and $u(x, y-h)$ imply that the expression

$$\frac{u(x-h, y) + u(x+h, y) + u(x, y-h) + u(x, y+h) - 4u(x, y)}{h^2}$$

approximates $\Delta u(x, y)$ with discretization error $O(h^2)$ for all $(x, y) \in \Omega$. Representing the numerical approximation to $u(x_m, y_n)$ by $u_{m,n}$ and letting $f_{m,n} = f(x_m, y_n)$, we obtain that

$$\forall (m, n) \in 1..M - 1 \times 1..N - 1 :$$
$$(4u_{m,n} - u_{m-1,n} - u_{m+1,n} - u_{m,n-1} - u_{m,n+1})/h^2 = f_{m,n}. \tag{8.3}$$

Equation 8.3 specifies $(M - 1)(N - 1)$ equations; this matches exactly the number of unknowns. Note that only values $u_{m,n}$ on *interior grid points* are unknown, say those values $u_{m,n}$ with $(m, n) \in 1..M - 1 \times 1..N - 1$. The rest are specified by the Dirichlet-boundary conditions.

The discretization of Equation 8.2 at grid point (m, n) involves u-values at grid points $(m - 1, n)$, $(m + 1, n)$, (m, n), $(m, n - 1)$, and $(m, n + 1)$. This set of grid points is called the *stencil* of the discretization. The second-order discretization of the Laplace operator leads to the classical *five-point stencil*, represented graphically in Figure 8.1.

Equation 8.3 can be rewritten as a system of linear equations $A\vec{u} = \vec{b}$. The dimension of the vector of unknowns $\vec{u}$ is $(M - 1)(N - 1)$. Each component u_ℓ of $\vec{u}$ is identified with a certain unknown value $u_{m,n}$. Referring to both u_ℓ and $u_{m,n}$ is not ambiguous: the quantity with two subscripts refers to function values on the $M \times N$ grid and the quantity with one subscript to a component of the vector $\vec{u}$. This slight abuse of notation is justified, because it reduces significantly the number of symbols.

The function values $u_{m,n}$ can be mapped to vector components u_ℓ in many ways. One possibility is the so-called *lexicographic ordering of un-knowns*:

$$u_{(n-1)(M-1)+m-1} = u_{m,n}. \tag{8.4}$$

This ordering maps u-values on grid line n to a contiguous block of $M-1$ components of the vector $\vec{u}$ such that

$$\begin{bmatrix} u_{(n-1)(M-1)} \\ u_{(n-1)(M-1)+1} \\ \vdots \\ u_{(n-1)(M-1)+M-2} \end{bmatrix} = \begin{bmatrix} u_{1,n} \\ u_{2,n} \\ \vdots \\ u_{M-1,n} \end{bmatrix}.$$

Using lexicographic ordering, Equation 8.3 for $(m,n) \in 2..M-2 \times 2..N-2$ can be rewritten in the form:

$$(4u_\ell - u_{\ell-1} - u_{\ell+1} - u_{\ell-M+1} - u_{\ell+M-1})/h^2 = f_{m,n}, \tag{8.5}$$

where $\ell = (n-1)(M-1)+m-1$. Equation 8.5 must be adapted if $m=1$, $m = M-1$, $n=1$, or $n = N-1$. In those cases, known boundary values must be transferred to the right-hand side. For example, consider the case $m=1$, $n \in 2..N-2$, and $\ell = (n-1)(M-1)+m-1 = (n-1)(M-1)$. Equation 8.5 then becomes

$$(4u_\ell - u_{\ell+1} - u_{\ell-M+1} - u_{\ell+M-1})/h^2 = f_{1,n} + u_{0,n}/h^2,$$

where $u_{0,n}$ is known because of the Dirichlet-boundary conditions. The other boundaries are treated analogously.

To complete the transformation of Equation 8.3 into a linear system $A\vec{u} = \vec{b}$, we must assign a certain order to the equations. Almost always, one chooses the order of the equations identical to the order of the unknowns. This corresponds to the intuition that equations, like unknowns, are tied to particular grid points. Using lexicographic ordering, the components of $\vec{b}$ are given by:

$$b_{(n-1)(M-1)+m-1} = f_{m,n} + \text{boundary terms}. \tag{8.6}$$

Let $\vec{f}$ be the vector of grid values $f_{m,n}$ mapped into vector components f_ℓ by lexicographic ordering, and let $\vec{g}$ be the vector of Dirichlet-boundary values at grid points on the boundary $\partial\Omega$. The boundary terms of Equation 8.6 are a correction on the vector $\vec{f}$ such that

$$\vec{b} = \vec{f} - B\vec{g} \tag{8.7}$$

is the right-hand side of the matrix formulation. The matrix B is a sparse matrix, whose structure is determined by the stencil of the discretization, the ordering of the boundary values as components of $\vec{g}$, and the ordering of the equations; see Exercise 36. The *discrete Poisson problem* on Ω is thus equivalent to the sparse system of equations

$$A\vec{u} = \vec{f} - B\vec{g}. \tag{8.8}$$

The coefficient matrix A is a representation of the discrete Laplace operator on Ω. Its classical structure is studied in Exercise 36.

We shall refer to Equation 8.3 as the *grid formulation* and to Equation 8.8 as the *matrix formulation* of the discrete Poisson problem. Both formulations contain identical information. The matrix formulation is used primarily in classical convergence theory of relaxation methods and other theoretical developments; see Section 3.2. The matrix formulation is notationally compact and hides the partial-differential-equation aspects of the original problem. What remains is the linear-algebra problem of solving a sparse system of equations. This system could be solved by any general sparse-system solver. For example, the conjugate-gradient method is applicable, because the matrix A is a symmetric positive-definite matrix; see Section 8.6 for implementation details.

For implementation purposes, the matrix formulation is often inconvenient and inefficient, and the grid formulation is almost always preferred. For example, the notational problems with the boundary terms are an artifact of the matrix formulation. In the remainder, we shall switch between grid and matrix formulations, and we shall continue to overload the notation by using the same symbol in different contexts, like u in $u_{m,n}$, u_ℓ, and $\vec{u}$.

8.2 Jacobi Relaxation

8.2.1 Specification

As mentioned in Section 3.2, Jacobi relaxation converges for linear systems with coefficient matrices that are strictly diagonally dominant. This result is not sufficient for the discrete Poisson problem. For a convergence argument that is applicable, see Section 9.1.

The grid formulation of the Jacobi relaxation is given by:

$$4u_{m,n}^{(k+1)} = h^2 f_{m,n} + u_{m-1,n}^{(k)} + u_{m+1,n}^{(k)} + u_{m,n-1}^{(k)} + u_{m,n+1}^{(k)}. \tag{8.9}$$

Program Jacobi-1 stores the unknown interior values $u_{m,n}$ and the known boundary values $g_{m,n}$ in array u. The right-hand-side values $f_{m,n}$ are stored in array f. Both arrays u and f are defined over the index set $0..M \times 0..N$. Program Jacobi-1 performs K Jacobi-relaxation steps.

```
program Jacobi-1
declare
    k, m, n : integer ;
    u, f : array[0..M × 0..N] of real
initially
    ⟨ ; m, n : (m, n) ∈ 0..M × 0..N :: f[m, n] = f̃_{m,n} ⟩ ;
    ⟨ ; m, n : (m, n) ∈ 1..M − 1 × 1..N − 1 :: u[m, n] = ũ_{m,n}^{(0)} ⟩ ;
    ⟨ ; n : n ∈ 0..N :: u[0, n], u[M, n] = g̃_{0,n}, g̃_{M,n} ⟩ ;
    ⟨ ; m : m ∈ 0..M :: u[m, 0], u[m, N] = g̃_{m,0}, g̃_{m,N} ⟩
assign
    ⟨ ; k : 0 ≤ k < K ::
        ⟨ ‖ m, n : (m, n) ∈ 1..M − 1 × 1..N − 1 ::
            u[m, n] := 0.25(h² f[m, n] + u[m − 1, n] + u[m, n − 1]
                                     + u[m + 1, n] + u[m, n + 1])
        ⟩
    ⟩
end
```

In the **initially** section, the right-hand-side values, the initial guess for u on the interior grid points, and the Dirichlet-boundary values are specified. In the **assign** section, the concurrent separator of the inner quantification guarantees that all u-values on the right-hand side of the assignment are those of the previous iteration step.

8.2.2 Implementation

We could use a distributed-matrix structure to store the values $u_{m,n}$ and $f_{m,n}$. However, the similarity between grids and matrices is quite superficial. Only two-dimensional grid problems could use matrix structures, and even in this case problems arise: whereas the first index referred to matrix rows, it refers to grid columns. In grid-oriented algorithms, we shall abandon the terms "row" and "column." Instead, we shall refer to grid lines in the x- and y-direction. (The latter have constant first index.) Because grids and matrices are different mathematical entities, their representations are different. Data distributions for grids must accommodate the increased emphasis on access to neighboring values.

We shall use modified linear distributions of both indices, m and n. Figure 8.2 illustrates the data distribution on a 3×4 process mesh, and Figure 8.3 details the ranges of local and global indices in process (p, q). The data distribution is defined by the sequence of global indices M_p with

$$1 = M_0 < M_1 < \ldots < M_p < M_{p+1} < \ldots < M_P = M$$

and the sequence N_q with

$$1 = N_0 < N_1 < \ldots < N_q < N_{q+1} < \ldots < N_Q = N.$$

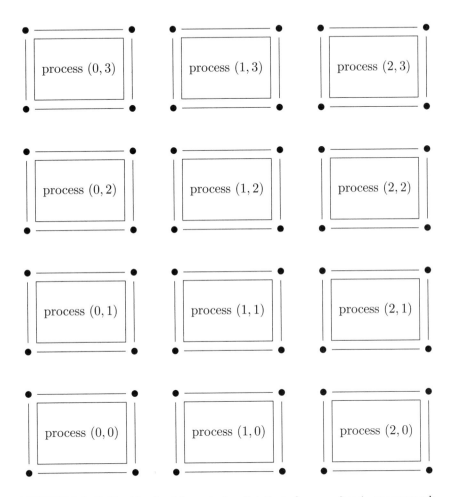

FIGURE 8.2. Grid with ghost boundaries distributed over a 3×4 process mesh.

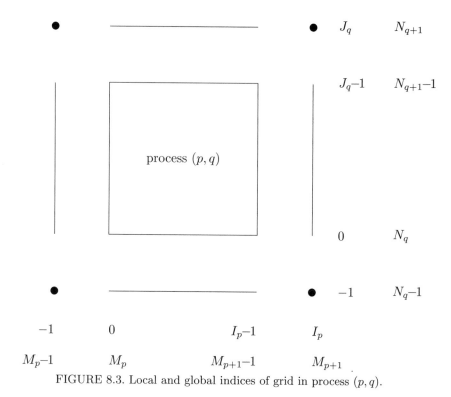

FIGURE 8.3. Local and global indices of grid in process (p, q).

A grid value $u_{m,n}$ with

$$(m, n) \in M_p..M_{p+1} - 1 \times N_q..N_{q+1} - 1$$

is mapped to process (p, q) and to local indices

$$(i, j) \in 0..I_p - 1 \times 0..J_q - 1.$$

The local grid in process (p, q) has dimensions $I_p \times J_q$, where

$$\begin{cases} I_p &= M_{p+1} - M_p \\ J_q &= N_{q+1} - N_q. \end{cases}$$

In Figures 8.2 and 8.3, this local grid is represented by rectangles. In process (p, q), the map between global indices (m, n) and local indices (i, j) is given by:

$$(m, n) = (M_p + i, N_q + j).$$

The *exterior boundary* is the set of grid points with global indices (m, n), such that $m = 0$, $m = M$, $n = 0$, or $n = N$. The data distribution as defined thus far does not include the exterior boundary. We shall also refer to the *interior boundaries* of a distributed grid. The interior boundary in process (p,q) is the set of grid points with local indices (i,j), such that $i = 0$, $i = I_p - 1$, $j = 0$, or $j = J_q - 1$.

The five-point stencil implies that the relaxation at a grid point requires access to values at four neighboring grid points. Such access is not immediately available to grid points on the interior boundaries. Therefore, we extend the data distribution with a *ghost boundary*, which contains duplicate values of neighbors residing in other processes. This expands the range of local indices (i,j) in process (p,q) from $0..I_p - 1 \times 0..J_q - 1$ to $-1..I_p \times -1..J_q$. Here, local index $i = -1$ corresponds to global index $m = M_p - 1$, $i = I_p$ to $m = M_{p+1}$, $j = -1$ to $n = N_q - 1$, and $j = J_q$ to $n = N_{q+1}$. In Figures 8.2 and 8.3, the ghost boundaries are represented by lines and dots, the dots representing the *ghost corners*. The exterior boundary of the grid is represented by parts of the ghost boundary in *boundary processes*, which are those processes (p,q) with $p = 0$, $p = P - 1$, $q = 0$, or $q = Q - 1$.

Duplicating program Jacobi-1, imposing the above data distribution, and subsequently converting to local indices leads to multicomputer program Jacobi-2.

$0..P - 1 \times 0..Q - 1 \parallel (p,q)$ **program** Jacobi-2
declare
      $k,\ i,\ j\ :\ $ integer ;
      $u\ :\ $ array$[-1..I_p \times -1..J_q]$ of real ;
      $f\ :\ $ array$[0..I_p - 1 \times 0..J_q - 1]$ of real ;
      $b_1,\ b_2\ :\ $ array$[-1..J_q - 1]$ of real
initially
    $\langle\ ;\ i,j\ :\ (i,j) \in 0..I_p - 1 \times 0..J_q - 1\ ::\ f[i,j] = \tilde{f}_{M_p+i,N_q+j}\ \rangle\ ;$
    $\langle\ ;\ i,j\ :\ (i,j) \in -1..I_p \times -1..J_q\ ::\ u[i,j] = \tilde{u}^{(0)}_{M_p+i,N_q+j}\ \rangle\ ;$
    if $p = 0$ **then** $\langle\ ;\ j\ :\ j \in -1..J_q\ ::\ u[-1,j] = \tilde{g}_{0,N_q+j}\ \rangle\ ;$
    if $p = P - 1$ **then** $\langle\ ;\ j\ :\ j \in -1..J_q\ ::\ u[I_p,j] = \tilde{g}_{M,N_q+j}\ \rangle\ ;$
    if $q = 0$ **then** $\langle\ ;\ i\ :\ i \in -1..I_p\ ::\ u[i,-1] = \tilde{g}_{M_p+i,0}\ \rangle\ ;$
    if $q = Q - 1$ **then** $\langle\ ;\ i\ :\ i \in -1..I_p\ ::\ u[i,J_q] = \tilde{g}_{M_p+i,N}\ \rangle$
assign
    $\langle\ ;\ k\ :\ 0 \le k < K\ ::$
        { Local Relaxation }
        $\langle\ ;\ j\ :\ j \in 0..J_q - 1\ ::\ b_2[j] := u[-1,j]\ \rangle\ ;$
        $\langle\ ;\ i\ :\ i \in 0..I_p - 1\ ::$
            $\langle\ ;\ j\ :\ j \in -1..J_q - 1\ ::\ b_1[j] := u[i,j]\ \rangle\ ;$
            $\langle\ ;\ j\ :\ j \in 0..J_q - 1\ ::$
                $u[i,j] := 0.25(h^2 f[i,j] + b_1[j-1] + b_2[j]$
                              $+ u[i,j+1] + u[i+1,j])$
            $\rangle\ ;$
            $\langle\ ;\ j\ :\ j \in 0..J_q - 1\ ::\ b_2[j] := b_1[j]\ \rangle$
        $\rangle\ ;$

```
{ Ghost-Boundary Exchange }
if q < Q − 1 then
    send {u[i, J_q − 1] : i ∈ 0..I_p − 1} to (p, q + 1) ;
if q > 0 then
    receive {u[i, −1] : i ∈ 0..I_p − 1} from (p, q − 1) ;
if q > 0 then
    send {u[i, 0] : i ∈ 0..I_p − 1} to (p, q − 1) ;
if q < Q − 1 then
    receive {u[i, J_q] : i ∈ 0..I_p − 1} from (p, q + 1) ;
if p < P − 1 then
    send {u[I_p − 1, j] : j ∈ −1..J_q} to (p + 1, q) ;
if p > 0 then
    receive {u[−1, j] : j ∈ −1..J_q} from (p − 1, q) ;
if p > 0 then
    send {u[0, j] : j ∈ −1..J_q} to (p − 1, q) ;
if p < P − 1 then
    receive {u[I_p, j] : j ∈ −1..J_q} from (p + 1, q)
⟩
end
```

Array f is declared without ghost boundaries. The **initially** section specifies the right-hand-side values, the initial guess for u on the interior grid points, and the Dirichlet-boundary values.

The **assign** section specifies K Jacobi-relaxation steps, each step consisting of two phases: the local relaxation and the ghost-boundary exchange. The local relaxation is equivalent to the concurrent quantification:

$$\langle \parallel i, j : (i, j) \in 0..I_p − 1 \times 0..J_q − 1 ::$$
$$u[i, j] := 0.25(h^2 f[i, j] + u[i−1, j] + u[i+1, j] + u[i, j−1] + u[i, j+1])$$
$$\rangle$$

Because individual multicomputer processes must be sequential, this concurrent quantification was converted into an equivalent sequential quantification by means of two temporary buffers, arrays b_1 and b_2.

The ghost-boundary exchange updates the values of the ghost boundaries. To study this part of the program, consider the code without the **if–then** statements, which prevent boundary processes from changing the exterior boundary.

The ghost-boundary exchange, represented graphically in Figure 8.4, consists of four stages. Each stage updates one part of the ghost boundary. In stage 1, for example, process (p, q) sends its north interior boundary to its north neighbor and receives the values for its south ghost boundary from its south neighbor. Only the message exchanges involving process (p, q) are pictured. Of course, all other processes simultaneously execute a similar exchange with their neighbors. Hence, stage 1 updates all south ghost boundaries. This is repeated for the three other ghost boundaries.

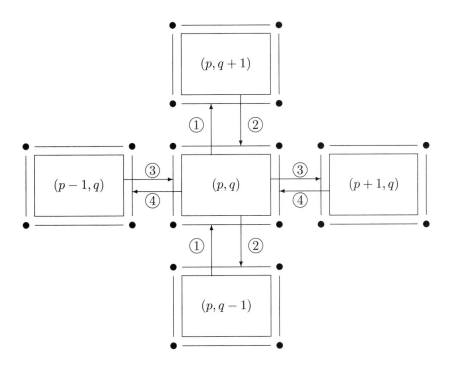

FIGURE 8.4. Ghost-boundary exchange for interior process (p, q) of program Jacobi-2.

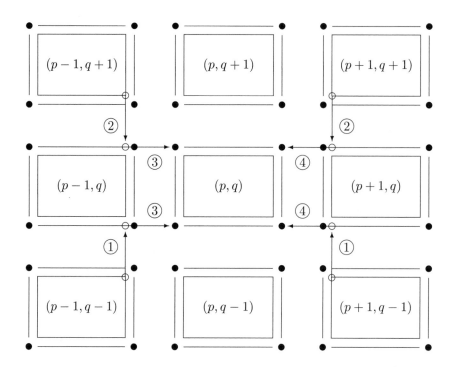

FIGURE 8.5. Updating the ghost corners.

The east and west exchanges differ somewhat from the north and south exchanges, because the former include the update of the ghost corners. Figure 8.5 shows how the four stages of the ghost-boundary exchange cooperate to initialize the ghost corners. Because the second-order discretization of the Laplace operator has a five-point stencil, the ghost corners are never used in program Jacobi-2. For other stencils, however, it is important that a correct update of the ghost corners be achieved at virtually no extra cost.

The ghost-boundary exchange of program Jacobi-2 is not the only possible exchange strategy. Because the ghost corners are not used, the following code may be substituted:

> **if** $q < Q - 1$ **then**
>   **send** $\{u[i, J_q - 1] : i \in 0..I_p - 1\}$ **to** $(p, q + 1)$;
> **if** $q > 0$ **then**
>   **send** $\{u[i, 0] : i \in 0..I_p - 1\}$ **to** $(p, q - 1)$;
> **if** $p < P - 1$ **then**
>   **send** $\{u[I_p - 1, j] : j \in 0..J_q - 1\}$ **to** $(p + 1, q)$;
> **if** $p > 0$ **then**
>   **send** $\{u[0, j] : j \in 0..J_q - 1\}$ **to** $(p - 1, q)$;
>
> **if** $q > 0$ **then**
>   **receive** $\{u[i, -1] : i \in 0..I_p - 1\}$ **from** $(p, q - 1)$ □
> **if** $q < Q - 1$ **then**
>   **receive** $\{u[i, J_q] : i \in 0..I_p - 1\}$ **from** $(p, q + 1)$ □
> **if** $p > 0$ **then**
>   **receive** $\{u[-1, j] : j \in 0..J_q - 1\}$ **from** $(p - 1, q)$ □
> **if** $p < P - 1$ **then**
>   **receive** $\{u[I_p, j] : j \in 0..J_q - 1\}$ **from** $(p + 1, q)$

In this code fragment, the asynchronous separator □ is introduced. Statements separated by □ may be evaluated in any order. In this case, it is natural to receive messages in the order of their arrival.

8.2.3 Performance Analysis

The performance of a concurrent relaxation method depends on three factors: the number of iteration steps necessary to achieve the required accuracy, the number of floating-point operations per iteration step, and the efficiency of the multicomputer program.

As discussed in Section 3.2, the number of iteration steps is determined by the convergence factor. The convergence factor for Jacobi relaxation applied to the discrete Poisson problem defined by Equation 8.3 is

$$\rho_J = 1 - \frac{h^2}{2} + O(h^4),$$

where $h = \pi/M$ is the grid spacing; see Section 9.1.3 for a derivation. The number of Jacobi-relaxation steps required to reduce an $O(1)$ error on the initial guess to $O(\epsilon)$ is estimated by

$$K_J \geq \frac{\log \epsilon}{\log \rho_J} \approx \frac{-2 \log \epsilon}{h^2}.$$

Typically, the solution of the discrete problem is computed with an accuracy of the same order of magnitude as the discretization error, which is $O(h^2)$ in this case. Hence, we expect that

$$K_J = O\left(\frac{-\log h^2}{h^2}\right) = O(M^2 \log M^2).$$

Each iteration step evaluates $6(M-1)(N-1)$ floating-point operations. In our simple performance model, there is a direct correspondence between the operation count and the sequential-execution time:

$$T_1 = 6(M-1)(N-1)\tau_A \approx 6MN\tau_A.$$

The multicomputer-execution time also depends on the process-mesh dimensions, the data distribution, and the ghost-boundary exchange. With

$$\bar{I} = \max_{0 \leq p < P} I_p$$
$$\bar{J} = \max_{0 \leq q < Q} J_q,$$

the estimated multicomputer-execution time per iteration step, still assuming one process per node, is

$$T_{P \times Q} \approx 6\bar{I}\bar{J}\tau_A + 2\tau_C(\bar{I}) + 2\tau_C(\bar{J}).$$

The first term is due to floating-point arithmetic, the second and third terms are due to communication. For coarse-grained load-balanced computations, we may assume that $\bar{I} \approx M/P$ and $\bar{J} \approx N/Q$, and

$$T_{P \times Q} \approx \frac{T_1}{PQ} + 4\tau_S + 2\beta\left(\frac{M}{P} + \frac{N}{Q}\right).$$

The performance of Jacobi relaxation on multicomputers is a manifestation of the so-called *area-perimeter law*. The arithmetic time is proportional to the area of the local grids $(\bar{I}\bar{J})$, and the communication time is proportional to their perimeter $(2(\bar{I} + \bar{J}))$. Note that this "law" neglects the latency τ_S. If $I = J$, communication time is a linear function of the problem dimension while arithmetic time is a quadratic function. As a result, communication time should be negligible for sufficiently coarse-grained computations. Whether or not this is the case in practice remains to be seen, and a discussion of area-perimeter concerns is postponed until Section 8.8.

How do the two ghost-boundary-exchange strategies discussed above stack up against each other? The answer to this question varies from one computer to the next. The asynchronous strategy has the advantage that the times spent waiting for messages overlap. However, the asynchronous ghost-boundary exchange injects all ghost-boundary values simultaneously into the communication network, which increases the probability of network contention (see Section 12.3). As long as the network is not saturated by the number and size of the messages involved, the asynchronous strategy is likely to be superior. Our simple performance model does not include this effect and cannot be used to decide the issue.

8.3   Gauss-Seidel Relaxation

8.3.1   Specification

Gauss-Seidel relaxation converges for systems with symmetric positive-definite coefficient matrices and, therefore, can be used to solve the discrete Poisson problem. The grid formulation of a relaxation step is given by:

$$4u_{m,n}^{(k+1)} = h^2 f_{m,n} + u_{m-1,n}^{(k+1)} + u_{m+1,n}^{(k)} + u_{m,n-1}^{(k+1)} + u_{m,n+1}^{(k)}. \qquad (8.10)$$

As in program Jacobi-1, the unknown interior values $u_{m,n}$ and known boundary values $g_{m,n}$ are stored in array u, and the right-hand-side values $f_{m,n}$ are stored in array f. Both arrays are defined over the index set $0..M \times 0..N$. Program Gauss-Seidel-1 performs K Gauss-Seidel-relaxation steps based on lexicographic ordering of unknowns and equations.

```
program Gauss-Seidel-1
declare
    k, m, n  :  integer ;
    u, f  :  array[0..M × 0..N] of real
assign
    ⟨ ; k  :  0 ≤ k < K  ::
        ⟨ ; n  :  n ∈ 1..N − 1  ::
            ⟨ ; m  :  m ∈ 1..M − 1  ::
                u[m, n] := 0.25(h² f[m, n] + u[m − 1, n] + u[m, n − 1]
                                          + u[m + 1, n] + u[m, n + 1])
            ⟩
        ⟩
    ⟩
end
```

The **initially** section is omitted, because it is identical to that of program Jacobi-1. In the **assign** section, the sequential quantifications over m and n ensure that the most recently computed u-values are always used in the right-hand side of the assignment.

8.3.2 Implementation

Although we use the same data distribution, the ghost-boundary exchanges of Section 8.2.2 are not applicable, because the Gauss-Seidel relaxation requires the current values of $u[m-1,n]$ and $u[m,n-1]$. To proceed with the local relaxation in process (p,q), its west and south ghost boundaries must be updated with the values of the east and north interior boundaries of processes $(p-1,q)$ and $(p,q-1)$, respectively. The next iteration step may proceed only after the east and north ghost boundaries are updated as well. These insights lead to multicomputer program Gauss-Seidel-2.

$0..P-1 \times 0..Q-1 \parallel (p,q)$ **program** Gauss-Seidel-2
declare
    $k,\ i,\ j\ :\ $integer ;
    $u\ :\ $array$[-1..I_p \times -1..J_q]$ of real ;
    $f\ :\ $array$[0..I_p-1 \times 0..J_q-1]$ of real
assign
    $\langle\ ;\ k\ :\ 0 \le k < K\ ::$
        if $p > 0$ **then**
            receive $\{u[-1,j] : j \in 0..J_q-1\}$ **from** $(p-1,q)$;
        if $q > 0$ **then**
            receive $\{u[i,-1] : i \in 0..I_p-1\}$ **from** $(p,q-1)$;
        $\langle\ ;\ j\ :\ j \in 0..J_q-1\ ::$
            $\langle\ ;\ i\ :\ i \in 0..I_p-1\ ::$
                $u[i,j] := 0.25(h^2 f[i,j] + u[i-1,j] + u[i,j-1]$
                                  $+ u[i+1,j] + u[i,j+1])$
            $\rangle$
        $\rangle$;
        if $p < P-1$ **then**
            send $\{u[I_p-1,j] : j \in 0..J_q-1\}$ **to** $(p+1,q)$;
        if $q < Q-1$ **then**
            send $\{u[i,J_q-1] : i \in 0..I_p-1\}$ **to** $(p,q+1)$;
        if $p > 0$ **then**
            send $\{u[0,j] : j \in 0..J_q-1\}$ **to** $(p-1,q)$;
        if $q > 0$ **then**
            send $\{u[i,0] : i \in 0..I_p-1\}$ **to** $(p,q-1)$;
        if $p < P-1$ **then**
            receive $\{u[I_p,j] : j \in 0..J_q-1\}$ **from** $(p+1,q)$;
        if $q < Q-1$ **then**
            receive $\{u[i,J_q] : i \in 0..I_p-1\}$ **from** $(p,q+1)$
    $\rangle$
end

The omitted **initially** section is identical to that of program Jacobi-2.

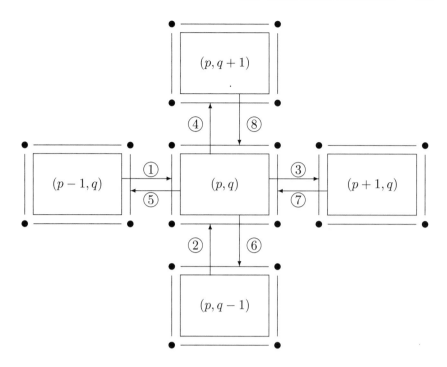

FIGURE 8.6. Ghost-boundary exchange for interior process (p, q) of program Gauss-Seidel-2.

Figure 8.6 pictures the ghost-boundary exchange of program Gauss-Seidel-2. The eight stages of the exchange correspond to the eight communication operations that take place in every iteration step. For example, stage number 4 corresponds to the fourth communication instruction of program Gauss-Seidel-2. As in program Jacobi-2, the **if–then** statements preceding the communication instructions prevent the boundary processes from changing the exterior boundary.

8.3.3 Performance Analysis

The convergence factor of Gauss-Seidel relaxation applied to the discrete Poisson problem is

$$\rho_{GS} = 1 - h^2 + O(h^4).$$

The number of Gauss-Seidel-relaxation steps to reduce the error on the solution estimate from $O(1)$ to $O(\epsilon)$ is, therefore, about one-half the number of Jacobi-relaxation steps. Per iteration step, both methods perform the same number of floating-point operations.

The increased numerical performance of Gauss-Seidel relaxation is offset, however, by unacceptable multicomputer performance. One multicomputer Gauss-Seidel-relaxation step starts with just process $(0,0)$ active and all other processes waiting for the updates of their west and south ghost boundaries. After process $(0,0)$ has sent out its interior-boundary values, processes $(1,0)$ and $(0,1)$ become active. Continuing in this fashion, processes become active in a wave propagating from the lower-left to the upper-right corner of the process mesh. Within one Gauss-Seidel-relaxation step, the maximum number of active processes at any one time is the minimum of P and Q.

Full concurrency is only achieved if the number of iteration steps is large. As the wave of processes working on iteration step k propagates, processes left in its wake may proceed with iteration step $k+1$ as soon as their ghost-boundary values are updated. If K is sufficiently large, say $K \gg \max(P, Q)$, many iteration steps are pipelined in this fashion, because many waves traverse the process mesh simultaneously. Once this pipeline is filled, all processes are fully active, except for the communication delay due to the ghost-boundary exchange. There are two other sources of multicomputer inefficiency. At the beginning of the iteration, all processes ahead of the first wave are idle. At the end of the iteration, the last wave leaves idle processes in its wake. A complete performance analysis of program Gauss-Seidel-2 is left as an exercise.

8.3.4   Gauss-Seidel Variants

Various modifications to basic Gauss-Seidel relaxation are possible. Successive over-relaxation, symmetric successive over-relaxation, and red-black Gauss-Seidel relaxation are the best known.

The matrix formulation of successive over-relaxation was discussed in Section 3.2. The grid formulation, when applied to the discrete Poisson problem, is given by:

$$4u_{m,n}^{(k+1)} = \omega(h^2 f_{m,n} + u_{m-1,n}^{(k+1)} + u_{m+1,n}^{(k)} + u_{m,n-1}^{(k+1)} + u_{m,n+1}^{(k)}) + 4(1-\omega)u_{m,n}^{(k)}. \quad (8.11)$$

From an algorithmic point of view, successive over-relaxation hardly differs from Gauss-Seidel relaxation.

The multicomputer program for symmetric successive over-relaxation based on a lexicographic ordering of unknowns and equations is even more problematic than program Gauss-Seidel-2. Because a forward sweep is followed by a backward sweep, at most one relaxation wave can be active in the process mesh. This severely limits the multicomputer efficiency of the method.

Red-black ordering is the most frequently used alternative ordering of unknowns and equations. Color the grid points (m, n) according to the parity of $m + n$: if $m + n$ is even, color the grid point red, and if $m + n$ is odd, color it black. In this way, the grid is colored in a checker-board pattern. When mapping the grid values $u_{m,n}$ to vector components u_ℓ, values corresponding to red grid points are listed first and are followed by values corresponding to black grid points. (The mapping of $f_{m,n}$ to f_ℓ is handled similarly.) Now apply standard Gauss-Seidel relaxation to the matrix formulation obtained by red-black ordering. For operators with a five-point stencil, grid points of the same color may be relaxed concurrently, because the u-value at a grid point depends only on u-values of grid points of a different color. The specification program is given by program Red-Black-1.

> **program** Red-Black-1
> **declare**
> > $k,\ m,\ n\ :\ $ integer ;
> > $u,\ f\ :\ $ array$[0..M \times 0..N]$ of real
>
> **assign**
> > $\langle\ ;\ k\ :\ 0 \le k < K\ ::$
> > > $\langle\ \|\ m,n\ :\ (m,n) \in 1..M-1 \times 1..N-1$ **and even**$(m+n)\ ::$
> > > > $u[m,n] := 0.25(h^2 f[m,n] + u[m-1,n] + u[m,n-1]$
> > > > $\qquad\qquad\qquad\qquad + u[m+1,n] + u[m,n+1])$
> > > $\rangle\ ;$
> > > $\langle\ \|\ m,n\ :\ (m,n) \in 1..M-1 \times 1..N-1$ **and odd**$(m+n)\ ::$
> > > > $u[m,n] := 0.25(h^2 f[m,n] + u[m-1,n] + u[m,n-1]$
> > > > $\qquad\qquad\qquad\qquad + u[m+1,n] + u[m,n+1])$
> > > $\rangle$
> > $\rangle$
>
> **end**

The omitted **initially** section is identical to that of program Jacobi-1.

The multicomputer implementation of a red-black Gauss-Seidel relaxation shares properties of programs Jacobi-2 and Gauss-Seidel-2. The local relaxation of the red (black) points proceeds as in program Gauss-Seidel-2; there is no need for buffer arrays as in program Jacobi-2. The ghost-boundary exchange after the relaxation of the red (black) grid points proceeds as in program Jacobi-2, but only the red (black) u-values are exchanged. One red-black-relaxation step exchanges twice as many messages as one Jacobi-relaxation step. Although messages are only one-half the size on average, the total communication time increases because of the latency τ_S incurred by each message exchange. Unlike lexicographic Gauss-Seidel relaxation, the multicomputer efficiency of red-black Gauss-Seidel relaxation does not depend on the number of iteration steps, because the iteration steps need not be pipelined.

Red-black coloring cannot always break the sequential nature of the Gauss-Seidel relaxation. In general, multicoloring schemes must be used. For example, just changing the boundary conditions from Dirichlet to Neumann type may invalidate the red-black coloring scheme. In this case, the appropriate coloring scheme depends on the discretization of the Neumann-boundary condition. For any coloring scheme, the method remains a Gauss-

Seidel relaxation, but each ordering results in a different matrix A, a different decomposition $A = L + D + U$, a different splitting $A = G - H$, and a different convergence factor $\rho(G^{-1}H)$.

8.4   Line Relaxation

8.4.1   Specification

Relaxation methods studied thus far are *point-relaxation methods*: in one elementary step, one unknown is updated, while all other unknowns are given a particular value. In one elementary step of a *group-relaxation method*, a subset of the unknowns is updated simultaneously by solving a reduced system of equations. *Line-relaxation methods* are the most common application of this idea. In one elementary line-relaxation step, one grid line of unknowns is updated simultaneously. For example, an elementary line-relaxation step in the x-direction updates all unknowns $u_{m,n}$ with $m \in 1..M - 1$ and n fixed. Similarly, an elementary line-relaxation step in the y-direction updates all unknowns $u_{m,n}$ with $n \in 1..N - 1$ and m fixed. In the Jacobi version of line relaxation, all other unknowns are given their value from the previous iteration step. In the Gauss-Seidel version of line relaxation, the most recently computed values are used.

The grid formulation of the *line-Jacobi relaxation method* for the discrete Poisson problem is given by

$$4u_{m,n}^{(k+1)} - u_{m-1,n}^{(k+1)} - u_{m+1,n}^{(k+1)} = h^2 f_{m,n} + u_{m,n-1}^{(k)} + u_{m,n+1}^{(k)}. \tag{8.12}$$

To update one line of unknowns simultaneously, a system of equations with tridiagonal coefficient matrix $T = \mathrm{trid}(-1, 4, -1)$ must be solved. In particular, line n in the x-direction is relaxed by solving

$$T \begin{bmatrix} u_{1,n}^{(k+1)} \\ u_{2,n}^{(k+1)} \\ u_{3,n}^{(k+1)} \\ \vdots \\ u_{M-2,n}^{(k+1)} \\ u_{M-1,n}^{(k+1)} \end{bmatrix} = \begin{bmatrix} h^2 f_{1,n} + u_{0,n} + u_{1,n-1}^{(k)} + u_{1,n+1}^{(k)} \\ h^2 f_{2,n} + u_{2,n-1}^{(k)} + u_{2,n+1}^{(k)} \\ h^2 f_{3,n} + u_{3,n-1}^{(k)} + u_{3,n+1}^{(k)} \\ \vdots \\ h^2 f_{M-2,n} + u_{M-2,n-1}^{(k)} + u_{M-2,n+1}^{(k)} \\ h^2 f_{M-1,n} + u_{M,n} + u_{M-1,n-1}^{(k)} + u_{M-1,n+1}^{(k)} \end{bmatrix}.$$

Program Line-Jacobi-1 is the corresponding specification program. The tridiagonal matrix T and its LU-decomposition are represented by three arrays a, b, and c. The entries of a, b, and c are initialized to represent the main, the lower, and the upper diagonal of T, respectively. The LU-decomposition of T is computed in place. This procedure slightly differs from program Tridiagonal-LU-1 of Section 6.1.2, because the indices start at 1 instead of 0.

Because the coefficient matrix does not depend on n, it suffices to perform the LU-decomposition just once. However, a new right-hand side must be constructed, and a tridiagonal-back-solve procedure must be performed for every n. The right-hand side is computed in the array of unknowns, and the back-solve procedure is performed in place.

program Line-Jacobi-1
declare
  $k,\ m,\ n\ :$ integer ;
  $u,\ f\ :\ $array$[0..M \times 0..N]$ of real ;
  $a,\ b,\ c\ :\ $array$[0..M - 1]$ of real
initially
  $\langle\ ;\ m, n\ :\ (m, n) \in 0..M \times 0..N\ ::\ f[m, n] = \tilde{f}_{m,n}\ \rangle$;
  $\langle\ ;\ n\ :\ n \in 0..N\ ::\ u[0, n], u[M, n] = \tilde{g}_{0,n}, \tilde{g}_{M,n}\ \rangle$;
  $\langle\ ;\ m\ :\ m \in 0..M\ ::\ u[m, 0], u[m, N] = \tilde{g}_{m,0}, \tilde{g}_{m,N}\ \rangle$;
  $\langle\ ;\ m\ :\ m \in 1..M - 1\ ::\ a[m] = 4\ \rangle$;
  $\langle\ ;\ m\ :\ m \in 2..M - 1\ ::\ b[m], c[m] = -1, -1\ \rangle$
assign
  { Tridiagonal LU-Decomposition }
  $\langle\ ;\ m\ :\ m \in 2..M - 1\ ::$
   $b[m] := b[m]/a[m - 1]$;
   $a[m] := a[m] - c[m]b[m]$
  $\rangle$;
  $\langle\ ;\ k\ :\ 0 \le k < K\ ::$
   $\langle\ \|\ n\ :\ n \in 1..N - 1\ ::$
    { Construct Right-Hand Side }
    $u[1, n] := h^2 f[1, n] + u[0, n] + u[1, n - 1] + u[1, n + 1]$;
    $\langle\ ;\ m\ :\ 2 \le m < M - 1\ ::$
     $u[m, n] := h^2 f[m, n] + u[m, n - 1] + u[m, n + 1]$
    $\rangle$;
    $u[M-1, n] := h^2 f[M-1, n] + u[M, n]$
       $+ u[M-1, n-1] + u[M-1, n+1]$;
    { Solve System for $u[m, n], m \in 1..M - 1$ }
    $\langle\ ;\ m\ :\ 2 \le m < M\ ::$
     $u[m, n] := u[m, n] - b[m]u[m - 1, n]$
    $\rangle$;
    $u[M - 1, n] := u[M - 1, n]/a[M - 1]$;
    $\langle\ ;\ m\ :\ M - 2 \ge m > 0\ ::$
     $u[m, n] := (u[m, n] - c[m + 1]u[m + 1, n])/a[m]$
    $\rangle$
   $\rangle$
  $\rangle$
end

The development of other line-relaxation variants is left as an exercise. In particular, derive the *line-Gauss-Seidel* and the *line-red-black relaxation*.

In a line-Jacobi relaxation, all lines must be relaxed concurrently. In a line-Gauss-Seidel relaxation, the quantification over n must be a sequential quantification. The line-red-black relaxation consists of two concurrent quantifications: one over the even and one over the odd values of n.

Line-relaxation methods treat grid lines in the x-direction differently from those in the y-direction. To rectify this, a line-relaxation step in the x-direction is usually followed by one in the y-direction. Such methods are called *alternating-direction line-relaxation methods*.

8.4.2 Implementation

As in Sections 8.2.2 and 8.3.2, we impose a two-dimensional data distribution over a $P \times Q$ process mesh.

A line-Gauss-Seidel-relaxation iteration in the x-direction propagates a wave of distributed tridiagonal solves in the y-direction. Such a program combines the concurrency of the tridiagonal solver in the x-direction and the concurrency of the pipeline in the y-direction. This method can be used with any distributed tridiagonal solver. Two distributed solvers were discussed in Chapter 6: a direct solver based on full recursive doubling and an iterative solver. In addition, the sequential solver can be adapted for use on distributed data (see Exercise 39).

The viability of using a distributed direct tridiagonal solver with multicomputer efficiency η_P follows from a back-of-the-envelope calculation. The maximal speed-up from applying this solver in the x-direction is $\eta_P P$. The speed-up from pipelining in the y-direction is at most Q, an estimate in which filling and emptying the pipeline and communication are ignored. The speed-up of the combined procedure is, therefore, at most $\eta_P P Q$. However, pipelining is impossible in an alternating-direction scheme. In this case, line relaxation in the x-direction has a speed-up of at most $\eta_P P$ and an efficiency of at most η_P / Q. Similarly, line relaxation in the y-direction has a speed-up of at most $\eta_Q Q$ and an efficiency of at most η_Q / P.

When using alternating-direction methods, one is usually restricted to line-Jacobi and line-red-black relaxation. In these cases, distributed tridiagonal solves in one direction are followed by a boundary-value exchange in the other. A back-of-the-envelope calculation, ignoring communication as above, leads to a maximal speed-up of $\frac{\eta_P + \eta_Q}{2} PQ$.

Distributed tridiagonal solvers can be avoided by using one-dimensional data distributions. If $P = 1$, line relaxations in the x-direction can be performed using a nondistributed sequential tridiagonal solver. Because the data for the line relaxations in the y-direction are still distributed over Q processes, the grid needs to be transposed before performing a line relaxation in the y-direction.

The grid-transpose strategy was already discussed in Section 7.5.6 for the multivariate fast Fourier transform. While concurrent and sequential fast Fourier transforms perform the same number of floating-point operations, concurrent tridiagonal solvers require considerably more floating-point operations than their sequential alternative. By using nondistributed sequential tridiagonal solvers and a grid transpose, the floating-point overhead is avoided, and only the communication overhead remains. The grid-transpose strategy has, therefore, an edge over the use of distributed tridiagonal solvers. However, memory must be set aside to store the transposed grid, unless the transpose can be performed in place. Another disadvantage of the grid-transpose strategy is that one-dimensional distributions lead to granularity problems for smaller numbers of processes. Implementation of the grid-transpose operation usually requires computer-dependent programming to avoid network contention; see Section 12.3.3.

The iterative tridiagonal solver of Section 6.4 is an easily implemented alternative to direct solvers. When using the iterative solver, it is important to limit the number of tridiagonal-iteration steps. This number determines whether or not the iterative is competitive with the direct tridiagonal solver. If the average number of tridiagonal-iteration steps per relaxation-iteration step is L, the effective speed-up of the concurrent iterative with respect to the sequential direct tridiagonal solver is approximately P/L. When this is compared with the speed-up S_P of the concurrent direct solver, the number of tridiagonal-iteration steps is limited by:

$$L < P/S_P = 1/\eta_P,$$

where η_P is the efficiency of the concurrent direct tridiagonal solver.

The number of tridiagonal-iteration steps depends on three factors: the accuracy of the initial guess, the tolerance on the solution of the tridiagonal system, and the spectral radius of the tridiagonal-iteration matrix. The latter is fixed once the problem and data distribution are defined. The values of the previous relaxation-iteration step are excellent initial guesses for the tridiagonal iteration. One should never solve the tridiagonal system more accurately than warranted by the discretization error. Moreover, it makes sense to relax the tolerance of the tridiagonal solver at the beginning of the relaxation iteration. This may reduce the average number of tridiagonal-iteration steps per relaxation-iteration step.

Program Line-Jacobi-2 is a multicomputer implementation of line-Jacobi relaxation based on the iterative tridiagonal solver. This program performs a fixed number of tridiagonal-iteration steps per relaxation step. In every tridiagonal-iteration step, the right-hand side is computed using new ghost-boundary values of the solution vector. Subsequently, the tridiagonal-back-solve procedure is performed, and the ghost boundaries are exchanged in the east-west direction. The array variable r holds partially computed right-hand-side values for the tridiagonal systems. This array is the line-relaxation equivalent of the buffer arrays b_1 and b_2 of program Jacobi-2. The buffer array r not only helps in sequentializing a concurrent quantification. It also eliminates some floating-point operations and allows us to bundle the messages of all J_q tridiagonal systems, hence avoiding multiple latencies.

$0..P - 1 \times 0..Q - 1 \parallel (p,q)$ **program** Line-Jacobi-2
declare
    k, ℓ, i, j : integer ;
    u : array$[-1..I_p \times -1..J_q]$ of real ;
    f, r : array$[0..I_p - 1 \times 0..J_q - 1]$ of real ;
    a, b, c : array$[0..I_p - 1]$ of real
initially
    $\langle \ ; i,j \ : \ (i,j) \in 0..I_p - 1 \times 0..J_q - 1 \ :: \ f[i,j] = \tilde{f}_{M_p+i,N_q+j} \ \rangle$;
    $\langle \ ; i,j \ : \ (i,j) \in -1..I_p \times -1..J_q \ :: \ u[i,j] = \tilde{u}^{(0)}_{M_p+i,N_q+j} \ \rangle$;
    if $p = 0$ **then** $\langle \ ; j \ : \ j \in -1..J_q \ :: \ u[-1,j] = \tilde{g}_{0,N_q+j} \ \rangle$;
    if $p = P - 1$ **then** $\langle \ ; j \ : \ j \in -1..J_q \ :: \ u[I_p,j] = \tilde{g}_{M,N_q+j} \ \rangle$;
    if $q = 0$ **then** $\langle \ ; i \ : \ i \in -1..I_p \ :: \ u[i,-1] = \tilde{g}_{M_p+i,0} \ \rangle$;
    if $q = Q - 1$ **then** $\langle \ ; i \ : \ i \in -1..I_p \ :: \ u[i,J_q] = \tilde{g}_{M_p+i,N} \ \rangle$;
    $\langle \ ; i \ : \ i \in 0..I_p - 1 \ :: \ a[i],b[i],c[i] = 4,-1,-1 \ \rangle$
assign
    { Tridiagonal LU-Decomposition }
    $\langle \ ; i \ : \ i \in 1..I_p - 1 \ ::$
        $b[i] := b[i]/a[i-1]$;
        $a[i] := a[i] - c[i]b[i]$
    $\rangle$;
    { Line-Relaxation Iteration in x-Direction }
    $\langle \ ; k \ : \ 0 \leq k < K \ ::$
        { Prepare Right-Hand Sides }
        $\langle \ ; i,j \ : \ (i,j) \in 0..I_p - 1 \times 0..J_q - 1 \ ::$
            $r[i,j] := h^2 f[i,j] + u[i,j-1] + u[i,j+1]$
        $\rangle$;
        { Iterate all J_q Tridiagonal Systems }
        $\langle \ ; \ell \ : \ 0 \leq \ell < L \ ::$
            $\langle \ ; j \ : \ j \in 0..J_q - 1 \ ::$
                { Construct Right-Hand Side }
                $u[0,j] := r[0,j] + u[-1,j]$;
                $\langle \ ; i \ : \ 0 < i < I_p - 1 \ :: \ u[i,j] := r[i,j] \ \rangle$;
                $u[I_p - 1,j] := r[I_p - 1,j] + u[I_p,j]$;
                { Solve Local Tridiagonal System }
                $\langle \ ; i \ : \ 1 \leq i < I_p \ ::$
                    $u[i,j] := u[i,j] - b[i]u[i-1,j]$
                $\rangle$;
                $u[I_p - 1,j] := u[I_p - 1,j]/a[I_p - 1]$;
                $\langle \ ; i \ : \ I_p - 2 \geq i \geq 0 \ ::$
                    $u[i,j] := (u[i,j] - c[i+1]u[i+1,j])/a[i]$
                $\rangle$
            $\rangle$;

{ East-West Ghost-Boundary Exchange }
if $p < P - 1$ then
    send $\{u[I_p - 1, j] : j \in 0..J_q - 1\}$ to $(p + 1, q)$;
if $p > 0$ then
    receive $\{u[-1, j] : j \in 0..J_q - 1\}$ from $(p - 1, q)$;
if $p > 0$ then
    send $\{u[0, j] : j \in 0..J_q - 1\}$ to $(p - 1, q)$;
if $p < P - 1$ then
    receive $\{u[I_p, j] : j \in 0..J_q - 1\}$ from $(p + 1, q)$
$\rangle$;

{ North-South Ghost-Boundary Exchange }
if $q < Q - 1$ then
    send $\{u[i, J_q - 1] : i \in 0..I_p - 1\}$ to $(p, q + 1)$;
if $q > 0$ then
    receive $\{u[i, -1] : i \in 0..I_p - 1\}$ from $(p, q - 1)$;
if $q > 0$ then
    send $\{u[i, 0] : i \in 0..I_p - 1\}$ to $(p, q - 1)$;
if $q < Q - 1$ then
    receive $\{u[i, J_q] : i \in 0..I_p - 1\}$ from $(p, q + 1)$

$\rangle$
end

8.5 Domain-Decomposed Relaxations

Until now, we focused on the multicomputer implementation of well-known relaxation methods. With this approach, no new numerical-convergence arguments must be developed, and all concurrency-related problems are purely algorithmic in nature. *Domain-decomposed relaxation methods* try to achieve a higher level of concurrency by changing the numerical method.

The most elementary example of a domain-decomposed relaxation arises when point-Gauss-Seidel relaxation is combined with the ghost-boundary exchange of concurrent Jacobi relaxation. This avoids the inefficient pipelining of program Gauss-Seidel-2 as well as the buffer arrays b_1 and b_2 of program Jacobi-2. On a local grid that is mapped to a particular process, this method performs a classical point-Gauss-Seidel relaxation. On the interior boundary, however, this method uses the neighboring values of the previous iteration step, just like in Jacobi relaxation. We call this a *domain-decomposed point-Gauss-Seidel relaxation*. In the coarse-grain limit, a computation with just one process, this method reduces to the point-Gauss-Seidel relaxation. In the fine-grain limit, a computation with one process per grid point, this method reduces to the point-Jacobi relaxation. For computations on a $P \times Q$ process mesh, we expect a convergence rate in between that for point-Jacobi and point-Gauss-Seidel relaxation. The actual convergence rate depends on P, Q, and the location of the interior boundaries.

Line-Gauss-Seidel relaxation can be adapted in a similar fashion to obtain *domain-decomposed line-Gauss-Seidel relaxation*. By not waiting for the neighboring values at interior boundaries, the inefficient pipeline of line-Gauss-Seidel relaxation is replaced by the more efficient line-Jacobi ghost-boundary exchange.

The implementation of line-Jacobi relaxation based on the concurrent iterative tridiagonal solver is the basis of another domain-decomposed relaxation method. In program Line-Jacobi-2, do not iterate the tridiagonal solver with the intent of solving the tridiagonal system. Instead, stop the tridiagonal iteration after a few steps, and accept the extra error on the computed solution of the tridiagonal system as part of the relaxation method. When only one tridiagonal-iteration step per line-relaxation step is performed, we obtain *segment relaxation*. It reduces to a classical line relaxation in each local grid combined with a Jacobi ghost-boundary exchange. Its multicomputer implementation is left as an exercise. Of course, segment relaxation has Jacobi, Gauss-Seidel, and alternating-direction variants.

Although it is possible to be very creative in the development of new relaxation methods, the realities of convergence, convergence rate, and robustness severely constrain the imagination. The convergence rate and, therefore, the number of iteration steps until convergence depend on the process-mesh dimensions and on the location of all interior boundaries. A convergence theory for these methods should be valid for as wide a range of computational parameters as possible. Until such convergence theories are established, one should take into consideration the risk that domain-decomposed relaxation methods may converge for some values of the computational parameters and diverge for other nearby values. This dependence on computational parameters makes these methods less robust. It also creates a computer dependency, because many computational parameters are a function of computer-dependent quantities.

Poorly understood convergence theories and a risk of nonconvergence and nonrobustness do not necessarily prevent the pragmatically minded from using these methods. After all, for realistic applications, the convergence of even the most classical relaxation methods often must be verified experimentally.

8.6   The Conjugate-Gradient Method

The conjugate-gradient method is applicable to the discrete Poisson problem, because the matrix formulation has a symmetric positive-definite coefficient matrix. The conjugate-gradient method is discussed at length in Section 3.3, where an implementation suitable for linear-algebra applications was proposed. That implementation is applicable to the matrix formulation of the discrete Poisson problem. However, implementations for the grid formulation are almost always preferred, because preconditioners specialized for the Poisson problem can be more easily integrated.

The main operation in the conjugate-gradient method is the matrix–vector-product assignment

$$\vec{w} := A\vec{v}.$$

Because the matrix A represents the discrete Laplace operator, it has the special sparsity structure of the five-point stencil, which enables us to use the distributed-grid structure of Section 8.2 for the vectors $\vec{v}$ and $\vec{w}$. Every component v_ℓ of the vector $\vec{v}$ corresponds to a value $v_{m,n}$ on the $M \times N$ grid. Similarly, every component w_ℓ of $\vec{w}$ corresponds to a value $w_{m,n}$. For $(m, n) \in 2..M - 2 \times 2..N - 2$, the matrix–vector-product assignment is equivalent to:

$$w_{m,n} := (4v_{m,n} - v_{m-1,n} - v_{m+1,n} - v_{m,n-1} - v_{m,n+1})/h^2.$$

Near the exterior boundary, this grid formulation of the matrix–vector product must be adapted to avoid including exterior-boundary values in the computation. However, if $\vec{v}$ vanishes on the exterior boundary, the above assignment is valid for all $(m, n) \in 1..M - 1 \times 1..N - 1$, and no special adaptations near the exterior boundary are required. This is the case for the conjugate-gradient method. Because $\vec{v}$ is a search direction and because boundary values of the solution are fixed by the Dirichlet-boundary conditions, the vector $\vec{v}$ always vanishes on the exterior boundary. The multicomputer implementation of the matrix–vector product in grid formulation only requires an update of the ghost boundaries.

The other important operation for the conjugate-gradient method is the inner product of two vectors. Again, the exterior-boundary points must be excluded. The implementation is straightforward.

For fast convergence, the matrix A must be preconditioned. Given a convergent relaxation method based on the splitting

$$A = G - H,$$

with G an invertible symmetric positive-definite matrix, we obtain a preconditioner for A by solving the system

$$G\vec{w} = \vec{v}.$$

For the discrete Poisson problem, only Jacobi relaxation and symmetric successive over-relaxation lead to symmetric positive-definite matrices G. It follows from Equation 3.16 that Jacobi preconditioning is equivalent to simple diagonal preconditioning. Equation 3.23 implies that symmetric successive over-relaxation leads to the preconditioning system

$$(D + \omega L)D^{-1}(D + \omega L^T)\vec{w} = \omega(2 - \omega)\vec{v}.$$

The solution of this system requires a forward and a backward sweep of successive over-relaxation. As pointed out in Section 8.3.4, lexicographic ordering is ill-suited for multicomputer implementation. However, symmetric successive over-relaxation based on red-black ordering is particularly successful. Not only is the multicomputer implementation efficient, red-black ordering also has the property of *computational collapse*. In principle, a forward sweep followed by a backward sweep consists of four stages performed in sequence: red, black, black, and red relaxation. After the first black relaxation, the residuals at the black grid points are zero. Because of this, the second black relaxation does not change any values. One step of symmetric successive over-relaxation based on red-black ordering consists, therefore, of just three stages performed in sequence: red, black, and red relaxation.

Most relaxation methods are based on a splitting $A = G - H$ in which G is not a symmetric positive-definite matrix. This precludes their use as preconditioners in the conjugate-gradient method. However, they can be used as preconditioners in other Krylov-space methods that do not rely on positive definiteness. For example, successive over-relaxation, red-black or lexicographic Gauss-Seidel, and line-relaxation methods are allowed as preconditioners for the quasi-minimal-residual method of Section 3.4.

8.7   FFT-Based Fast Poisson Solvers

"Fast" solvers are *direct* solution methods with an operation count that is strictly less than quadratic in the number of unknowns. The operation count of "fast" solvers may or may not be lower than the operation count of some iterative solvers. Direct solvers always compute the solution to maximal accuracy (except for round-off error), while iterative solvers can be tuned to fit the need. Moreover, the operation count of iterative solvers can be very low if a good initial guess is available. Such information cannot be used by direct solvers. Finally, the term "fast" only refers to the operation count, not to actual performance: some low-operation-count direct solvers are difficult to implement efficiently. FFT-Based fast Poisson solvers have a low operation count and have efficient implementations on multicomputers. However, even these solvers are often outperformed by advanced iterative methods like the multigrid methods of Chapter 9.

8.7.1   The Poisson Problem in Fourier Space

The solution $u(x, y)$ of the Poisson problem with Dirichlet-boundary conditions (Equation 8.2) can be written as $u(x, y) = v(x, y) + w(x, y) + t(x, y)$. The functions $v(x, y)$ and $w(x, y)$ are solutions of the Laplace equation on Ω. Function $v(x, y)$ satisfies the Dirichlet-boundary conditions

$$\begin{aligned} \forall x \in (0, \pi): \quad & v(x, 0) = v(x, \pi) = 0 \\ \forall y \in (0, \pi): \quad & \begin{cases} v(0, y) = g(0, y) \\ v(\pi, y) = g(\pi, y), \end{cases} \end{aligned} \tag{8.13}$$

and $w(x, y)$ satisfies

$$\begin{aligned} \forall x \in (0, \pi): \quad & \begin{cases} w(x, 0) = g(x, 0) \\ w(x, \pi) = g(x, \pi) \end{cases} \\ \forall y \in (0, \pi): \quad & w(0, y) = w(\pi, y) = 0. \end{aligned} \tag{8.14}$$

Finally, function $t(x, y)$ satisfies the Poisson equation

$$\forall (x, y) \in (0, \pi) \times (0, \pi): \quad -\Delta t = f(x, y) \tag{8.15}$$

with homogeneous (vanishing) Dirichlet-boundary conditions.

First, we construct $v(x, y)$. Because it vanishes for $y = 0$ and $y = \pi$, the function $v(x, y)$ for fixed x has a Fourier-sine-series expansion

$$v(x, y) = \sum_{\ell=1}^{\infty} \hat{v}_{\ell}(x) \sin(\ell y).$$

Assuming uniform convergence of the Fourier-sine series and using that $v(x, y)$ is a solution of the Laplace equation, we have that

$$\frac{\partial^2 v}{\partial x^2} + \frac{\partial^2 v}{\partial x^2} = \sum_{\ell=1}^{\infty} \left(\frac{d^2 \hat{v}_{\ell}(x)}{dx^2} - \ell^2 \hat{v}_{\ell}(x) \right) \sin(\ell y) = 0$$

and, because the coefficients of this expansion must vanish, that

$$\forall \ell > 0 \text{ and } \forall x \in (0, \pi): \quad \frac{d^2 \hat{v}_{\ell}(x)}{dx^2} - \ell^2 \hat{v}_{\ell}(x) = 0. \tag{8.16}$$

The Fourier-sine-series expansions of $g(0, y)$ and $g(\pi, y)$ and the second part of Equation 8.13 imply that

$$v(0, y) = g(0, y) = \sum_{\ell=1}^{\infty} \hat{g}_{\ell}^{(0, \cdot)} \sin(\ell y)$$

$$v(\pi, y) = g(\pi, y) = \sum_{\ell=1}^{\infty} \hat{g}_{\ell}^{(\pi, \cdot)} \sin(\ell y)$$

and, therefore, that

$$\forall \ell > 0: \quad \hat{v}_{\ell}(0) = \hat{g}_{\ell}^{(0, \cdot)} \quad \text{and} \quad \hat{v}_{\ell}(\pi) = \hat{g}_{\ell}^{(\pi, \cdot)}. \tag{8.17}$$

The solution of Equation 8.16 with the boundary conditions of Equation 8.17 is given by

$$\forall \ell > 0 : \ \hat{v}_\ell(x) = \frac{\hat{g}_\ell^{(0,\cdot)} \sinh(\ell(\pi - x)) + \hat{g}_\ell^{(\pi,\cdot)} \sinh(\ell x)}{\sinh(\ell \pi)}. \qquad (8.18)$$

The construction of $w(x, y)$ proceeds analogously and is given by

$$w(x, y) = \sum_{k=1}^{\infty} \hat{w}_k(y) \sin(kx),$$

where

$$\forall k > 0 : \ \hat{w}_k(y) = \frac{\hat{g}_k^{(\cdot,0)} \sinh(k(\pi - y)) + \hat{g}_k^{(\cdot,\pi)} \sinh(ky)}{\sinh(k\pi)}. \qquad (8.19)$$

Equation 8.15 with homogeneous boundary conditions is solved by performing a Fourier-sine transform in both the x- and y-directions:

$$t(x, y) = \sum_{k=1}^{\infty} \sum_{\ell=1}^{\infty} \hat{t}_{k,\ell} \sin(kx) \sin(\ell y)$$

$$f(x, y) = \sum_{k=1}^{\infty} \sum_{\ell=1}^{\infty} \hat{f}_{k,\ell} \sin(kx) \sin(\ell y).$$

The unknown coefficients $\hat{t}_{k,\ell}$ are easily computed from

$$\forall (k, \ell) \in \mathbb{N}_0^2 : \ (k^2 + \ell^2)\hat{t}_{k,\ell} = \hat{f}_{k,\ell}, \qquad (8.20)$$

which was obtained by applying the multivariate Fourier-sine transform to Equation 8.15.

8.7.2   Fast Poisson Solvers

FFT-Based fast Poisson solvers apply the procedure of the previous section to the discrete Poisson problem. Here, we only compute the discrete approximation of $t(x, y)$. The procedures for $v(x, y)$ and $w(x, y)$ should be obvious.

Application of the multivariate discrete Fourier-sine transform to the second-order discretization of Equation 8.15 results in

$$\forall (k, \ell) \in 1..M{-}1 \times 1..N{-}1 : \ 4(\sin^2(\frac{kh}{2}) + \sin^2(\frac{\ell h}{2}))\hat{t}_{k,\ell} = h^2 \hat{f}_{k,\ell}, \qquad (8.21)$$

where

$$t_{m,n} = \sum_{k=1}^{M-1} \sum_{\ell=1}^{N-1} \hat{t}_{k,\ell} \sin(mkh) \sin(n\ell h).$$

To compute $t_{m,n}$, the multivariate discrete Fourier-sine transform is applied to $f_{m,n}$, the coefficients $\hat{f}_{k,\ell}$ are multiplied by the amplification factor that follows from Equation 8.21, and an inverse multivariate discrete Fourier-sine transform is applied. The implementation of the multivariate discrete Fourier-sine transform is obtained by combining Sections 7.5.4 and 7.5.6.

The above procedure computes the solution of a second-order discretization of the Poisson problem. With *less* effort, the solution of a spectrally-accurate discretization can be computed: reduce the infinite-dimensional system of Equation 8.20 to a finite-dimensional system, and replace Equation 8.21 by

$$\forall (k,\ell) \in 1..M - 1 \times 1..N - 1 : \ (k^2 + \ell^2)\hat{t}_{k,\ell} = \hat{f}_{k,\ell}.$$

The amplification factors associated with this set of equations require significantly less floating-point operations than those defined by Equation 8.21.

The discrete Fourier-sine transform and its inverse have an operation count of $O(MN \log(MN))$. The multiplication by the amplification factors and inclusion of the boundary terms do not alter the order of this estimate. The resulting direct solver easily qualifies as a fast solver.

8.7.3 *Variants*

Transform Equation 8.15 by means of a univariate Fourier-sine transform in the x-direction, and obtain the systems

$$\forall k > 0 : \ \frac{d^2 \tilde{t}_k(y)}{dy^2} - k^2 \tilde{t}_k(y) = \tilde{f}_k(y).$$

One can solve these without performing a Fourier transform in the y-direction. In the discrete case, each of these one-dimensional boundary-value problems reduces to a tridiagonal system of equations. The operation count of the fast Fourier transform in one direction and the nondistributed sequential tridiagonal solver in the other is actually lower than the operation count of the multivariate fast Fourier transform. On multicomputers, however, any operation-count advantage is wiped out if distributed tridiagonal solvers are required. This method is, therefore, restricted to one-dimensional data distributions and nondistributed sequential tridiagonal solvers.

FFT-Based fast Poisson solvers can also be developed for other than Dirichlet-boundary conditions. Periodic-boundary conditions require a real Fourier transform, and Neumann-boundary conditions require a Fourier-cosine transform. However, when boundary conditions are arbitrarily mixed, Dirichlet for $x = 0$ and Neumann for $x = \pi$, for example, the Fourier approach is inappropriate. It may still be possible to apply a Fourier transform in one direction and to solve the one-dimensional boundary-value problems in the other direction.

8.8 Three-Dimensional Problems

To obtain programs for the three-dimensional Poisson problem, one must merely replace two-dimensional by three-dimensional array indices in comparable programs for the two-dimensional problem. Numerically and computationally, however, three-dimensional problems are substantially harder. Even for coarse discretizations, the number of unknowns is large. This leads to an increased operation count per iteration step and to an increased number of iteration steps until convergence. Here, we show that it is also more difficult to reach high multicomputer performance for three-dimensional than for two-dimensional problems.

"The area-perimeter law," which was introduced in Section 8.2.3, states that floating-point arithmetic, communication, and memory use due to the ghost boundaries can be neglected for sufficiently coarse-grained calculations, because those resources are proportional to the perimeter of the local grid, while resources needed by the local grid itself are proportional to its area. This "law" extends to three-dimensional computations if perimeter is replaced by surface area and area by volume. To examine this "law" in greater detail, assume that all computing resources are proportional to memory use.

Consider a two-dimensional computation in which a local grid of size $I \times I$ is mapped to each process. The local grid requires I^2 floating-point words and the ghost boundary $4I + 4$ words. In the limit of I approaching ∞ (coarse granularity), the memory for the ghost boundary is indeed negligible. For finite values of I, a measure is needed to judge whether the ghost-boundary resources can or cannot be neglected for a particular computation. One such measure is the *break-even point*: that value of I for which ghost boundary and local grid require roughly an equal amount of memory. Here, the break-even point is $I = 5$ and the number of unknowns per process at the break-even point is 25. If the number of unknowns per process is less than or comparable to 25, the resources needed by the ghost boundary are comparable to those needed by the local grid and, by definition, cannot be neglected. Note that the ratio of ghost-boundary to local-grid memory is approximately given by the slowly decaying function $4/I$. It follows that practical values of I for which the ghost-boundary resources are negligible are substantially higher.

Consider now the situation for three-dimensional computations. For a local grid of size $I \times I \times I$, the local grid requires I^3 words and the ghost boundary $6I^2 + 12I + 8$ words. The break-even point is $I = 8$, and the number of unknowns per process at the break-even point is $8^3 = 512$. The ghost-boundary resources are negligible only if the number of unknowns per process is considerably larger than 512.

Attaching the term "law" to the "area-perimeter law" is obviously an overstatement. In fact, the details of area-perimeter considerations show that values of I where this "law" is valid are unrealistically high. To stress this point, let us compute when ghost-boundary resources are truly negligible. In a three-dimensional calculation, the ratio of ghost-boundary to local-grid memory is approximately given by the function $6/I$. If we accept a ratio of 1% as negligible, we need that $6/I \leq 1\% = 0.01$ or $I \geq 600$. An $I \times I \times I$ local grid with $I = 600$ contains $600^3 = 2.16 \; 10^8$ grid points! Clearly, "the area-perimeter law" is valid only for problems that are far too large. Practical three-dimensional computations necessarily operate in near-fine-grain conditions, where ghost-boundary resources are not at all negligible.

8.9 Notes and References

The Poisson problem is an easy test problem for most convergence theories of iterative methods. As a result, most papers referred to in Section 3.5 are also relevant for this chapter. Ortega and Voigt [68] survey concurrent solution methods for partial-differential equations. Harrar [44] discusses and surveys coloring methods that can introduce concurrency in relaxation steps.

Dorr [26] and Temperton [79] survey and compare fast Poisson solvers, including some that are not based on the fast Fourier transform. The FFT-based fast solvers have an important practical advantage: all concurrency-related problems are hidden in the fast Fourier transform. The multicomputer implementation of other fast solvers, on the other hand, requires a redevelopment of the complete program.

Exercises

Exercise 36 *Derive the precise structure of the matrices A and B of the matrix formulation for the discrete Poisson problem with Dirichlet-boundary conditions. First, use lexicographic ordering of the grid points. Subsequently, repeat the exercise for red-black ordering.*

Exercise 37 *Rewrite program Jacobi-2 such that it uses only one buffer array. Is the reduction in memory use significant?*

Exercise 38 *Analyze the performance of program Gauss-Seidel-2 on a $P \times Q$ process mesh.*

Exercise 39 *Examine how a distributed sequential tridiagonal solver can be used to implement a line-Gauss-Seidel-relaxation method on a $P \times Q$ process mesh. How about line-Jacobi relaxation?*

Exercise 40 *Develop a multicomputer implementation of segment relaxation.*

Exercise 41 *Modify program Line-Jacobi-2 to obtain*

- *an alternating-direction line-Gauss-Seidel relaxation,*

- *a domain-decomposed alternating-direction line-Gauss-Seidel, and*

- *an alternating-direction segment relaxation.*

Exercise 42 *Develop a fast solver for:*

$$\forall (x, y) \in (0, \pi) \times (0, 2\pi) : \quad -\Delta u = f(x, y)$$

with boundary conditions:

$$\forall x \in (0, \pi) : \quad u(x, 0) = u(x, 2\pi)$$
$$\forall y \in (0, 2\pi) : \quad \begin{cases} u(0, y) & = & g(y) \\ u(\pi, y) & = & h(y). \end{cases}$$

9
Multigrid Methods

The multigrid methodology has been successfully applied to a wide range of problems and applications. The design of a successful multigrid program usually requires advanced theoretical insight as well as practical experience in the particular application. Such discussions are clearly beyond the scope of this book. On the other hand, the wide applicability of the multigrid idea justifies devoting a separate chapter to an introductory study of multigrid methods and their multicomputer implementation.

We shall develop a multigrid theory, specification program, and multicomputer program for the one-dimensional Poisson equation with Dirichlet-boundary conditions. This simple problem is a perfect vehicle to study the algorithmic aspects of concurrent multigrid methods on regular grids. The generalization to regular-grid problems in two or more dimensions is straightforward. Although the numerical aspects will be oversimplified, this chapter is an important first step in developing an intuitive and fundamental understanding of the multigrid methodology on multicomputers.

9.1 Relaxation Methods

In model-problem analysis, one studies a solution method by applying it to a simple problem of similar type as the problem one really wants to solve. Whether or not specific results carry over to the general case depends on the particular circumstances. Here, the insights gained from a model-problem analysis of a relaxation method lead to an intuitive development of two-grid and multigrid methods. To confirm our intuition, we shall also examine the two-grid method by means of a model-problem analysis.

9.1.1 Model Problem

The one-dimensional Poisson problem with Dirichlet-boundary conditions,

$$\forall x \in (0, \pi) : \ -\frac{d^2 u}{dx^2} = f(x)$$
$$\begin{cases} u(0) = g_0 \\ u(\pi) = g_\pi, \end{cases} \tag{9.1}$$

is discretized on an equidistant grid with grid spacing $h = \pi/M$ to obtain

$$\forall m \in 1..M - 1 : \ \frac{1}{h^2}(-u_{m-1} + 2u_m - u_{m+1}) = f_m$$
$$\begin{cases} u_0 = g_0 \\ u_M = g_\pi. \end{cases} \tag{9.2}$$

In matrix formulation, this discrete problem is represented by

$$A_h \vec{u}_h = \vec{b}_h, \tag{9.3}$$

where $A_h = \text{trid}(-\frac{1}{h^2}, \frac{2}{h^2}, -\frac{1}{h^2})$, $\vec{u}_h = [u_m]_{m=1}^{M-1}$, and

$$\vec{b}_h = \begin{bmatrix} f_1 + (u_0/h^2) \\ f_2 \\ \vdots \\ f_{M-2} \\ f_{M-1} + (u_M/h^2) \end{bmatrix}.$$

The eigenvalues and eigenvectors of the $(M - 1) \times (M - 1)$ tridiagonal matrix A_h are given by

$$\forall k \in 1..M - 1 : \ \begin{cases} \lambda_k^h = \left(\frac{\sin(kh/2)}{(h/2)} \right)^2 \\ \vec{s}_k^h = [\sin(kmh)]_{m=1}^{M-1}. \end{cases} \tag{9.4}$$

The spectral representation of Equation 9.3 is the system $\hat{A}_h \hat{\vec{u}}_h = \hat{\vec{b}}_h$, where

$$\begin{aligned} \hat{A}_h &= S_h^{-1} A_h S_h = \text{diag}(\lambda_1^h, \lambda_2^h, \dots, \lambda_{M-1}^h) \\ \vec{u}_h &= S_h \hat{\vec{u}}_h \\ \vec{b}_h &= S_h \hat{\vec{b}}_h \\ S_h &= [\vec{s}_1^h, \vec{s}_2^h, \dots, \vec{s}_{M-1}^h]. \end{aligned}$$

Because the eigenvectors of A_h are the vectors of Lemma 20, the spectral representation of the discrete model problem is its discrete Fourier-sine transform.

9.1.2 Model-Problem Analysis

We now apply model-problem analysis to a Jacobi-relaxation method that is modified to include an over-relaxation parameter ω. When applied to Equation 9.2, the relaxation method is defined by

$$\forall m \in 1..M - 1: \ u_m^{(t+1)} = \frac{\omega}{2}(f_m h^2 + u_{m+1}^{(t)} + u_{m-1}^{(t)}) + (1 - \omega)u_m^{(t)}. \quad (9.5)$$

For reasons to become clear later, we call this *Jacobi under-relaxation*.

Like other relaxation methods (see Section 3.2), Jacobi under-relaxation corresponds to a splitting $G_h - H_h$ of the matrix A_h, and its convergence properties are determined by the iteration matrix $R_h = G_h^{-1}H_h$. The error vectors of successive iteration steps t and $t + 1$ satisfy

$$\vec{v}_h^{(t+1)} = R_h \vec{v}_h^{(t)}, \quad (9.6)$$

where $\vec{v}_h^{(t)} = \vec{u}_h^* - \vec{u}_h^{(t)}$ and $\vec{u}_h^*$ is the exact solution of Equation 9.3. The spectral representation of R_h is obtained by writing the error vectors as linear combinations of the eigenvectors of A_h. Substituting $\vec{v}_h^{(t)} = S_h \hat{\vec{v}}_h^{(t)}$ and $\vec{v}_h^{(t+1)} = S_h \hat{\vec{v}}_h^{(t+1)}$ into Equation 9.6, we obtain that

$$\hat{R}_h = S_h^{-1} R_h S_h. \quad (9.7)$$

In practice, $\hat{R}_h$ is computed by a discrete Fourier-sine transform. The grid formulation of Equation 9.6 follows from Equation 9.5 and is given by

$$\forall m \in 1..M - 1: \ v_m^{(t+1)} = \frac{\omega}{2}(v_{m+1}^{(t)} + v_{m-1}^{(t)}) + (1 - \omega)v_m^{(t)}. \quad (9.8)$$

Substitute the discrete Fourier-sine expansion

$$\forall m \in 1..M - 1: \ v_m^{(t)} = \sum_{k=1}^{M-1} \hat{v}_k^{(t)} \sin kmh$$

into Equation 9.8, and obtain that

$$\forall k \in 1..M - 1: \ \hat{v}_k^{(t+1)} = (1 - 2\omega \sin^2 \frac{kh}{2})\hat{v}_k^{(t)},$$

which is equivalent to

$$\hat{R}_h = \text{diag}(1 - 2\omega \sin^2 \frac{kh}{2}). \quad (9.9)$$

The model-problem analysis is simplified because of a fortunate coincidence: the matrices R_h and A_h have the same eigenvectors. This is why $\hat{R}_h$
is diagonal and why Jacobi under-relaxation does not introduce any coupling between spectral components of the error. One iteration step merely
multiplies spectral component number k of the error by the *amplification
factor* $1 - 2\omega \sin^2 \frac{kh}{2}$.

9.1.3 Convergence Factor

The convergence factor of Jacobi under-relaxation equals the spectral radius of R_h. Because the matrices $\hat{R}_h$ and R_h are similar (Equation 9.7),
their spectral radii are identical, and Equation 9.9 implies that

$$\rho_J(\omega) = \max_{0 < k < M} |1 - 2\omega \sin^2 \frac{kh}{2}|.$$

For convergence, it is necessary that $\rho_J(\omega) < 1$ or, equivalently, that

$$\forall k \in 1..M - 1 : \ -1 < 1 - 2\omega \sin^2 \frac{kh}{2} < 1.$$

This requirement implies that ω must satisfy

$$\forall k \in 1..M - 1 : \ 0 < \omega < \frac{1}{\sin^2 \frac{kh}{2}}.$$

The second inequality is most restrictive for $k = M - 1$. After taking the
limit for $h \to 0$, we find the necessary condition for convergence:

$$0 < \omega \leq 1.$$

This justifies calling the method Jacobi *under*-relaxation.

We now compute the convergence factor $\rho_J(\omega)$ for values of $\omega \in (0, 1]$.
As k takes on all values in the range $1..M - 1$, the value $y = 2\omega \sin^2 \frac{kh}{2}$
ranges over the interval $[2\omega \sin^2 \frac{h}{2}, 2\omega \cos^2 \frac{h}{2}]$, and

$$\rho_J(\omega) \leq \max_{2\omega \sin^2 \frac{h}{2} < y < 2\omega \cos^2 \frac{h}{2}} |1 - y| = 1 - 2\omega \sin^2 \frac{h}{2} \approx 1 - \omega \frac{h^2}{2}.$$

Fastest convergence is obtained for $\omega = 1$, which corresponds to classical
Jacobi relaxation.

9.1.4 Smoothing Factor

The convergence factor is a worst-case measure. In the case of Jacobi under-
relaxation, error components with lowest frequency ($k = 1$) are reduced
the least. High-frequency components are attenuated much faster. This is

indicated by the *smoothing factor* $\mu_J(\omega)$, which measures by how much the top half of the error spectrum is reduced every relaxation step. For the model problem, the smoothing factor is given by

$$\mu_J(\omega) = \max_{M/2 \leq k < M} |1 - 2\omega \sin^2 \frac{kh}{2}|.$$

Considering the range of k, it follows that

$$\mu_J(\omega) \leq \max_{\omega \leq y \leq 2\omega} |1 - y|.$$

The convex and positive function $|1 - y|$ must attain its maximum at one of the edges of the domain, and

$$\mu_J(\omega) \leq \max(|1 - \omega|, |1 - 2\omega|).$$

The right-hand side is minimized for $\omega = 2/3$, and the smoothing factor of Jacobi under-relaxation with $\omega = 2/3$ is $\mu_J(2/3) \leq 1/3$.

9.1.5 Smoothing vs. Convergence

For fast convergence, one should always choose the parameter ω equal to one, because this minimizes the convergence factor of Jacobi under-relaxation. Even with this choice, the convergence factor remains $1 - O(h^2)$. Therefore, at least $O(h^{-2}|\log \epsilon|)$ relaxation steps are necessary to reduce the error from $O(1)$ to $O(\epsilon)$. Because the convergence factor depends on h, finer grids not only require more work per relaxation step, they also require more relaxation steps.

The analysis of Section 9.1.3 reveals that convergence is limited by the low-frequency errors, which are reduced the least by relaxation. High-frequency errors, on the other hand, are reduced effectively. In Section 9.1.4, it was shown that merely three iteration steps of Jacobi under-relaxation with $\omega = 2/3$ reduce the high-frequency errors by a factor of at least 27.

When iterating until convergence, the first few relaxation steps eliminate the high-frequency errors. The bulk of the work, the remaining $O(1/h^2)$ relaxation steps, is necessary to reduce low-frequency errors to acceptable levels. There is an intuitive explanation for the difference in convergence rate between high- and low-frequency errors. The low-frequency errors have a long wave length, which spans many grid points. Because one relaxation step involves only the neighboring grid points of its stencil, information can travel over just one grid spacing per relaxation step. Merely to reach distances comparable to the wave length, one must evaluate a number of relaxation steps proportional to the wave length and inversely proportional to the grid spacing.

9.2   Two-Grid Methods

9.2.1   Intuitive Development

The division line between high and low frequency depends on the grid: frequencies that are low on a fine grid may be high on a coarser grid. In multigrid methods, a combination of relaxations on grids of varying coarseness reduces errors of all frequencies.

As a first step in the development of this idea, we shall consider a problem discretized on one fine and one coarse grid. The resulting algorithm will use coarse-grid solutions to improve the convergence rate of fine-grid relaxations. The initial derivation will be based more on intuition than on hard fact. In Section 9.2.2, our intuition will be backed up by a rigorous treatment of the model problem.

The matrix formulation of the fine and coarse discretizations are given by $A_h \vec{u}_h = \vec{b}_h$ and $A_H \vec{u}_H = \vec{b}_H$, respectively. We assume that both A_h and A_H are invertible. The convergence and smoothing properties of the fine-grid relaxation are determined by the iteration matrix $R_h = G_h^{-1} H_h$. We assume that this relaxation method, like Jacobi under-relaxation, has a smoothing factor $\mu < 1$ that is independent of the grid spacing. For now, we assume that it is feasible to solve coarse-grid problems exactly. This last assumption will be removed in Section 9.3.

Given an initial guess $\vec{u}_h^0$ for the fine-grid solution, apply K_1 relaxation steps to obtain a new estimate $\vec{u}_h^1$. The error vectors $\vec{v}_h^0 = \vec{u}_h^* - \vec{u}_h^0$ and $\vec{v}_h^1 = \vec{u}_h^* - \vec{u}_h^1$, where $\vec{u}_h^*$ is the unknown exact fine-grid solution, satisfy

$$\vec{v}_h^1 = R_h^{K_1} \vec{v}_h^0, \tag{9.10}$$

and the smoothing property ensures that the high-frequency components of $\vec{v}_h^1$ are negligible compared to its low-frequency components. Our goal is to compute an estimate for $\vec{v}_h^1$, which can be used to improve the solution estimate $\vec{u}_h^1$.

The defect associated with $\vec{u}_h^1$,

$$\vec{d}_h^1 = \vec{b}_h - A_h \vec{u}_h^1,$$

is easily computed and is related to $\vec{v}_h^1$ via

$$A_h \vec{v}_h^1 = \vec{d}_h^1. \tag{9.11}$$

Because $\vec{v}_h^1$ is smooth with respect to the fine grid, it can be represented accurately as the interpolation of a coarse-grid vector $\vec{v}_H^1$:

$$\vec{v}_h^1 \approx I_H^h \vec{v}_H^1.$$

The operator I_H^h is a suitable *interpolation operator* from the coarse grid to the fine grid. We introduce this approximate equality into Equation 9.11

and obtain an approximate system for $\vec{v}_H^1$:

$$A_h I_H^h \vec{v}_H^1 \approx \vec{d}_h^1.$$

Assuming also that the defect vector $\vec{d}_h^1$ is smooth, this system can be transferred to the coarse grid by means of a *restriction operator* I_h^H:

$$(I_h^H A_h I_H^h)\vec{v}_H^1 \approx I_h^H \vec{d}_h^1.$$

Usually, the interpolation and restriction operators are such that

$$I_h^H A_h I_H^h \approx A_H.$$

To obtain a numerical method, reverse the above and take

$$A_H \vec{v}_H^1 = I_h^H \vec{d}_h^1$$

as a defining statement for $\vec{v}_H^1$. Having computed $\vec{v}_H^1$ by solving this coarse-grid problem, estimate the fine-grid error $\vec{v}_h^1$ by interpolating $\vec{v}_H^1$. This estimate is added to $\vec{u}_h^1$ to obtain a new fine-grid-solution estimate $\vec{u}_h^2$. More formally, the *coarse-grid correction* is given by

$$
\begin{aligned}
\vec{d}_h^1 &= \vec{b}_h - A_h \vec{u}_h^1 \\
\vec{d}_H^1 &= I_h^H \vec{d}_h^1 \\
\vec{v}_H^1 &= A_H^{-1} \vec{d}_H^1 \\
\vec{v}_h^1 &= I_H^h \vec{v}_H^1 \\
\vec{u}_h^2 &= \vec{u}_h^1 + \vec{v}_h^1,
\end{aligned}
$$

or, in matrix form, by

$$\vec{u}_h^2 = \vec{u}_h^1 + I_H^h A_H^{-1} I_h^H (\vec{b}_h - A_h \vec{u}_h^1).$$

The error vectors with respect to the exact fine-grid solution $\vec{u}_h^*$ satisfy

$$\vec{v}_h^2 = (I_h - I_H^h A_H^{-1} I_h^H A_h)\vec{v}_h^1 = C_h \vec{v}_h^1. \qquad (9.12)$$

The matrix C_h is called the *coarse-grid-correction matrix*.

Each *two-grid step* consists of a *fine-grid-smoothing iteration* of K_1 relaxation steps and a *coarse-grid correction*. Equations 9.10 and 9.12 imply that the convergence factor of the two-grid method is the spectral radius of the iteration matrix $C_h R_h^{K_1}$. We know that the smoothing eliminates the high-frequency components, and we hope that the coarse-grid correction reduces the low-frequency components without introducing new high-frequency components.

9.2.2 Model-Problem Analysis

We now use the model problem to confirm our intuition that the two-grid method reduces errors of all frequencies by a factor independent of the grid spacing. The model problem is discretized on a coarse grid with N grid cells and grid spacing $H = \pi/N$ and on a fine grid with $M = 2N$ grid cells and grid spacing $h = \pi/M = H/2$. All spectral properties of A_h are known from Equation 9.4. The eigenvalues of R_h are given by Equation 9.9. The columns of S_h are the eigenvectors of both R_h and A_h. Similarly, all spectral properties of A_H are known, and the columns of the matrix

$$S_H = [\vec{s}_1^H, \vec{s}_2^H, \ldots, \vec{s}_{N-1}^H]$$

are the eigenvectors of A_H. We shall also use unit vectors of $\mathbb{R}^{M-1}$ and $\mathbb{R}^{N-1}$, which we denote by $\vec{e}_k^h$ and $\vec{e}_k^H$, respectively.

The spectral representation of $C_h R_h^{K_1}$ is given by

$$\hat{C}_h \hat{R}_h^{K_1} = S_h^{-1}(C_h R_h^{K_1})S_h = (S_h^{-1}C_h S_h)\hat{R}_h^{K_1}.$$

We already know $\hat{R}_h$ from Equation 9.9. In the computation of

$$
\begin{aligned}
\hat{C}_h &= S_h^{-1}(I_h - I_H^h A_H^{-1} I_h^H A_h)S_h \\
&= (S_h^{-1} I_h S_h) - (S_h^{-1} I_H^h S_H)(S_H^{-1} A_H^{-1} S_H)(S_H^{-1} I_h^H S_h)(S_h^{-1} A_h S_h) \\
&= I_h - \hat{I}_H^h \hat{A}_H^{-1} \hat{I}_h^H \hat{A}_h,
\end{aligned}
\tag{9.13}
$$

only the spectral representations of the interpolation and restriction operators are unknown.

The *linear-interpolation operation* $\vec{w}_h = I_H^h \vec{w}_H$ is defined by

$$\forall n \in 0..N-1: \left\{ \begin{array}{ll} w_{2n+1}^h &= 0.5(w_n^H + w_{n+1}^H) \\ w_{2n}^h &= w_n^H. \end{array} \right.$$

When this is applied to the coarse-grid eigenvectors, we find that

$$\forall k \in 1..N-1: I_H^h \vec{s}_k^H = \cos^2 \frac{kh}{2}\vec{s}_k^h - \sin^2 \frac{kh}{2}\vec{s}_{M-k}^h. \tag{9.14}$$

Column number k of $\hat{I}_H^h$ is given by

$$\hat{I}_H^h \vec{e}_k^H = S_h^{-1} I_H^h S_H \vec{e}_k^H = S_h^{-1} I_H^h \vec{s}_k^H.$$

With Equation 9.14, we find that

$$\forall k \in 1..N-1: \hat{I}_H^h \vec{e}_k^H = \cos^2 \frac{kh}{2}\vec{e}_k^h - \sin^2 \frac{kh}{2}\vec{e}_{M-k}^h. \tag{9.15}$$

The interpolation operator, therefore, maps coarse-grid component k into fine-grid components k and $M - k$.

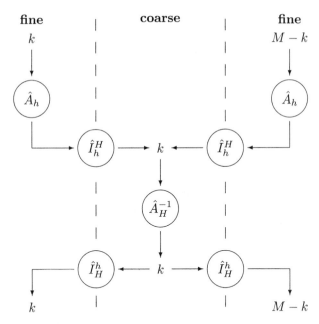

FIGURE 9.1. Frequency coupling of the coarse-grid-correction operation.

The *full-weight-restriction operation* $\vec{w}_H = I_h^H \vec{w}_h$ is defined by

$$\forall n \in 1..N-1 : \quad w_n^H = 0.25(w_{2n-1}^h + 2w_{2n}^h + w_{2n+1}^h).$$

From $\hat{I}_h^H = S_H^{-1} I_h^H S_h$, $\hat{I}_h^H \vec{e}_k^h = S_H^{-1} I_h^H \vec{s}_k^h$, and

$$\begin{cases} \forall k \in 1..N-1 : & I_h^H \vec{s}_k^h = \cos^2 \frac{kh}{2} \vec{s}_k^H \\ k = N : & I_h^H \vec{s}_N^h = \vec{0} \\ \forall k \in N+1..M-1 : & I_h^H \vec{s}_k^h = -\sin^2 \frac{(M-k)h}{2} \vec{s}_{M-k}^H, \end{cases} \qquad (9.16)$$

it follows that $\hat{I}_h^H = (\hat{I}_H^h)^T$. Full-weight restriction maps fine-grid component $k \in 1..N-1$ into coarse-grid component k. However, fine-grid component $k \in N+1..M-1$ is mapped into coarse-grid component $M-k$.

Only the nondiagonal matrices $\hat{I}_h^H$ and $\hat{I}_H^h$ introduce any coupling between the spectral components of the error. This coupling of the frequencies is displayed in Figure 9.1. The fine-grid error after smoothing, $\vec{v}_h^1$, contains spectral components $k \in 1..M-1$. When multiplied by the diagonal matrix $\hat{A}_h$, each component is amplified or reduced individually. Coupling occurs when the error is multiplied by the nondiagonal matrix $\hat{I}_h^H$, the restriction operator. This operator couples fine-grid components k and $M-k$ into one coarse-grid component $k \in 1..N-1$. The multiplication by the diagonal matrix $\hat{A}_H^{-1}$ does not couple any coarse-grid components. However, coarse-grid component k is split back into fine-grid components k and $M-k$ by the interpolation operator $\hat{I}_H^h$.

Because the coarse-grid-correction operator only couples fine-grid components k and $M - k$, the two-dimensional subspace of $\mathbb{R}^{M-1}$ spanned by $\vec{e}_k^h$ and $\vec{e}_{M-k}^h$ is an invariant subspace of $\hat{C}_h$. There are $N-1$ such invariant subspaces, one for each $k \in 1..N - 1$.

The matrix $\hat{C}_h$ is thus a sparse matrix with nonzeros only in the positions indicated by the indices of the elements of the 2×2 submatrices

$$\forall k \in 1..N - 1 : \begin{bmatrix} c_{k,k} & c_{k,M-k} \\ c_{M-k,k} & c_{M-k,M-k} \end{bmatrix}.$$

Submatrix number k maps invariant subspace number k unto itself; this submatrix is computed by combining Equations 9.13, 9.15, and 9.16. Using the short-hand notation $s = \sin \frac{kh}{2}$ and $c = \sin \frac{(M-k)h}{2} = \cos \frac{kh}{2}$, the coarse-grid correction in subspace number k is given by

$$\begin{bmatrix} 1 & 0 \\ 0 & 1 \end{bmatrix} - \begin{bmatrix} c^2 \\ -s^2 \end{bmatrix} \left[\left(\frac{\sin \frac{kH}{2}}{\frac{H}{2}} \right)^2 \right]^{-1} \begin{bmatrix} c^2 & -s^2 \end{bmatrix} \begin{bmatrix} \frac{4}{h^2}s^2 & 0 \\ 0 & \frac{4}{h^2}c^2 \end{bmatrix},$$

which simplifies to

$$\begin{bmatrix} s^2 & c^2 \\ s^2 & c^2 \end{bmatrix}.$$

This and Equation 9.9 imply that two-grid iteration based on Jacobi under-relaxation with $\omega = 2/3$ and acting on subspace number k has the spectral representation

$$\begin{bmatrix} s^2 & c^2 \\ s^2 & c^2 \end{bmatrix} \begin{bmatrix} 1 - \frac{4}{3}s^2 & 0 \\ 0 & 1 - \frac{4}{3}c^2 \end{bmatrix}^{K_1} = \begin{bmatrix} s^2(1 - \frac{4}{3}s^2)^{K_1} & c^2(1 - \frac{4}{3}c^2)^{K_1} \\ s^2(1 - \frac{4}{3}s^2)^{K_1} & c^2(1 - \frac{4}{3}c^2)^{K_1} \end{bmatrix}.$$

The spectral radius of this matrix is

$$\rho_k = |s^2(1 - \frac{4}{3}s^2)^{K_1} + c^2(1 - \frac{4}{3}c^2)^{K_1}|. \tag{9.17}$$

The definitions of s, c, k, and h imply that

$$\frac{1}{3} < 1 - \frac{4}{3}s^2 < 1 \tag{9.18}$$

$$-\frac{1}{3} < 1 - \frac{4}{3}c^2 < \frac{1}{3} \tag{9.19}$$

$$0 < s^2 < \frac{1}{2}. \tag{9.20}$$

Equation 9.18 guarantees that, whatever the value of K_1, the first term of the right-hand side of Equation 9.17 will not blow up. Equation 9.19 guarantees that the second term of the right-hand side of Equation 9.17 is negligible if K_1 is sufficiently large. This and Equation 9.20 implies that

$$\rho_k \leq s^2 < \frac{1}{2}.$$

Because it is independent of the particular subspace, this bound is valid for all of $\mathbb{R}^{M-1}$, and the convergence factor of the two-grid iteration satisfies

$$\rho_{TGM} = \rho(C_h R_h^{K_1}) = \rho(\hat{C}_h \hat{R}_h^{K_1}) = \max_{0<k<M} \rho_k < \frac{1}{2}.$$

The model-problem analysis of the two-grid method shows that all error components are reduced by a factor independent of the grid spacing h. As a result, the number of two-grid steps necessary to reduce the error from $O(1)$ to $O(\epsilon)$ is independent of h. Of course, the amount of work per step increases as h decreases.

9.3   Multigrid Methods

The problem is discretized on a sequence of $L{+}1$ grids numbered 0 through L. Grid ℓ has a grid spacing h_ℓ, and $h_0 > h_1 > \ldots > h_L$. The matrix formulation of the level-ℓ problem is given by

$$A_\ell \vec{u}_\ell = \vec{b}_\ell. \tag{9.21}$$

We assume the following:

- The matrix A_ℓ is invertible for all $\ell \in 0..L$.

- It is feasible to compute exact solutions of Equation 9.21 only for $\ell = 0$. That solution method is called the *coarsest-grid solver*.

- On every level ℓ, there exists a *smoothing operator* with smoothing factor μ independent of the grid spacing. The smoothing operator is obtained by splitting the matrix A_ℓ into $G_\ell - H_\ell$, and its convergence and smoothing properties are determined by the iteration matrix $R_\ell = G_\ell^{-1} H_\ell$.

- There exist *interpolation operators* $I_\ell^{\ell+1}$ that transfer grid functions from level ℓ to the next finer level.

- There exist *restriction operators* $I_\ell^{\ell-1}$ that transfer grid functions from level ℓ to the next coarser level.

9.3.1   Elementary Multigrid

Program Two-Grid-0 is a preliminary specification of the two-grid method on level ℓ. This specification uses data of the undefined type "vector." The precise data structure depends on whether the problem is one-, two-, or three-dimensional. It also depends on whether the discrete problems are defined on a regular or on an irregular grid. Because the top-level structure of multigrid programs is independent of these considerations, we postpone these low-level implementation decisions until Section 9.4, where the multicomputer programs will be developed.

program Two-Grid-0$(K, \ell, \vec{u}_\ell, \vec{b}_\ell, \vec{u}_{\ell-1}, \vec{b}_{\ell-1})$
declare
    K, ℓ, k, k_1 : integer ;
    $\vec{u}_\ell, \vec{b}_\ell, \vec{u}_{\ell-1}, \vec{b}_{\ell-1}$: vector
assign
    $\langle\ ;\ k\ :\ 0 \le k < K\ ::$
        $\langle\ ;\ k_1\ :\ 0 \le k_1 < K_1\ ::\ \vec{u}_\ell := G_\ell^{-1}(\vec{b}_\ell + H_\ell \vec{u}_\ell)\ \rangle\ ;$
        $\vec{b}_{\ell-1} := I_\ell^{\ell-1}(\vec{b}_\ell - A_\ell \vec{u}_\ell)\ ;$
        $\vec{u}_{\ell-1} := A_{\ell-1}^{-1} \vec{b}_{\ell-1}\ ;$
        $\vec{u}_\ell := \vec{u}_\ell + I_{\ell-1}^{\ell} \vec{u}_{\ell-1}$
    $\rangle$
end

The parameter K is the number of two-grid steps to be performed. When program Two-Grid-0 is called, vector $\vec{u}_\ell$ is an estimate for the solution and vector $\vec{b}_\ell$ is the right-hand side of the problem on level ℓ. The parameters $\vec{u}_{\ell-1}$ and $\vec{b}_{\ell-1}$ are used in the coarse-grid-correction step. The quantification over k_1 is the smoothing iteration. The undeclared operators $A_{\ell-1}^{-1}$, G_ℓ^{-1}, and H_ℓ and the undeclared constant K_1 are assumed known.

Program Two-Grid-0 applied to a problem on level ℓ uses exact solutions of problems on level $\ell - 1$. This is infeasible, unless $\ell = 1$. To remedy this, replace the exact coarse-grid solver by an approximate iterative solver. A natural choice is to use the same solver on level $\ell - 1$ as on level ℓ. Program Multigrid-0 is a specification of this idea.

program Multigrid-0$(K, \ell, \vec{u}_\ell, \vec{b}_\ell, \ldots, \vec{u}_0, \vec{b}_0)$
declare
    K, ℓ, k, k_1 : integer ;
    $\vec{u}_0, \vec{b}_0, \ldots, \vec{u}_\ell, \vec{b}_\ell$: vector
assign
    if $\ell = 0$ **then**
        $\vec{u}_0 := A_0^{-1} \vec{b}_0$
    else
    $\langle\ ;\ k\ :\ 0 \le k < K\ ::$
        $\langle\ ;\ k_1\ :\ 0 \le k_1 < K_1\ ::\ \vec{u}_\ell := G_\ell^{-1}(\vec{b}_\ell - H_\ell \vec{u}_\ell)\ \rangle\ ;$
        $\vec{b}_{\ell-1} := I_\ell^{\ell-1}(\vec{b}_\ell - A_\ell \vec{u}_\ell)\ ;$
        $\vec{u}_{\ell-1} := \vec{0}\ ;$
        Multigrid-0$(K_2, \ell - 1, \vec{u}_{\ell-1}, \vec{b}_{\ell-1}, \ldots, \vec{u}_0, \vec{b}_0)\ ;$
        $\vec{u}_\ell := \vec{u}_\ell + I_{\ell-1}^{\ell} \vec{u}_{\ell-1}$
    $\rangle$
end

On level $\ell = 0$, program Multigrid-0 applies the coarsest-grid solver and returns. On levels $\ell > 0$, the program performs K multigrid-iteration steps. Each multigrid-iteration step consists of K_1 smoothing steps and a coarse-grid-correction step. The latter recursively calls program Multigrid-0 to perform K_2 multigrid steps on level $\ell - 1$.

9.3.2 Performance Analysis

For the purpose of performance analysis, we make a number of additional assumptions:

- the partial-differential operator is an operator in D space dimensions, say the Laplace operator on a D-dimensional domain Ω,

- the grid spacings are given by $h_\ell = 2^{-\ell} h_0$,

- the partial-differential operator is discretized to second order,

- the number of floating-point operations performed by the smoothing, restriction, and interpolation operators on level ℓ is proportional to the number of unknowns on that level, which is $O(h_\ell^{-D})$,

- the operation count of the coarsest-grid solver is W_0.

The performance of any iterative method depends on the convergence factor and on the operation count per iteration step.

The convergence factor ρ of any iterative method determines the number of iteration steps to reach a certain accuracy. Typically, one wishes to reduce the error of the initial guess, which is $O(1)$, to the order of magnitude of the discretization error. For second-order discretization on grid L, the discretization error is $O(h_L^2)$. In this case, one must perform at least

$$O\left(\frac{\log h_L^2}{\log \rho}\right)$$

iteration steps. For a D-dimensional problem, the number of unknowns on level L is $O(h_L^{-D})$. The number of iteration steps until convergence is, therefore, proportional to the logarithm of the number of unknowns, provided ρ is less than 1 and independent of the grid size. We already know that the convergence factor of the two-grid method applied to the model problem satisfies the latter requirements. The question is whether there exist values of K_1 and K_2 such that the multigrid-convergence factor is less than 1 and independent of the grid size.

Intuitively, one might conclude that K_1 must be large enough to eliminate all high-frequency errors. It is actually preferable to use a small value for K_1. For the model problem, the relation between the two-grid-convergence factor and the number smoothing steps follows from Equations 9.17 through 9.20:

$$\rho_{TGM} < \frac{1}{2} + \frac{1}{3^{K_1}}.$$

The marginal reduction in ρ_{TGM} by performing one additional smoothing step is negligible if $K_1 > 2$.

If the number of coarse-grid-iteration steps K_2 is large, then the approximate iterative coarse-grid solver of the multigrid method is equivalent to the exact coarse-grid solver of the two-grid method. If K_2 is large, the multigrid-convergence factor ρ_{MG} and the two-grid-convergence factor ρ_{TGM} are identical. In practice, it is preferable to choose a small value for K_2 and to accept a ρ_{MG} that is somewhat larger than the corresponding ρ_{TGM}. The number of coarse-grid-iteration steps K_2 almost never exceeds 2. The case $K_2 = 1$ is called *V-cycle multigrid*, and the case $K_2 = 2$ is called *W-cycle multigrid*.

Tests and analyses that are not covered here show that a multigrid-convergence factor less than 1 and independent of the grid spacing can be achieved for small values of K_1 and K_2. Occasionally, K_1 and K_2 are chosen adaptively. Such adaptive strategies increase robustness but rely on various error estimates, which usually increase the operation count per iteration step.

The operation count W_ℓ of one multigrid-iteration step on level ℓ satisfies the recursion:

$$\forall \ell \in 2..L : \quad W_\ell = 2^{D\ell}C + K_2 W_{\ell-1}$$
$$\ell = 1 : \quad W_1 = 2^D C + W_0.$$

The term $2^{D\ell}C$ includes all floating-point operations performed by the smoothing, restriction, and interpolation operators on grid level ℓ. This term is dominated by the smoothing iteration, and the constant C depends linearly on the constant K_1. By solving the above recursion, we obtain that

$$
\begin{aligned}
W_L &= 2^{DL}C\frac{1-\left(\frac{K_2}{2^D}\right)^L}{1-\frac{K_2}{2^D}} + K_2^{L-1}W_0 \\
&< 2^{DL}C\frac{2^D}{2^D-K_2} + K_2^{L-1}W_0.
\end{aligned}
\tag{9.22}
$$

The inequality is valid if $1 \le K_2 < 2^D$ and is usually a tight upper bound, because $(K_2/2^D)^L \ll 1$.

For V-cycle multigrid, the upper bound is further simplified to

$$W_L < 2^{DL}C\frac{2^D}{2^D-1} + W_0.$$

The first term is proportional to the number of unknowns on the finest level. If W_0 is negligible, the operation count of the coarser levels is only a small fraction of the operation count of the finest level. Whether or not W_0 is negligible depends on L and the coarsest-grid solver. The number of unknowns on the coarsest level is a factor of 2^{DL} less than the number of unknowns on the finest level. For V-cycle multigrid with many levels, it is likely that W_0 is negligible. However, for moderate values of L, which are more prevalent, W_0 might not be negligible, because the coarsest-grid solver may be an expensive direct solver.

For W-cycle multigrid, the upper bound for the number of floating-point operations per multigrid step is

$$W_L < 2^{DL} C \frac{2^D}{2^D - 2} + 2^{L-1} W_0.$$

The coarsest-grid solver plays a more prominent role here, and its operation count is often not negligible.

We may summarize multigrid performance as follows. Let $N = 2^{DL}$ be the number of unknowns on the finest grid. Assume that the convergence factor ρ_{MG} is less than one and independent of the grid size for small values of K_1 and K_2. In this case, just $(\log N)$ multigrid-iteration steps reduce the $O(1)$ error of the initial guess to the size of the discretization error. Neglecting W_0, the operation count of one multigrid-iteration step is $O(N)$. Solving the problem up to discretization accuracy via multigrid, therefore, requires an operation count proportional to $N \log N$.

9.3.3 Full Multigrid

The operation count can be reduced further by starting the multigrid iteration from a better initial guess. Program Full-Multigrid-0 calls program Multigrid-0 to compute solutions on coarser levels as good initial guesses for the finer levels.

> **program** Full-Multigrid-0$(L, \vec{u}_L, \vec{b}_L, \ldots, \vec{u}_0, \vec{b}_0)$
> **declare**
> k, ℓ, L : integer ;
> $\vec{u}_0, \vec{b}_0, \ldots, \vec{u}_L, \vec{b}_L$: vector
> **assign**
> $\vec{u}_0 := A_0^{-1} \vec{b}_0$;
> $\langle \; ; \; \ell \; : \; 1 \le \ell \le L \; ::$
> $\vec{u}_\ell := I_{\ell-1}^\ell \vec{u}_{\ell-1}$;
>   Multigrid-0$(N_\ell, \ell, \vec{u}_\ell, \vec{b}_\ell, \ldots, \vec{u}_0, \vec{b}_0)$
> $\rangle$
> **end**

The vector $\vec{b}_\ell$ must be initialized with values obtained from the level-ℓ problem. None of the vectors $\vec{u}_\ell$ need be initialized.

The success of full multigrid depends on the values for N_ℓ, the number of multigrid steps on every level ℓ. Again, assume that the tolerance should be consistent with the discretization error. On level ℓ, the incoming initial guess was determined by interpolation from a level $\ell - 1$ solution and has an error $O(h_{\ell-1}^2)$. After the multigrid iteration on level ℓ, the error should be $O(h_\ell^2)$. As a result, we have that

$$N_\ell = O\left(-\log \frac{h_\ell^2}{h_{\ell-1}^2}\right) = O(\log 4) = O(1).$$

It is, therefore, sufficient to apply the same number of multigrid steps on every level, say $N_\ell = K_3$. The full-multigrid method requires W_{FMG} floating-point operations with

$$W_{FMG} = W_0 + K_3 \sum_{\ell=1}^{L} W_\ell.$$

Neglecting W_0 here and in Equation 9.22, the following operation count for full multigrid is obtained:

$$W_{FMG} \approx K_3 C \frac{2^D}{2^D - K_2} 2^D \frac{2^{DL} - 1}{2^D - 1} \approx \frac{2^D}{2^D - K_2} K_3 C 2^{DL}.$$

The factor 2^{DL} is proportional to the number of unknowns on the finest level. Because all other factors are constants, full multigrid solves the problem, up to discretization accuracy, with an operation count that is proportional to the number of unknowns!

9.4   Multicomputer Multigrid

9.4.1   Specification Programs

The specification programs of Section 9.3 did not specify the data type "vector" and used operator notation for the smoothing, restriction, interpolation, and coarsest-grid solver. Such abstractions are convenient for developing general specifications, but specific representations are required to derive multicomputer programs and to discuss data distributions.

The model problem of Equation 9.1 is discretized on $L + 1$ grids numbered 0 through L. Grid 0 is the coarsest grid and has grid spacing $h_0 = \pi/M$. Grid ℓ has grid spacing $h_\ell = 2^{-\ell} h_0 = \pi/(2^\ell M)$. Grid functions on level ℓ are represented as real arrays over the index set $0..2^\ell M$.

Programs that specify the elementary multigrid operations are easily obtained. Given a solution estimate u and a right-hand side f on level ℓ, program Relax-1 performs K iteration steps of Jacobi under-relaxation with optimal smoothing parameter $\omega = 2/3$. Program Compute-Defect-1 computes the defect associated with a given solution estimate u and right-hand side f. Program Restrict-1 performs a full-weight restriction from level ℓ to level $\ell - 1$. Program Interpolate-1 maps a level $\ell - 1$ grid function to level ℓ by linear interpolation.

program Relax-1(K, ℓ, u, f)
declare
    $K,\ k,\ \ell,\ m\ :\ $ integer ;
    $u,\ f\ :\ $ array$[0..2^\ell M]$ of real
assign
    $\langle\ ;\ k\ :\ 0 \le k < K\ ::$
        $\langle\ \|\ m\ :\ 0 < m < 2^\ell M\ ::$
        $u[m] := (h_\ell^2 f[m] + u[m+1] + u[m-1] + u[m])/3$
        $\rangle$
    $\rangle$
end

program Compute-Defect-1(ℓ, u, f, y)
declare
    $\ell,\ m\ :\ $ integer ;
    $u,\ f,\ y\ :\ $ array$[0..2^\ell M]$ of real
assign
    $\langle\ \|\ m\ :\ 0 < m < 2^\ell M\ ::$
        $y[m] := f[m] + (u[m+1] - 2u[m] + u[m-1])/h_\ell^2$
    $\rangle$
end

program Restrict-1(ℓ, y, f)
declare
    $\ell,\ n\ :\ $ integer ;
    $y\ :\ $ array$[0..2^\ell M]$ of real ;
    $f\ :\ $ array$[0..2^{\ell-1} M]$ of real
assign
    $\langle\ \|\ n\ :\ 0 < n < 2^{\ell-1} M\ ::\ f[n] := (y[2n-1] + 2y[2n] + y[2n+1])/4\ \rangle$
end

program Interpolate-1(ℓ, f, y)
declare
    $\ell,\ n\ :\ $ integer ;
    $f\ :\ $ array$[0..2^{\ell-1} M]$ of real ;
    $y\ :\ $ array$[0..2^\ell M]$ of real
assign
    $\langle\ \|\ n\ :\ 0 \le 2n \le 2^\ell M\ ::\ y[2n] := f[n]\ \rangle\ \|$
    $\langle\ \|\ n\ :\ 0 \le 2n+1 \le 2^\ell M\ ::\ y[2n+1] := (f[n] + f[n+1])/2\ \rangle$
end

On the coarsest grid, we use Jacobi relaxation until convergence, but any other solver could easily be substituted. As discussed in Section 9.1.5, under-relaxation is not appropriate for iteration until convergence, and we set $\omega = 1$ in program Coarsest-Grid-Solver-1. It is assumed that the number of iteration steps K is large, and a convergence test is omitted.

> **program** Coarsest-Grid-Solver-1(ℓ, u, f)
> **declare**
> k, ℓ, m : integer ;
> u, f : array$[0..2^{\ell}M]$ of real
> **assign**
> $\langle$; k : $0 \le k < K$ ::
> $\langle \parallel m$: $0 < m < 2^{\ell}M$ :: $u[m] := (h^2 f[m] + u[m+1] + u[m-1])/2 \rangle$
> $\rangle$
> **end**

In the multigrid procedure, the vectors $\vec{u}_0, \dots, \vec{u}_\ell, \dots, \vec{u}_L$ can be considered parts of one global data structure U, an array of arrays with varying dimensions. Such a data structure is declared with the statement

$$U : \text{array}[j \in 0..L] \text{ of array}[0..2^j M] \text{ of real}$$

The variable j in this declaration is used to specify how the second array dimension varies as a function of the first index; it is not a program variable. The array $U[\ell]$ represents the vector $\vec{u}_\ell$ of the more abstract formulation, and the entry $U[\ell][m]$ represents the m-th component of $\vec{u}_\ell$. Similar data structures represent the right-hand side and an auxiliary grid function. The auxiliary grid function Y simplifies the programs by providing space for the defect and the interpolated coarse-grid correction.

> **program** Multigrid-1(K, ℓ, U, F, Y)
> **declare**
> K, ℓ, k, m, n : integer ;
> U, F, Y : array$[j \in 0..L]$ of array$[0..2^j M]$ of real
> **assign**
>   **if** $\ell = 0$ **then** Coarsest-Grid-Solver-1$(0, U[0], F[0])$
>   **else**
> $\langle$; k : $0 \le k < K$ ::
>   Relax-1$(K_1, \ell, U[\ell], F[\ell])$;
>   Compute-Defect-1$(\ell, U[\ell], F[\ell], Y[\ell])$;
>   Restrict-1$(\ell, Y[\ell], F[\ell-1])$;
> $\langle \parallel n$: $0 \le n \le 2^{\ell-1}M$ :: $U[\ell-1][n] := 0 \rangle$;
>   Multigrid-1$(K_2, \ell-1, U, F, Y)$;
>   Interpolate-1$(\ell, U[\ell-1], Y[\ell])$;
> $\langle \parallel m$: $0 < m < 2^{\ell}M$ :: $U[\ell][m] := U[\ell][m] + Y[\ell][m] \rangle$
> $\rangle$
> **end**

When program Multigrid-1 is called for level ℓ, the array $F[\ell]$ must be initialized with the discretized right-hand-side function of the continuous problem. The entries of $U[\ell]$ that represent the Dirichlet-boundary conditions must be initialized with exact values. The entries of $U[\ell]$ that represent interior grid points must contain initial guesses. No initialization is required for the Y arrays or for the U and F arrays on levels below ℓ.

9.4.2 Multicomputer Relaxation

The multicomputer versions of programs Relax-1, Compute-Defect-1, and Coarsest-Grid-Solver-1 are an immediate application of Section 8.2.

On level ℓ, the index set $0..2^\ell M$ is distributed over P processes. The global indices mapped to process p are given by the range $M_p^\ell..M_{p+1}^\ell - 1$. The indices $M_p^\ell - 1$ and M_{p+1}^ℓ point to the left and right ghost boundaries, respectively. For all $\ell \in 0..L$, the indices M_p^ℓ satisfy:

$$1 = M_0^\ell < M_1^\ell < \ldots < M_p^\ell < \ldots < M_P^\ell = 2^\ell M. \qquad (9.23)$$

The global-index range $M_p^\ell..M_{p+1}^\ell - 1$ is mapped into the local-index range $0..I_p^\ell - 1$ with $I_p^\ell = M_{p+1}^\ell - M_p^\ell$. The local indices of left and right ghost boundaries are given by -1 and I_p^ℓ, respectively.

As in the two-dimensional case, the multicomputer version of Jacobi relaxation relies on a boundary exchange. Programs Boundary-Exchange-3 and Relax-3 are easily obtained.

```
0..P − 1 ∥ p program Boundary-Exchange-3(ℓ, u)
declare
    ℓ : integer ;
    u : array[−1..I_p^ℓ] of real
assign
    if p < P − 1 then send u[I_p^ℓ − 1] to p + 1 ;
    if p > 0 then receive u[−1] from p − 1 ;
    if p > 0 then send u[0] to p − 1 ;
    if p < P − 1 then receive u[I_p^ℓ] from p + 1
end
```

```
0..P − 1 ∥ p program Relax-3(K, ℓ, u, f)
declare
    K, k, ℓ, i : integer ;
    u, f : array[−1..I_p^ℓ] of real
assign
    ⟨ ; k : 0 ≤ k < K ::
        Boundary-Exchange-3(ℓ, u) ;
        ⟨ ∥ i : 0 ≤ i < I_p^ℓ :: u[i] := (h_ℓ^2 f[i] + u[i+1] + u[i−1] + u[i])/3 ⟩
    ⟩
end
```

The transformations of programs Compute-Defect-1 and Coarsest-Grid-Solver-1 into multicomputer programs Compute-Defect-3 and Coarsest-Grid-Solver-3, respectively, are similar. These multicomputer programs are not displayed. Throughout this chapter, program names with suffix "−3" are reserved for fully developed multicomputer programs. Program names with suffix "−2" are used for intermediate versions that use global indices.

9.4.3 Multicomputer Restriction

The key data-distribution issue of multicomputer multigrid methods is addressed in the intergrid transfers. The principal question is whether new communication operations need to be developed.

The level-ℓ data available in process p are entries with global index in the range $M_p^\ell-1..M_{p+1}^\ell$. This range includes the ghost boundaries, which are assumed initialized by a boundary exchange. In program Restrict-1, the computation of an entry of f on level $\ell-1$ with global index $n \in M_p^{\ell-1}..M_{p+1}^{\ell-1}-1$ (this range does not include the ghost boundaries) needs the values stored in $y[2n-1]$, $y[2n]$, and $y[2n+1]$. These values are available in process p if and only if

$$\forall n \in M_p^{\ell-1}..M_{p+1}^{\ell-1} - 1 :\ 2n-1, 2n, 2n+1 \in M_p^\ell - 1..M_{p+1}^\ell.$$

This is satisfied if

$$\forall p \in 0..P-1 : \left\{ \begin{array}{ll} 2M_p^{\ell-1} - 1 & \geq\ M_p^\ell - 1 \\ 2M_{p+1}^{\ell-1} - 1 & \leq\ M_{p+1}^\ell. \end{array} \right.$$

This set of conditions is equivalent to

$$\forall p \in 1..P-1 :\ M_p^{\ell-1} = \left\lceil \frac{M_p^\ell}{2} \right\rceil. \tag{9.24}$$

The data distribution of the finest grid is usually chosen to achieve load balance. Once all indices M_0^L, M_1^L, ..., and M_P^L are known, Equations 9.23 and 9.24 uniquely determine the data distribution of the coarser levels, say all indices M_p^ℓ, where $0 \leq \ell < L$ and $0 \leq p \leq P$.

For multilevel data distributions that satisfy Equation 9.24, we obtain program Restrict-2 with global indices.

$P-1 \parallel p$ **program** Restrict-2(ℓ, y, f)
declare
    ℓ, n : integer ;
    y : array$[M_p^\ell - 1..M_{p+1}^\ell]$ of real ;
    f : array$[M_p^{\ell-1} - 1..M_{p+1}^{\ell-1}]$ of real
assign
    Boundary-Exchange-2(ℓ, y) ;
    $\langle\ \parallel\ n\ :\ M_p^{\ell-1} \leq n < M_{p+1}^{\ell-1}\ ::\ f[n] := (y[2n-1]+y[2n]+y[2n+1])/4\ \rangle$
end

After a boundary exchange on the fine grid, the restriction can be computed on all coarse-grid points except those on ghost boundaries.

To introduce local indices, we need to derive the relation between the local indices of the fine grid and those of the coarse grid. For global indices, fine-grid index m and coarse-grid index n represent the same grid point if $m = 2n$. To find the correspondence in local indices, substitute

$$\begin{cases} m &= i + M_p^\ell \\ n &= j + M_p^{\ell-1} \end{cases}$$

into $m = 2n$, and find that local fine-grid index i and local coarse-grid index j represent the same grid point if

$$i = m - M_p^\ell = 2n - M_p^\ell = 2j + 2M_p^{\ell-1} - M_p^\ell = 2j + \delta_p^\ell.$$

Equation 9.24 implies that

$$\delta_p^\ell = 2M_p^{\ell-1} - M_p^\ell = \begin{cases} 0 & \text{if } M_p^\ell \text{ is even} \\ 1 & \text{if } M_p^\ell \text{ is odd.} \end{cases} \tag{9.25}$$

Program Restrict-3 is the resulting local-index version of program Restrict-2.

```
P − 1 ∥ p program Restrict-3(ℓ, y, f)
declare
    ℓ, j : integer ;
    y : array[−1..Iₚˡ] of real ;
    f : array[−1..Iₚˡ⁻¹] of real
assign
    Boundary-Exchange-3(ℓ, y) ;
    if δₚˡ = 0 then
        ⟨ ; j : 0 ≤ j < Iₚˡ⁻¹ ::
            f[j] := (y[2j − 1] + 2y[2j] + y[2j + 1])/4
        ⟩
    else
        ⟨ ; j : 0 ≤ j < Iₚˡ⁻¹ ::
            f[j] := (y[2j] + 2y[2j + 1] + y[2j + 2])/4
        ⟩
end
```

Within an individual multicomputer process, computations are sequential. Therefore, the concurrent was replaced by the sequential separator in the appropriate quantifications.

9.4.4 Multicomputer Interpolation

The multicomputer implementation of the linear-interpolation operator, which was specified by program Interpolate-1, must not impose any additional constraints on the data distribution, because the latter was completely specified by the implementation of the restriction operator. Fortunately, level ℓ quantities, including the ghost boundaries, only require level $\ell - 1$ quantities available in each process. The global-index version is given by program Interpolate-2.

$0..P - 1 \parallel p$ **program** Interpolate-2(ℓ, f, y)
declare
    ℓ, n : integer ;
    f : array$[M_p^{\ell-1} - 1..M_{p+1}^{\ell-1}]$ of real ;
    y : array$[M_p^{\ell} - 1..M_{p+1}^{\ell}]$ of real
assign
    Boundary-Exchange-2$(\ell - 1, f)$;
    $\langle \parallel n : M_p^{\ell}-1 \leq 2n \leq M_{p+1}^{\ell} :: y[2n] := f[n] \rangle \parallel$
    $\langle \parallel n : M_p^{\ell}-1 \leq 2n+1 \leq M_{p+1}^{\ell} :: y[2n+1] := (f[n]+f[n+1])/2 \rangle$
end

The local-index ranges for the two quantifications are obtained by substituting $n = j + M_p^{\ell-1}$ into the global-index ranges and by using Equations 9.24 and 9.25. For the first quantification, we obtain that

$$-\delta_p^{\ell} \leq j \leq I_p^{\ell-1} - \delta_{p+1}^{\ell}$$

and, for the second quantification, that

$$-1 \leq j < I_p^{\ell-1}.$$

After conversion of all global into local indices and converting concurrent into sequential quantifications, program Interpolate-3 is obtained.

$0..P - 1 \parallel p$ **program** Interpolate-3(ℓ, f, y)
declare
    ℓ, j : integer ;
    f : array$[-1..I_p^{\ell-1}]$ of real ;
    y : array$[-1..I_p^{\ell}]$ of real
assign
    Boundary-Exchange-3$(\ell - 1, f)$;
    if $\delta_p^{\ell} = 0$ **then**
        begin
            $\langle ; j : 0 \leq j \leq I_p^{\ell-1} - \delta_{p+1}^{\ell} :: y[2j] := f[j] \rangle$;
            $\langle ; j : -1 \leq j < I_p^{\ell-1} :: y[2j + 1] := (f[j] + f[j + 1])/2 \rangle$
        end
    else
        begin
            $\langle ; j : -1 \leq j \leq I_p^{\ell-1} - \delta_{p+1}^{\ell} :: y[2j + 1] := f[j] \rangle$;
            $\langle ; j : -1 \leq j < I_p^{\ell-1} :: y[2j + 2] := (f[j] + f[j + 1])/2 \rangle$
        end
end

9.4.5   Multicomputer Multigrid

To obtain multicomputer program Multigrid-3, assemble the parts.

$0..P-1 \parallel p$ **program** Multigrid-3(K, ℓ, U, F, Y)
declare
$\quad K,\ \ell,\ k,\ i,\ j\ :\ $ integer ;
$\quad U,\ F,\ Y\ :\ $ array$[j \in 0..L]$ of array$[-1..I_p^j]$ of real
assign
$\quad$ **if** $\ell = 0$ **then**
$\quad\quad$ Coarsest-Grid-Solver-3$(0, U[0], F[0])$
$\quad$ **else**
$\quad\quad \langle\ ;\ k\ :\ 0 \le k < K\ ::$
$\quad\quad\quad$ Relax-3$(K_1, \ell, U[\ell], F[\ell])$;
$\quad\quad\quad$ Compute-Defect-3$(\ell, U[\ell], F[\ell], Y[\ell])$;
$\quad\quad\quad$ Restrict-3$(\ell, Y[\ell], F[\ell - 1])$;
$\quad\quad\quad \langle\ ;\ j\ :\ -1 \le j \le I_p^{\ell-1}\ ::\ U[\ell - 1][j] := 0\ \rangle$;
$\quad\quad\quad$ Multigrid-3$(K_2, \ell - 1, U, F, Y)$;
$\quad\quad\quad$ Interpolate-3$(\ell, U[\ell - 1], Y[\ell])$;
$\quad\quad\quad \langle\ ;\ i\ :\ 0 \le i < I_p^{\ell}\ ::\ U[\ell][i] := U[\ell][i] + Y[\ell][i]\ \rangle$
$\quad\quad \rangle$
end

9.5   Notes and References

The multigrid idea was proposed by Brandt [11]. For an introductory tutorial to the multigrid method, consult Briggs [12]. Hackbusch and Trottenberg [43] and Hackbusch [42] offer more rigor, greater detail, and generality.

Recent numerical investigations into concurrent multigrid methods have centered around adapting the basic multigrid method to increase its concurrency. We mention just two of these efforts.

Frederickson and McBryan [33] introduce more than one coarse grid for each fine grid. In the coarse-grid correction, multiple coarse-grid solutions are combined, and the resulting multigrid method has an improved convergence factor over standard multigrid. By increasing the number of coarse grids, one also increases the amount of work per multigrid step. Is the improvement in the convergence factor worth the extra work? If the number of levels is large, computing time on the coarse levels can be dominated by the communication time. In this case, the extra arithmetic work contributes a negligible amount of computing time, and the improved convergence factor is likely to help. Whether or not multiple coarse grids are helpful depends on the actual improvement in convergence factor, the number of levels, the size of the finest grid, and computer-dependent characteristics. This leaves plenty of room for debate; see Decker [20].

Although the operation count of the coarse-grid computations is often negligible, they have a measurable impact on the efficiency of the multigrid algorithm, because coarse-grid computations are less efficient and because they form synchronization points. Hart and McCormick [45] proposed asynchronous multigrid methods. Here, all levels are active simultaneously, each level occasionally communicating with the next finest and the next coarsest level. The number of processes allocated to each level is proportional to the size of the problem. In fully asynchronous methods, the arithmetic depends on execution times of individual processes. This is an inherent computer dependency, which substantially complicates guaranteeing robustness of the method.

Exercises

Exercise 43 *Compute the convergence and smoothing factor of Jacobi relaxation for*

$$\forall x \in (0, 2\pi) : \quad -\frac{d^2 u}{dx^2} + cu = f(x)$$

with periodic-boundary conditions. What happens if $c = 0$?

Exercise 44 *Compute the convergence and smoothing factor of Jacobi relaxation for the Poisson problem with Dirichlet-boundary conditions in a two- and three-dimensional domain.*

Exercise 45 *Develop a multicomputer multigrid program on a $P \times Q$ process mesh for the two-dimensional Poisson problem with Dirichlet-boundary conditions. Analyze the performance of the program.*

10
Domain Decomposition

In the concurrent-computing literature, there is ambiguity over the use of the terms "domain decomposition" and "data distribution." Many authors use the term domain decomposition for data distributions in which the data local to each process topologically correspond to a subdomain of the computational domain. We shall not use the term domain decomposition in this sense. We reserve the term domain decomposition for certain numerical methods that split the computational domain into two or more subdomains. Some elementary domain-decomposition methods were already encountered in Section 8.5, where domain-decomposed relaxation methods were discussed.

Domain decomposition is a concept from numerical analysis. Although many domain-decomposition methods were developed for the purpose of achieving concurrency, it is reasonable to use them in sequential as well as in concurrent computations. Domain-decomposition issues are resolved when developing the numerical method and the specification program.

Data distribution, on the other hand, is a technique to derive multicomputer programs from specification programs. Data-distribution issues arise only if the specification program is transformed into a multicomputer program.

In part because of its promise for concurrent computing, domain decomposition is a very active area of research. We limit our study to the algorithmic aspects of just one approach to domain decomposition. As in previous chapters, we do this by developing a Poisson solver. The approach is applicable to more general problems, although significant mathematical obstacles arise.

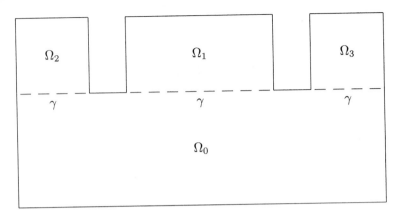

FIGURE 10.1. Irregular domain that is a composite of four regular subdomains.

10.1 Composite Domains

If an irregular domain is the union of regular subdomains, the Poisson problem on the composite domain can be reduced to a combination of several Poisson problems on the subdomains. In this fashion, one can avoid complicated irregular grids. The regular subproblems can be solved by any standard technique; in particular, fast Poisson solvers are applicable on the subdomains.

10.1.1 The Capacitance System

Consider Poisson problems of the form

$$\begin{cases} \forall (x,y) \in \Omega : & -\Delta u = f(x,y) \\ \forall (x,y) \in \partial\Omega : & u(x,y) = g(x,y) \end{cases} \qquad (10.1)$$

on a domain Ω that is the union of an arbitrary number of non-overlapping subdomains, say R subdomains Ω_r, where $0 \leq r < R$. An example of such a domain is pictured in Figure 10.1. A regular grid is introduced on each subdomain Ω_r. We make the simplifying assumption that grid points of two different subdomains coincide on shared subdomain boundaries. The discrete Poisson problem on the composite domain Ω is given by

$$A\vec{u} = \vec{f} - B\vec{g}, \qquad (10.2)$$

where the matrices A and B and the vectors $\vec{u}$, $\vec{f}$, and $\vec{g}$ have their usual interpretation. The matrix A represents the discrete Laplace operator on Ω. Although its structure is more complicated than before because of the irregularity of Ω, we may assume that A is a symmetric positive-definite matrix. In the context of domain-decomposition methods, the vector $\vec{g}$ is called an *exterior-boundary vector*.

The vector $\vec{u}$, which contains all unknowns on the composite grid, is split into $R+1$ subvectors. The first R subvectors $\vec{u}_r$ with $0 \leq r < R$ contain the interior unknowns of subdomain Ω_r; these vectors are called *interior vectors*. The subvector $\vec{u}_R$ is an *interior-boundary vector*, and its components are the unknowns on the interior boundary γ, which is the union of all subdomain boundaries inside Ω. In Figure 10.1, the interior boundary is represented by a collection of dashed lines. We split the matrices A and B and the vector $\vec{f}$ in a fashion compatible with the splitting of the vector $\vec{u}$. Because there is no direct coupling between interior unknowns of two different subdomains (between $\vec{u}_i$ and $\vec{u}_j$ with $i \neq j$ and $0 \leq i, j < R$), the structure of the discrete composite-domain problem is given by

$$
\begin{bmatrix}
A_{0,0} & 0 & 0 & & A_{0,R} \\
0 & A_{1,1} & 0 & & A_{1,R} \\
0 & 0 & A_{2,2} & & A_{2,R} \\
& & & \ddots & \vdots \\
A_{0,R}^T & A_{1,R}^T & A_{2,R}^T & \cdots & A_{R,R}
\end{bmatrix}
\begin{bmatrix}
\vec{u}_0 \\ \vec{u}_1 \\ \vec{u}_2 \\ \vdots \\ \vec{u}_R
\end{bmatrix}
=
\begin{bmatrix}
\vec{f}_0 \\ \vec{f}_1 \\ \vec{f}_2 \\ \vdots \\ \vec{f}_R
\end{bmatrix}
-
\begin{bmatrix}
B_0 \\ B_1 \\ B_2 \\ \vdots \\ B_R
\end{bmatrix}
\vec{g}. \quad (10.3)
$$

Except for row and column permutations, Equations 10.2 and 10.3 are identical. However, we shall be able to exploit the special structure exhibited by the more detailed form of Equation 10.3.

The substructures of Equation 10.3 have an interpretation in terms of subdomain problems. For $r \in 0..R - 1$, the symmetric positive-definite submatrix $A_{r,r}$ represents the discrete Laplace operator on subdomain Ω_r. The vector $\vec{f}_r$ is a discretization of the right-hand-side function $f(x, y)$ restricted to Ω_r. Therefore, the subsystem

$$
A_{r,r}\vec{u}_r = \vec{f}_r - B_r\vec{g} - A_{r,R}\vec{u}_R \quad (10.4)
$$

is the matrix formulation of a discrete Poisson problem on Ω_r with Dirichlet-boundary conditions. Its complicated structure arises, because the boundary values for Ω_r are a subset of the components of $\vec{g}$ and $\vec{u}_R$, vectors which contain the boundary values for all subdomains.

For now, assume that $\vec{u}_R$ is a known quantity. The corrections on the right-hand side of the equation arise from the boundary conditions. The two terms $-B_r\vec{g}$ and $-A_{r,R}\vec{u}_R$ are due to the exterior- and interior-boundary conditions, respectively. The exterior-boundary vector $\vec{g}$ defines the discrete Dirichlet-boundary conditions on the exterior boundary of the composite problem. If part of the boundary of Ω_r is an exterior boundary, then some columns of B_r do not vanish. The components of $\vec{g}$ that correspond to nonvanishing columns of B_r define the Dirichlet conditions on the exterior part of the boundary of Ω_r. Similarly, the interior-boundary vector $\vec{u}_R$ defines values on the interior boundary. Components of $\vec{u}_R$ that correspond to nonvanishing columns of $A_{r,R}$ define the Dirichlet-boundary conditions on the interior part of the boundary of Ω_r. One must keep in mind that this complicated structure is an artifact of the matrix formulation, and that Equation 10.4 can be solved with any available Poisson solver.

Of course, one must know the interior-boundary vector $\vec{u}_R$ in order to solve Equation 10.4. We obtain a reduced system of equations in $\vec{u}_R$ from Equation 10.3 by eliminating the unknown vectors $\vec{u}_0$ through $\vec{u}_{R-1}$. Substitute

$$\vec{u}_r = A_{r,r}^{-1}(\vec{f}_r - B_r\vec{g} - A_{r,R}\vec{u}_R)$$

into block number R of Equation 10.3 to obtain the *capacitance system*

$$C\vec{u}_R = \vec{s}_R, \tag{10.5}$$

with the *capacitance matrix*

$$C = A_{R,R} - \sum_{r=0}^{R-1} A_{r,R}^T A_{r,r}^{-1} A_{r,R} \tag{10.6}$$

as coefficient matrix and with right-hand-side vector given by

$$\vec{s}_R = \vec{f}_R - B_R\vec{g} - \sum_{r=0}^{R-1} A_{r,R}^T \vec{v}_r, \tag{10.7}$$

where

$$\forall r \in 0..R - 1 : \quad A_{r,r}\vec{v}_r = \vec{f}_r - B_r\vec{g}. \tag{10.8}$$

Comparison of these systems and Equation 10.4 shows that $\vec{v}_r$ solves the discrete Poisson problem on Ω_r with homogeneous interior-boundary conditions and exterior-boundary conditions given by the components of $\vec{g}$.

The proposed domain-decomposition method proceeds as follows. First, the capacitance system is solved to find the interior-boundary vector $\vec{u}_R$. With known interior-boundary conditions, the R systems defined by Equation 10.4 can be solved. The solution of the composite problem is then given by the combination of the subdomain solutions.

10.1.2   Capacitance-System Solvers

The proposed domain-decomposition method is feasible only if an effective solver for the capacitance system can be developed.

Because the dimension of the capacitance system is much less than that of the original problem, one possible route is to use direct solution methods. These require that the capacitance matrix C and the right-hand-side vector $\vec{s}_R$ be computed explicitly. To compute $\vec{s}_R$, one must solve the R Poisson problems of Equation 10.8. To compute C, one must solve a Poisson problem on Ω_r for each nonzero column of $A_{r,R}$ or, in other words, one Poisson problem for each interior-boundary point on the subdomain boundary $\partial\Omega_r$.

The explicit calculation of C and $\vec{s}_R$ is avoided by using the conjugate-gradient method to solve the capacitance system. The conjugate-gradient method is applicable, because the capacitance matrix C is a symmetric positive-definite matrix; see Exercise 46. To achieve an acceptable convergence rate, a preconditioner is usually required. Because the development of good preconditioners for the capacitance matrix is outside the scope of this text, we shall use the basic conjugate-gradient method. Program Conjugate-Gradient, which was derived in Section 3.3.4, is displayed with line numbers for use in the discussion that follows.

```
0    program Conjugate-Gradient
1    declare
2        k  :  integer ;
3        ξ, β, π⁰ᵣᵣ, π¹ᵣᵣ, πₚw  :  real
4        p⃗, r⃗, w⃗, x⃗  :  vector ;
5    assign
6        r⃗ := r⃗ − Ax⃗ ;
7        p⃗ := r⃗ ;
8        π⁰ᵣᵣ := r⃗ᵀr⃗ ;
9        ⟨ ; k  :  0 ≤ k < K  ::
10            w⃗ := Ap⃗ ;
11            πₚw := p⃗ᵀw⃗ ;
12            ξ := π⁰ᵣᵣ/πₚw ;
13            x⃗ := x⃗ + ξp⃗ ;
14            r⃗ := r⃗ − ξw⃗ ;
15            π¹ᵣᵣ := r⃗ᵀr⃗ ;
16            β := π¹ᵣᵣ/π⁰ᵣᵣ ;
17            p⃗ := r⃗ + βp⃗ ;
18            π⁰ᵣᵣ := π¹ᵣᵣ
19        ⟩
20    end
```

When applied to the capacitance system, the four vectors $\vec{x}$, $\vec{p}$, $\vec{r}$, and $\vec{w}$ of program Conjugate-Gradient are interior-boundary vectors, and the matrix A of the program is the capacitance matrix. The key problems are the computation of the residual (line 6) and of the matrix–vector product (line 10).

The residual of the capacitance system for a solution estimate $\vec{v}_R$ follows from Equations 10.6, 10.7, and 10.8. It is given by

$$\vec{s}_R - C\vec{v}_R = \vec{f}_R - B_R\vec{g} - A_{R,R}\vec{v}_R - \sum_{r=0}^{R-1} A_{r,R}^T A_{r,r}^{-1}(\vec{f}_r - B_r\vec{g} - A_{r,R}\vec{v}_R).$$

As a first step in the computation of this residual, solve the systems

$$\forall r \in 0..R-1 :\ A_{r,r}\vec{v}_r = \vec{f}_r - B_r\vec{g} - A_{r,R}\vec{v}_R. \qquad (10.9)$$

Of the same form as Equation 10.4, these systems are discrete Poisson problems on the subdomains, for which solvers are available. It remains to compute the sum

$$\vec{f}_R - B_R\vec{g} - A_{R,R}\vec{v}_R - \sum_{r=0}^{R-1} A_{r,R}^T\vec{v}_r. \tag{10.10}$$

This is the residual vector of block number R of Equation 10.3 for the solution estimate defined by $\vec{v}_0$ through $\vec{v}_R$. The components of Equation 10.10 are, therefore, residuals of the discrete Poisson problem at interior-boundary grid points.

Based on Equation 10.6, the matrix–vector product of the capacitance matrix and an interior-boundary vector expands to:

$$C\vec{v}_R = A_{R,R}\vec{v}_R - \sum_{r=0}^{R-1} A_{r,R}^T A_{r,r}^{-1} A_{r,R}\vec{v}_R.$$

It requires the solution of the R subdomain problems

$$\forall r \in 0..R - 1 : \ A_{r,r}\vec{v}_r = -A_{r,R}\vec{v}_R, \tag{10.11}$$

each being the matrix formulation of the Laplace equation on Ω_r with interior-boundary conditions given by $\vec{v}_R$ and exterior-boundary conditions that vanish. Again, the vectors $\vec{v}_r$ are easily computed, and the sum

$$A_{R,R}\vec{v}_R + \sum_{r=0}^{R-1} A_{r,R}^T\vec{v}_r \tag{10.12}$$

is block number R of Equation 10.3. It is, therefore, the discrete Laplace operator on Ω applied to the grid points of the interior boundary.

10.2 Domain Decomposition for Concurrency

For the composite-domain problem, domain decomposition was introduced, because it is easier and/or faster to solve problems on regular domains than problems on irregular domains. Other applications of domain decomposition include discontinuous-coefficient problems and embedding of irregular into regular domains. These examples show that there can be compelling numerical reasons to use domain-decomposition methods, whether one considers sequential or concurrent implementation. *The interesting concurrent-computing question is whether domain decomposition is an effective tool to introduce concurrency into a computation, even when there is no a priori numerical motivation to use domain decomposition.* To examine this question, we study in detail the application of domain decomposition to the Poisson problem on a rectangular domain.

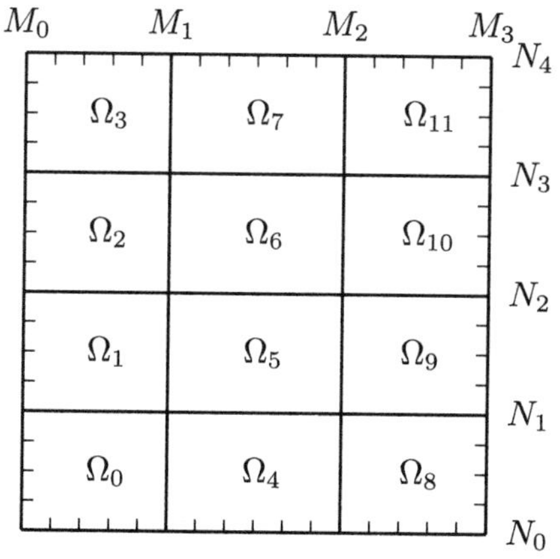

FIGURE 10.2. Regular domain decomposition into 3×4 subdomains on an underlying 16×16 grid.

The composite-grid example of Section 10.1 splits one Poisson problem on Ω into a combination of independent Poisson problems on R subdomains Ω_0 through Ω_{R-1}. Domain decomposition and concurrency are linked by the observation that the subdomain problems may be solved concurrently. To explore this possibility, we derive a multicomputer Poisson solver in which each process is responsible to solve Poisson problems on one subdomain. This one-to-one correspondence between processes and subdomains is by no means the only possible route to achieve concurrency.

10.2.1 Domain Decomposition

The Poisson problem with Dirichlet-boundary conditions, Equation 10.1, on the two-dimensional rectangular domain $\Omega = (0, \pi) \times (0, \pi)$ is discretized on an $M \times N$ regular grid. We always choose $M = N$ but retain the two symbols M and N for the number of grid cells in the x- and y-direction, respectively. The grid cells are squares with sides of length $h = \pi/M$.

For a computation on a $P \times Q$ process grid, define a decomposition of the domain Ω into $R = PQ$ non-overlapping subdomains Ω_r, where $0 \le r < R$. For the remainder of this chapter, we choose $r = Qp + q$. This defines a one-to-one correspondence between the subdomains and the processes. Domain Ω_r is given by

$$\Omega_r = (M_p h, M_{p+1} h) \times (N_q h, N_{q+1} h),$$

where $0 = M_0 < M_1 < \ldots < M_P = M$ and $0 = N_0 < N_1 < \ldots < N_Q = N$. Such a domain decomposition is illustrated in Figure 10.2, where $12 = 3 \times 4$ subdomains are constructed on top of a 16×16 global grid. The indices M_0 through M_3 of this domain decomposition are given by 0, 5, 11, and 16, and N_0 through N_4 are given by 0, 4, 8, 12, and 16.

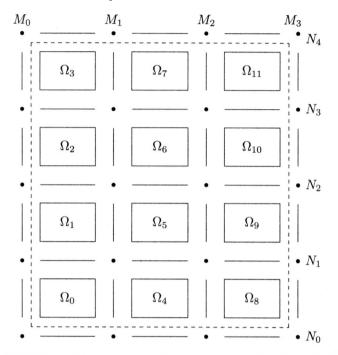

FIGURE 10.3. Grid decomposed into 3×4 nonoverlapping subgrids.

On each subdomain Ω_r, we use the local grid that is implied by the global $M \times N$ grid. This construction guarantees that grid points of two neighboring local grids coincide on shared subdomain boundaries.

As is usual, we represent the functions $u(x, y)$ and $f(x, y)$ of Equation 10.1 by their (approximate) values on grid points $(x_m, y_n) = (mh, nh)$, where $0 \leq m \leq M$ and $0 \leq n \leq N$. The declaration

$u, \; f \; : \; \text{array}[0..M \times 0..N]$ of real ;

represents all vectors $\vec{u}_0, \ldots, \vec{u}_{R-1}, \vec{u}_R$, and $\vec{g}$ and all vectors $\vec{f}_0, \ldots, \vec{f}_{R-1}$ and $\vec{f}_R$. This is illustrated in Figure 10.3. The interior grid points of sub-domains are represented by rectangles. They correspond to array entries $u[m, n]$ and $f[m, n]$ with indices that satisfy

$$\begin{cases} M_p & < & m & < & M_{p+1} \\ N_q & < & n & < & N_{q+1}. \end{cases}$$

These array entries store components of the interior vectors $\vec{u}_r$ and $\vec{f}_r$. Entries of u and f that correspond to the collection of lines and dots inside the dashed box store components of the interior-boundary vectors $\vec{u}_R$ and $\vec{f}_R$, respectively. Entries of u that correspond to lines and dots outside the dashed box store components of the exterior-boundary vector $\vec{g}$. Entries of f outside the dashed box are not used.

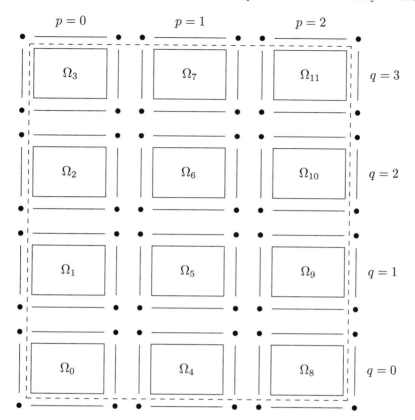

FIGURE 10.4. Grid decomposed into 3×4 nonoverlapping subgrids and distributed over a 3×4 process mesh.

The implementation of the conjugate-gradient method for the capacitance system must use this unconventional storage scheme of the interior-boundary vectors $\vec{u}_R$ and $\vec{f}_R$. For brevity, we skip the development of a specification program and proceed immediately to the construction of a multicomputer program.

10.2.2 Data Distribution

To solve discrete Poisson problems on Ω_r, where $r = Qp + q$, process (p, q) needs all data associated with grid points $(x_m, y_n) \in \Omega_r$. These must have indices $(m, n) \in M_p..M_{p+1} \times N_q..N_{q+1}$. It follows that grid line $m = M_{p+1}$ must be duplicated in processes (p, q) and $(p + 1, q)$. Similarly, grid line $n = N_{q+1}$ must be duplicated in processes (p, q) and $(p, q + 1)$. In Figure 10.4, the decomposed grid of Figure 10.3 is distributed over a 3×4 process grid. Again, the lines and dots outside the dashed box represent exterior-boundary vectors.

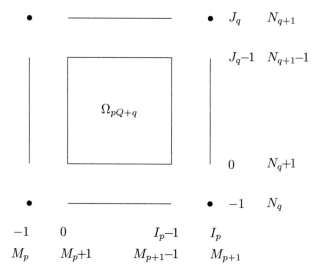

FIGURE 10.5. Local and global indices of the grid in process (p, q). This grid corresponds to domain Ω_r with $r = pQ + q$.

As is customary, the global indices (m, n) are mapped onto local indices $(i, j) \in -1..I_p \times -1..J_q$. In this case, the map between local and global indices is defined by

$$\begin{cases} i &=& m - M_p - 1 \\ j &=& n - N_q - 1, \end{cases}$$

and the ranges of local indices are determined by

$$\begin{cases} I_p &=& M_{p+1} - M_p - 1 \\ J_q &=& N_{q+1} - N_q - 1. \end{cases}$$

Figure 10.5 illustrates the data of process (p, q) with global and local indices. The variables $u[i, j]$ with $i = -1$, $i = I_p$, $j = -1$, or $j = J_q$ are represented by lines and dots and are called the ghost-boundary variables.

Figure 10.5 should be compared with Figure 8.3, which defines the data distribution used in the multicomputer programs for Jacobi and Gauss-Seidel relaxation. The two data distributions differ in their interpretation of the ghost boundaries. In Section 8.2.2, ghost boundaries contained duplicates of interior quantities. Ghost boundaries were mere message buffers for the boundary-exchange operation. In the present distribution, however, the ghost boundaries contain information not found anywhere else: without the ghost boundaries, it would be impossible to reconstruct the complete grid function. In spite of this significant difference, we shall retain the term ghost boundary. (Some duplication remains within the set of ghost boundaries, because ghost boundaries of neighboring processes are each other's duplicate.)

In multicomputer programs, two-dimensional arrays distributed over the $P \times Q$ process grid according to the data distribution of Figure 10.5 are declared by statements like

$$u, f \; : \; \text{array}[-1..I_p \times -1..J_q] \text{ of real} ;$$

Such arrays define all quantities on the computational grid.

For $r = Qp + q$, the interior vectors $\vec{u}_r$ and $\vec{f}_r$ are represented, respectively, by the variables $u[i, j]$ and $f[i, j]$ of process (p, q), where $0 \leq i < I_p$ and $0 \leq j < J_q$. The representation of the interior-boundary vector $\vec{u}_R$ is more complicated, because its components are distributed as well as partially duplicated. The set of all ghost-boundary variables of all processes contains all components of the interior-boundary vector. However, edge components (the lines in Figures 10.3, 10.4, and 10.5) are duplicated in up to two neighboring processes and corner components (the dots) in up to four processes. In boundary processes, one must exclude the exterior boundary, which is represented by the lines and dots outside the dashed box in Figures 10.3 and 10.4. In distributed array u, these excluded variables contain the components of the exterior-boundary vector $\vec{g}$.

10.2.3   Vector Operations

The vector operations used in lines 7, 8, 11, 13, 14, 15, and 17 of program Conjugate-Gradient act on interior-boundary vectors, which are stored in ghost boundaries of distributed arrays. Program Conjugate-Gradient requires four interior-boundary vectors: $\vec{p}$, $\vec{r}$, $\vec{w}$, and $\vec{x}$. In the ghost boundaries of the two distributed arrays u and f, on the other hand, there is room for only two interior-boundary vectors. To resolve this memory problem, we introduce a second ghost boundary around each distributed array. For reasons to be explained later, the array u actually needs a third ghost boundary. To avoid having more than one type of array, we declare both u and f with three ghost boundaries. This explains the declaration

$$u, f \; : \; \text{array}[-3..I_p + 2 \times -3..J_q + 2] \text{ of real} ;$$

For now, we wish to remain flexible about which interior-boundary vector is stored in which ghost boundary. Precise assignments will be made in Section 10.2.6. In programs, we shall identify ghost boundaries by means of a pair of variables: an array variable identifies the distributed array, and an integer, whose value is 0, 1, or 2, identifies the inner, middle, or outer ghost boundary, respectively.

For simplicity, the operations of this section are applied to the whole ghost boundary, not just to those segments that represent interior-boundary vectors. In Section 10.2.6, the ghost-boundary operations will be used to implement interior-boundary-vector operations. There, one must take appropriate steps to protect and/or to exclude the associated exterior-boundary vectors from the ghost-boundary operations.

Program Assign implements line 7 of program Conjugate-Gradient and copies ghost boundary g_v of array v to ghost boundary g_u of u. The values of g_u and g_v are either 0, 1, or 2.

$0..P - 1 \times 0..Q - 1 \parallel (p, q)$ **program** Assign(u, g_u, v, g_v)
declare

$\quad u, v \ : \ \text{array}[-3..I_p + 2 \times -3..J_q + 2] \text{ of real} \ ;$
$\quad g_u, u_n, u_e, u_s, u_w \ : \ \text{integer} \ ;$
$\quad g_v, v_n, v_e, v_s, v_w \ : \ \text{integer} \ ;$
$\quad i, j \ : \ \text{integer}$

assign

$\quad u_n, u_e, u_s, u_w := J_q + g_u, I_p + g_u, -1 - g_u, -1 - g_u \ ;$
$\quad v_n, v_e, v_s, v_w := J_q + g_v, I_p + g_v, -1 - g_v, -1 - g_v \ ;$
$\quad \langle \ ; \ i \ : \ 0 \le i < I_p \ :: \ u[i, u_s] := v[i, v_s] \ \rangle \ ;$
$\quad \langle \ ; \ i \ : \ 0 \le i < I_p \ :: \ u[i, u_n] := v[i, v_n] \ \rangle \ ;$
$\quad \langle \ ; \ j \ : \ 0 \le j < J_q \ :: \ u[u_w, j] := v[v_w, j] \ \rangle \ ;$
$\quad \langle \ ; \ j \ : \ 0 \le j < J_q \ :: \ u[u_e, j] := v[v_e, j] \ \rangle \ ;$
$\quad u[u_w, u_s] := v[v_w, v_s] \ ;$
$\quad u[u_w, u_n] := v[v_w, v_n] \ ;$
$\quad u[u_e, u_s] := v[v_e, v_s] \ ;$
$\quad u[u_e, u_n] := v[v_e, v_n]$

end

The indices u_n, u_e, u_s, and u_w are locations of the ghost-boundary segments of u. Those of v are identified analogously. Each quantification corresponds to a ghost edge (north, east, south, and west). The four assignments that follow correspond to ghost corners.

Program Vector-Sum adds a multiple of one ghost boundary to another and stores the result in a third. This operation is used to implement lines 13, 14, and 17 of program Conjugate-Gradient. Its calling sequence is

$0..P - 1 \times 0..Q - 1 \parallel (p, q)$ **program** Vector-Sum$(u, g_u, v, g_v, \alpha, w, g_w)$
declare

$\quad \alpha \ : \ \text{real} \ ;$
$\quad u, v, w \ : \ \text{array}[-3..I_p + 2 \times -3..J_q + 2] \text{ of real} \ ;$
$\quad g_u, g_v, g_w \ : \ \text{integer} \ ;$

The remainder of the program is similar to program Assign.

In program Inner-Product, each process computes the contributions of the north and east ghost boundaries to the inner product. These local contributions are summed by program Recursive-Double-Sum, which is a minor adaptation of program Recursive-Doubling-3 of Section 1.7 and is not displayed. To compute inner products of interior-boundary vectors, those segments of the ghost boundary that represent an exterior-boundary vector should be zero in either u or v.

$0..P - 1 \times 0..Q - 1 \parallel (p, q)$ **program** Inner-Product$(\pi_{uv}, u, g_u, v, g_v)$
declare
 π_{uv} : real ;
 $u,\ v$: array$[-3..I_p + 2 \times -3..J_q + 2]$ of real ;
 $g_u,\ u_n,\ u_e$: integer ;
 $g_v,\ v_n,\ v_e$: integer ;
 $i,\ j$: integer
assign
 $u_n, u_e := J_q + g_u, I_p + g_u$;
 $v_n, v_e := J_q + g_v, I_p + g_v$;
 $\pi_{uv} := u[u_e, u_n]v[v_e, v_n]$;
 $\pi_{uv} := \pi_{uv} + \langle\ +\ i\ :\ 0 \le i < I_p\ ::\ u[i, u_n]v[i, v_n]\ \rangle$;
 $\pi_{uv} := \pi_{uv} + \langle\ +\ j\ :\ 0 \le j < J_q\ ::\ u[u_e, j]v[v_e, j]\ \rangle$;
 Recursive-Double-Sum(π_{uv})
end

A ghost boundary represents an interior-boundary vector and an exterior-boundary vector. Operations on the interior-boundary vector must exclude the exterior-boundary vector. For this purpose, it is useful to have available program Zero-Exterior, which zeroes those ghost-boundary entries that correspond to the exterior boundary.

$0..P - 1 \times 0..Q - 1 \parallel (p, q)$ **program** Zero-Exterior(u, g_u)
declare
 u : array$[-3..I_p + 2 \times -3..J_q + 2]$ of real ;
 $g_u,\ u_n,\ u_e,\ u_s,\ u_w$: integer ;
 $i,\ j$: integer
assign
 $u_n, u_e, u_s, u_w := J_q + g_u, I_p + g_u, -1 - g_u, -1 - g_u$;
 if $q = 0$ **then**
  $\langle\ ;\ i\ :\ 0 \le i < I_p\ ::\ u[i, u_s] := 0\ \rangle$;
 if $p = 0$ **then**
  $\langle\ ;\ j\ :\ 0 \le j < J_q\ ::\ u[u_w, j] := 0\ \rangle$;
 if $q = Q - 1$ **then**
  $\langle\ ;\ i\ :\ 0 \le i < I_p\ ::\ u[i, u_n] := 0\ \rangle$;
 if $p = P - 1$ **then**
  $\langle\ ;\ j\ :\ 0 \le j < J_q\ ::\ u[u_e, j] := 0\ \rangle$;
  if $p = 0$ **and** $q = 0$ **then** $u[u_w, u_s] := 0$;
  if $p = 0$ **and** $q = Q - 1$ **then** $u[u_w, u_n] := 0$;
  if $p = P - 1$ **and** $q = 0$ **then** $u[u_e, u_s] := 0$;
  if $p = P - 1$ **and** $q = Q - 1$ **then** $u[u_e, u_n] := 0$
end

10.2.4   Sequential Poisson Solver

The data distribution of Section 10.2.2 was constructed such that the R discrete Poisson problems of Equation 10.9 can be solved independently, say the problem on Ω_r by process (p, q). The Poisson solver itself is not much of an issue: since the problem on Ω_r is defined on a regular grid and since all data associated with subdomain Ω_r is local to process (p, q), any sequential Poisson solver is applicable. It is even possible to use a different Poisson solver for each subdomain. For simplicity, we use a Gauss-Seidel-relaxation iteration with a fixed number of iteration steps for every subdomain. Program Poisson-Solver solves Poisson problems on all subdomains simultaneously; there is no communication between the processes. Program Poisson-Solver requires that the Dirichlet-boundary values of the subdomain problems be stored in the inner ghost boundary of u.

```
0..P − 1 × 0..Q − 1 ‖ (p, q) program Poisson-Solver(u, f)
declare
      u, f  :  array[−3..I_p + 2 × −3..J_q + 2] of real ;
      i, j, k  :  integer
assign
      ⟨ ; k  :  0 ≤ k < K ::
            ⟨ ; i,j  :  0 ≤ j < J_q and 0 ≤ i < I_p ::
                  u[i, j] := 0.25(h² f[i, j] + u[i − 1, j] + u[i + 1, j]
                                            + u[i, j − 1] + u[i, j + 1])
            ⟩
      ⟩
end
```

The discrete Laplace problems of Equation 10.11 are solved by program Laplace-Solver. This program is derived from program Poisson-Solver by setting all $f[i, j]$ to zero. It has the calling sequence

```
0..P − 1 × 0..Q − 1 ‖ (p, q) program Laplace-Solver(u)
declare
      u  :  array[−3..I_p + 2 × −3..J_q + 2] of real ;
```

10.2.5   Matrix Operations

Lines 6 and 10 of program Conjugate-Gradient involve a matrix–vector product of the capacitance matrix and an interior-boundary vector. The computation of the residual vector (line 6) begins with a ghost-boundary exchange. Here, values from neighboring processes are copied to the middle ghost boundary of u. Figure 10.6 illustrates which values are copied into which locations of process (p, q). In the figure, only the inner and middle ghost boundaries are drawn. A boundary-exchange operation overwrites all values stored in the middle ghost boundary. The code corresponding to Figure 10.6 is given by program Update-Ghost-Boundary.

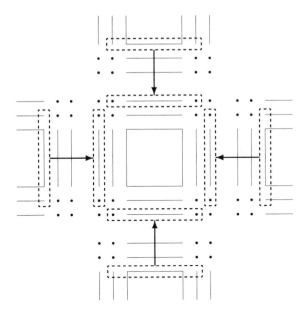

FIGURE 10.6. Boundary exchange for multicomputer Poisson solver based on domain decomposition.

$0..P - 1 \times 0..Q - 1 \parallel (p, q)$ **program** Update-Ghost-Boundary(u)
declare
      u : array$[-3..I_p + 2 \times -3..J_q + 2]$ of real ;
      i, j : integer
assign
      if $q < Q - 1$ **then**
          send $\{u[i, J_q - 1] : i \in -1..I_p\}$ **to** $(p, q + 1)$;
      if $q > 0$ **then**
          receive $\{u[i, -2] : i \in -1..I_p\}$ **from** $(p, q - 1)$;
      if $q > 0$ **then**
          send $\{u[i, 0] : i \in -1..I_p\}$ **to** $(p, q - 1)$;
      if $q < Q - 1$ **then**
          receive $\{u[i, J_q + 1] : i \in -1..I_p\}$ **from** $(p, q + 1)$;
      if $p < P - 1$ **then**
          send $\{u[I_p - 1, j] : j \in -1..J_q\}$ **to** $(p + 1, q)$;
      if $p > 0$ **then**
          receive $\{u[-2, j] : j \in -1..J_q\}$ **from** $(p - 1, q)$;
      if $p > 0$ **then**
          send $\{u[0, j] : j \in -1..J_q\}$ **to** $(p - 1, q)$;
      if $p < P - 1$ **then**
          receive $\{u[I_p + 1, j] : j \in -1..J_q\}$ **from** $(p + 1, q)$
end

After calling programs Poisson-Solver and Update-Ghost-Boundary, program Residual computes the expression of Equation 10.10 to complete the residual calculation. It is assumed that the inner ghost boundary of array u stores the exterior-boundary vector $\vec{g}$ and the interior-boundary vector $\vec{u}_R$. The array f is initialized with values of the right-hand-side function at all grid points.

The call to program Poisson-Solver computes the vectors $\vec{v}_r$ and stores them in the interiors of the distributed array u. The call to program Update-Ghost-Boundary makes available the neighboring values of ghost boundaries. Subsequently, the residuals are computed and stored in the inner ghost boundary of f. The residuals vanish on the exterior boundary, because the Dirichlet-boundary conditions give the exact values. Entries of f corresponding to exterior-boundary points are zeroed by the call to program Zero-Exterior.

$0..P-1 \times 0..Q-1 \parallel (p,q)$ **program** Residual(u,f)
declare
$\quad u,\ f\ :\ \text{array}[-3..I_p+2 \times -3..J_q+2]$ of real ;
$\quad i,\ j\ :\ \text{integer}$
assign
$\quad$ Poisson-Solver(u,f) ;
$\quad$ Update-Ghost-Boundary(u) ;
$\quad \langle\ ;\ i\ :\ -1 \leq i \leq I_p\ ::$
$\qquad f[i,-1] := f[i,-1] - h^{-2}(4u[i,-1]$
$\qquad\qquad\qquad\qquad\qquad\quad - u[i-1,-1] - u[i,-2]$
$\qquad\qquad\qquad\qquad\qquad\quad - u[i+1,-1] - u[i,0])$;
$\qquad f[i,J_q] := f[i,J_q] - h^{-2}(4u[i,J_q]$
$\qquad\qquad\qquad\qquad\qquad\quad - u[i-1,J_q] - u[i,J_q-1]$
$\qquad\qquad\qquad\qquad\qquad\quad - u[i+1,J_q] - u[i,J_q+1])$
$\quad \rangle\ ;$
$\quad \langle\ ;\ j\ :\ 0 \leq j < J_q\ ::$
$\qquad f[-1,j] := f[-1,j] - h^{-2}(4u[-1,j]$
$\qquad\qquad\qquad\qquad\qquad\quad - u[-1,j-1] - u[-2,j]$
$\qquad\qquad\qquad\qquad\qquad\quad - u[-1,j+1] - u[0,j])$;
$\qquad f[I_p,j] := f[I_p,j] - h^{-2}(4u[I_p,j]$
$\qquad\qquad\qquad\qquad\qquad\quad - u[I_p,j-1] - u[I_p-1,j]$
$\qquad\qquad\qquad\qquad\qquad\quad - u[I_p,j+1] - u[I_p+1,j])$
$\quad \rangle\ ;$
$\quad$ Zero-Exterior$(f,0)$
end

Line 9 of program Conjugate-Gradient implements Equation 10.12, which can be derived from Equation 10.10 by ignoring the terms in $\vec{f}_R$ and $\vec{g}$ and by performing a sign change. Program Apply-Capacitance-Matrix is thus a simplification of program Residual. It is assumed that the inner ghost boundary of u is initialized with zero exterior-boundary conditions and interior-boundary conditions given by $\vec{v}_R$. The computed vector sum, which corresponds to Equation 10.12, is stored in the inner ghost boundary of array f.

$0..P-1 \times 0..Q-1 \parallel (p,q)$ **program** Apply-Capacitance-Matrix(u,f)
declare
    $u,\ f\ :\ $array$[-3..I_p + 2 \times -3..J_q + 2]$ of real ;
    $i,\ j\ :\ $integer
assign
    Laplace-Solver(u) ;
    Update-Ghost-Boundary(u) ;
    $\langle\ ;\ i\ :\ -1 \leq i \leq I_p\ ::$
        $f[i, -1] := h^{-2}(4u[i, -1]$
                        $-\ u[i-1, -1] - u[i, -2]$
                        $-\ u[i+1, -1] - u[i, 0])$;
        $f[i, J_q] := h^{-2}(4u[i, J_q]$
                        $-\ u[i-1, J_q] - u[i, J_q - 1]$
                        $-\ u[i+1, J_q] - u[i, J_q + 1])$
    $\rangle$;
    $\langle\ ;\ j\ :\ 0 \leq j < J_q\ ::$
        $f[-1, j] := h^{-2}(4u[-1, j]$
                        $-\ u[-1, j-1] - u[-2, j]$
                        $-\ u[-1, j+1] - u[0, j])$;
        $f[I_p, j] := h^{-2}(4u[I_p, j]$
                        $-\ u[I_p, j-1] - u[I_p - 1, j]$
                        $-\ u[I_p, j+1] - u[I_p + 1, j])$
    $\rangle$;
    Zero-Exterior$(f, 0)$
end

When using programs Residual and Apply-Capacitance-Matrix, one must keep in mind that only the inner ghost boundary of u can be used to store the subdomain-boundary conditions and that values stored in the middle ghost boundary of u are lost due to the ghost-boundary exchange.

10.2.6 The Domain-Decomposition Program

Program Domain-Decomposition is essentially a line-by-line translation of program Conjugate-Gradient. The vectors $\vec{x}$, $\vec{p}$, $\vec{r}$, and $\vec{w}$ are stored in the ghost boundaries of the two distributed arrays u and f. In addition, program Domain-Decomposition includes the initialization of all quantities. Line numbers in program Domain-Decomposition refer to the corresponding lines in program Conjugate-Gradient.

```
0    0..P − 1 × 0..Q − 1 ∥ (p, q) program Domain-Decomposition
1    declare
2        k : integer ;
3        ξ, β, π⁰_rr, π¹_rr, π_pw : real
4        u, f : array[−3..I_p + 2 × −3..J_q + 2] of real ;
     initially
         ⟨ ; i, j : (i, j) ∈ −3..I_p + 2 × −3..J_q + 2 ::
             u[i, j], f[i, j] = 0, 0
         ⟩ ;
         ⟨ ; i, j : (i, j) ∈ −1..I_p × −1..J_q ::
             f[i, j] = f̃((i + M_p + 1)h, (j + N_q + 1)h) ;
             u[i, j] = ũ⁽⁰⁾_{i+M_p+1,j+N_q+1}
         ⟩ ;
         if q = 0 then
             ⟨ ; i : 0 ≤ i ≤ I_p :: u[i, −1] = g̃((i + M_p + 1)h, 0) ⟩ ;
         if q = Q − 1 then
             ⟨ ; i : 0 ≤ i ≤ I_p :: u[i, J_q] = g̃((i + M_p + 1)h, π) ⟩ ;
         if p = 0 then
             ⟨ ; j : 0 ≤ j ≤ J_q :: u[−1, j] = g̃(0, (j + N_q + 1)h) ⟩ ;
         if p = P − 1 then
             ⟨ ; j : 0 ≤ j ≤ J_q :: u[I_p, j] = g̃(π, (j + N_q + 1)h) ⟩
5    assign
6        Residual(u, f) ;
6a       Assign(u, 2, u, 0) ;
6b       Assign(f, 1, f, 0) ;
7        Assign(u, 0, f, 0) ;
8        Inner-Product(π⁰_rr, f, 0, f, 0) ;
9        ⟨ ; k : 0 ≤ k < K ::
10           Apply-Capacitance-Matrix(u, f) ;
11           Inner-Product(π_pw, u, 0, f, 0) ;
12           ξ := π⁰_rr/π_pw ;
13           Vector-Sum(u, 2, u, 2, ξ, u, 0) ;
14           Vector-Sum(f, 1, f, 1, −ξ, f, 0) ;
15           Inner-Product(π¹_rr, f, 1, f, 1) ;
16           β := π¹_rr/π⁰_rr ;
17           Vector-Sum(u, 0, f, 1, β, u, 0) ;
18           π⁰_rr := π¹_rr
19       ⟩ ;
19a      Assign(u,0,u,2) ;
19b      Poisson-Solver(u, f)
20   end
```

The first quantification of the **initially** section sets all entries of u and f equal to zero. The second quantification initializes the interior part and the inner ghost boundary of arrays f and u with values obtained from the right-hand-side function and with an initial guess, respectively. The part of the inner ghost boundary of u corresponding to exterior-boundary points is initialized with the Dirichlet-boundary values. The functions $\tilde{f}(x, y)$ and $\tilde{g}(x, y)$ refer, respectively, to the right-hand-side function and the function returning the Dirichlet-boundary values on the exterior boundary. The tilde notation avoids ambiguity with program variables.

In the following discussion of the **assign** section, the vectors $\vec{x}$, $\vec{r}$, $\vec{p}$, and $\vec{w}$ refer to the quantities of program Conjugate-Gradient. In program Domain-Decomposition, these quantities are stored in inner, middle, and outer ghost boundaries of arrays u and f.

In line 6, the residual vector $\vec{r}$ is computed and stored in the inner ghost boundary of f. In line 6a, the interior-boundary vector $\vec{x}$ and associated exterior-boundary vector $\vec{g}$, which were stored in the inner ghost boundary of u, are copied to the outer ghost boundary of u. In line 6b, the residual vector is copied into the middle ghost boundary of f. Note that the ghost-boundary entries corresponding to the exterior boundary were set equal to zero by program Residual. In line 7, the vector $\vec{p}$ is set equal to $\vec{r}$ and stored in the inner ghost boundary of u. Lines 7 through 19 are a line-by-line translation of program Conjugate-Gradient with the following assignments:

- $\vec{p}$ in the inner ghost boundary of u,

- $\vec{w}$ in the inner ghost boundary of f,

- $\vec{r}$ in the middle ghost boundary of f,

- $\vec{x}$ in the outer ghost boundary of u.

In line 19a, the outer ghost boundary of u, which now contains the interior-boundary vector $\vec{u}_R$ computed to a certain tolerance and the exterior-boundary vector $\vec{g}$, is copied back to the inner ghost boundary. In line 19b, the solution to the composite problem is computed using these boundary conditions.

A review of program Domain-Decomposition shows that u required three ghost boundaries, because the boundary exchange overwrites the middle ghost boundary. The outer ghost boundary of f is not used. However, if the capacitance-system solver were based on the preconditioned-conjugate-gradient method, this ghost boundary would be required for the vector $\vec{z}$ (see program Preconditioned-Conjugate-Gradient of Section 3.3.5).

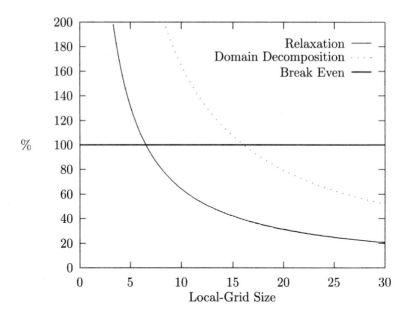

FIGURE 10.7. Ghost-boundary overhead as a function of local-grid size.

10.3   Performance Analysis

As discussed in Section 8.8, memory required for ghost boundaries often
cannot be neglected, because the area-perimeter law is often not valid.
This is particularly the case for three-dimensional problems. This problem is exacerbated here, because there are three ghost boundaries per grid
function. Figure 10.7 plots the ghost-boundary overhead as a function of
the local-grid size I. Ghost-boundary overhead is the memory required for
the ghost boundaries as a fraction of the memory required to represent a
grid function. This ratio is 100% at the break-even point. We assume that
our three-dimensional computation has a local grid consisting of $I \times I \times I$
grid points, not including any ghost boundaries. To compare relaxation
methods and the presented domain-decomposition method, we must make
two adjustments. As mentioned in Section 10.2.2, the ghost boundaries of
domain decomposition are more than message buffers; they contain information required to represent the grid functions. To compensate for this, we
count half of the inner ghost boundary as memory required to represent the
grid function. The other half of the inner ghost boundary and the middle
and outer ghost boundaries are counted as overhead. The formula for the
ghost-boundary overhead is then given by:

$$\frac{15I^2 + 66I + 86}{I^3 + 3I^2 + 6I + 4}.$$

In order to compare problems that have the same number of unknowns per process, one must compare a local $I \times I \times I$ grid of the domain-decomposition method with a local $J \times J \times J$ grid of relaxation, where

$$J^3 = I^3 + 3I^2 + 6I + 4.$$

The overhead ratio for relaxation method is then given by:

$$\frac{6J^2 + 12J + 8}{J^3}.$$

Figure 10.7 makes it obvious that ghost-boundary overhead for domain decomposition is significantly larger. Even more worrisome is that the break-even point for domain decomposition is around $I = 16$, which translates to more than 4,000 unknowns per process!

One conjugate-gradient-iteration step performs several vector operations and one matrix operation. The ghost-boundary overhead is directly related to the fraction of time spent on vector operations. If the computation is sufficiently coarse-grained, the execution time of one iteration step is dominated by the Poisson solver called when computing the matrix–vector product, and the execution time of the vector operations is negligible. In finer-grained computations, this may not be the case. The break-even point for the ghost-boundary overhead gives an indication of the dividing line between coarse- and fine-grained computations. In three dimensions, the vector operations are almost never negligible. This is a problem, because the multicomputer implementation of the vector operations is the least efficient: they duplicate most floating-point operations in at least two processes and, therefore, have an efficiency of at most 50%.

The ratio of communication over arithmetic time is small for the domain-decomposition method, since a boundary exchange occurs only between completed solution procedures on subdomain problems. Considerably more arithmetic is performed between two successive boundary-exchange operations in the domain-decomposition method than, say, in a relaxation method. This decreased ratio does not guarantee that the method is faster. Intuitively, one might even argue that it is a wasted effort to solve subdomain problems to high accuracy at the beginning of the conjugate-gradient iteration, when the Dirichlet-boundary values on the subdomains are still inaccurate. If one accepts this intuitive argument, then the decreased ratio is the result of performing superfluous floating-point operations, not the result of reducing communication overhead.

The total operation count depends on the number of conjugate-gradient-iteration steps until convergence and the number of floating-point operations per iteration step. The number of iteration steps is determined by the convergence rate, which depends on the condition of the capacitance matrix and on the preconditioner. A discussion of these factors is beyond the scope of this text.

It is an unresolved question as to which domain decomposition leads to an optimal operation count. The optimal number of subdomains is unknown, as is the configuration of such an optimal decomposition. Whatever the optimal decomposition is, it depends only on the partial-differential equation and its boundary conditions, not on the characteristics of the multicomputer that one happens to be using. In practical applications of domain decomposition as a tool to introduce concurrency, it is almost certain that the number and configuration of the subdomains is far from optimal.

Speed-up is often a misleading measure in a domain-decomposition context. Strictly speaking, a multicomputer-execution time obtained with R independent processes and R subdomains must be compared with the sequential-execution time of the same domain decomposition into R subdomains. Usually, however, one increases the number of subdomains together with the number of processes. This practice does not conform with the definition of speed-up and sometimes results in anomalous behavior. Suppose, for example, that a certain decomposition with R_0 subdomains minimizes the total operation count. As the number of processes and subdomains is increased from 1 to R_0, the total operation count decreases. If communication overhead is sufficiently small, one might actually observe superlinear speed-up. As is usual, the superlinear speed-up that is observed is due to a nonoptimal sequential computation: if one were to compare with a sequential computation based on the same domain decomposition of R_0 subdomains, the superlinear speed-up would disappear. Moreover, when increasing the number of processes and subdomains beyond the optimal number R_0 or when using another configuration of R_0 subdomains, the total operation count will increase and any performance advantage will be lost.

The domain-decomposition method we studied is significantly more difficult, numerically and algorithmically, than the multicomputer Poisson solvers obtained by pure data distribution. The convergence of the domain-decomposition method is difficult to understand and control, because it depends on the number of subdomains (processes) and on the configuration of the subdomains. When either are changed, one might have to switch preconditioners. Other computational characteristics depend on the number of processes. In coarse-grained computations, the cost of a conjugate-gradient-iteration step is dominated by the subdomain solver. In fine-grained computations, the subdomain problems are trivial (in the limit, only one grid point per subdomain), and the vector operations of the conjugate-gradient procedure dominate the computation. These practical problems severely constrain the application of this implementation of this domain-decomposition method as a tool to introduce concurrency.

As pointed out before, there are often compelling numerical arguments to use domain decomposition. As in Section 10.1, an irregular domain can be reduced to the union of several rectangular subdomains. An irregular domain can also be embedded into a rectangular one, which allows one to convert a solver for a rectangular domain into a solver for arbitrary geometry. Finally, some problems have interior boundaries in their formulation, which makes domain decomposition an obvious choice. In these numerical applications of domain decomposition, the number, size, and configuration of the subdomains are determined by the problem itself. When domain decomposition is used for the purpose of concurrency, on the other hand, the number of subdomains is determined by the number of processes, their size by load-balance requirements, and their configuration by communication-time considerations (communication is proportional to perimeter). Whether the numerical and concurrent applications of domain decomposition can be successfully combined remains a topic for further research.

10.4   Notes and References

The domain-decomposition method implemented and analyzed in this chapter was developed by Bjørstad and Widlund [9]. This is only one example of many existing domain-decomposition methods. The development of new domain-decomposition methods is a very active research area. Since there are no introductory texts, the proceedings of the international symposium on domain-decomposition methods for partial-differential equations [14] is a good starting point for further study of this topic.

Exercises

Exercise 46 *Prove that the capacitance matrix C of Equation 10.6 is a symmetric positive-definite matrix if A and $A_{r,r}$ for $r \in 0..R-1$ are symmetric positive-definite matrices.*

Exercise 47 *Develop a complete specification program for a Poisson problem on a composite domain consisting of two subdomains.*

Exercise 48 *Develop the specification program of Exercise 47 into a multicomputer program on a $P \times Q$ process mesh.*

Exercise 49 *Develop a domain-decomposed multicomputer Poisson solver based on the quasi-minimal-residual method.*

11
Particle Methods

The motion of stars in the universe is determined by a gravitational field, which itself depends on the position and mass of all stars. We shall use this interesting astrophysical problem, called *the N-body problem*, to introduce particle methods. Algorithmically, these methods are substantially different from grid-oriented computations for two reasons. First, grid operators are short-range operators: they act only on neighboring grid points. Standard particle methods, on the other hand, feature long-range interactions: all particles interact with all other particles. Second, the load balance of multicomputer computations based on particle methods depends on computed values. This contrasts with grid-oriented computations, where the load balance depends only on the amount of data. From our experience with LU-decomposition, which is based on long-range operators and whose load balance is data-dependent through pivoting, we already know that data-distribution strategies play an important role for particle methods on multicomputers. However, as for grid-oriented computations, the data distribution is also tied to the geometry of the problem.

The astrophysical N-body problem allows us to develop the main algorithmic ideas behind particle methods without requiring an extensive mathematical background. However, particle methods have uses far beyond this application. Fluid-dynamical calculations based on *vortex methods* are algorithmically similar to the particle methods of this chapter, but the particles in the computation are vectors. Combinations of grid and particle methods known as *particle-in-cell methods* are used in magneto-hydrodynamics.

The success of particle methods hinges on the ability to compute problems with many particles. It is, therefore, crucial to limit the growth of

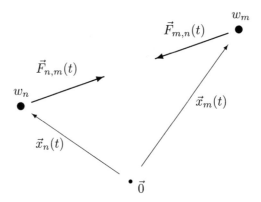

FIGURE 11.1. The field at $\vec{x}$ generated by a point mass of size w_m at $\vec{x}_m(t)$.

the operation count as the number of particles is increased. We are mainly interested in multicomputer implementations of particle methods with an acceptable operation count. We begin, however, with a study of the simplest algorithm.

11.1   The Astrophysical N-Body Problem

We use N bodies to model the universe. Each body is a point mass and represents a star or a group of stars. The time-dependent position of body number m, where $0 \leq m < N$, is given by $\vec{x}_m(t)$ and its mass by w_m. The force exerted by body number m on body number n, where $n \neq m$, follows from Newton's law of gravity:

$$\vec{F}_{n,m}(t) = -Gw_nw_m \frac{\vec{x}_n(t) - \vec{x}_m(t)}{\| \vec{x}_n(t) - \vec{x}_m(t) \|^3}.$$

Figure 11.1 is a graphical representation of this law. Of course, a force of equal magnitude in the opposite direction $\vec{F}_{m,n}(t) = -\vec{F}_{n,m}(t)$ is exerted by body number n on body number m. The acceleration of body number n is the sum of all its external forces divided by its mass:

$$\vec{a}_n(t) = \frac{1}{w_n} \sum_{m \neq n} \vec{F}_{n,m}(t)$$

$$= -G \sum_{m \neq n} w_m \frac{\vec{x}_n(t) - \vec{x}_m(t)}{\| \vec{x}_n(t) - \vec{x}_m(t) \|^3}. \tag{11.1}$$

Let $\vec{v}_n(t)$ be the velocity of body number n at time t. Then, we have that

$$\begin{cases} \vec{v}_n(t) = \vec{v}_n(0) + \int_0^t \vec{a}_n(\tau)\,d\tau \\ \vec{x}_n(t) = \vec{x}_n(0) + \int_0^t \vec{v}_n(\tau)\,d\tau. \end{cases} \tag{11.2}$$

Equations 11.1 and 11.2 are all we need to proceed. However, there is a somewhat hidden connection between this and previous chapters. Given a potential field $\phi(\vec{x}, t)$, the acceleration of a body at point $\vec{x}$ in space is given by

$$\vec{a}(\vec{x}, t) = -\nabla\phi(\vec{x}, t). \tag{11.3}$$

The differential formulation of Newton's law of gravity implies that the gravitational potential solves the Poisson equation

$$\Delta\phi = 4\pi G\rho(\vec{x}, t). \tag{11.4}$$

Here, the function $\rho(\vec{x}, t)$ is the mass density. In the astrophysical N-body problem, it is given by

$$\rho(\vec{x}, t) = \sum_{m=0}^{N-1} w_m \delta(\vec{x} - \vec{x}_m(t)),$$

where δ is the delta function. For mass densities of this form, Equation 11.4 can be solved explicitly, and

$$\phi(\vec{x}, t) = -G \sum_{m=0}^{N-1} \frac{w_m}{\| \vec{x} - \vec{x}_m(t) \|}. \tag{11.5}$$

Equations 11.3 and 11.5 imply Equation 11.1.

It follows from Equation 11.4 that we are developing yet another Poisson solver. The domain is unbounded, and the boundary conditions specify that the solution must vanish at infinity. Because the particles move to where the action is, particle methods are well suited for unbounded-domain problems. Grid-oriented methods need artificial boundary conditions at some finite boundaries that simulate true behavior at infinity. It should be noted that particle methods can be used for continuous mass densities, because it is possible to approximate (in a distribution sense) continuous functions by sums of delta functions.

11.2   The N^2 Algorithm

A simple time-step method, Euler's method, reduces the system consisting of Equations 11.1 and 11.2 to:

$$\begin{aligned}
\vec{a}_n(t) &= -G \sum_{m \neq n} w_m \frac{\vec{x}_n(t) - \vec{x}_m(t)}{\| \vec{x}_n(t) - \vec{x}_m(t) \|^3} \\
\vec{v}_n(t + \Delta t) &= v_n(t) + \Delta t \vec{a}_n(t) \\
\vec{x}_n(t + \Delta t) &= x_n(t) + \Delta t \vec{v}_n(t).
\end{aligned}$$

Euler's method is only first-order accurate in time. Only slightly more complicated, leap-frog time stepping achieves second-order accuracy. The particle positions and accelerations are computed at identical time levels, while the particle velocities are computed at time levels staggered by one-half time step:

$$\vec{a}_n(t) \;=\; -G \sum_{m \neq n} w_m \frac{\vec{x}_n(t) - \vec{x}_m(t)}{\| \vec{x}_n(t) - \vec{x}_m(t) \|^3} \tag{11.6}$$

$$\vec{v}_n(t + \tfrac{1}{2}\Delta t) \;=\; v_n(t - \tfrac{1}{2}\Delta t) + \Delta t \vec{a}_n(t) \tag{11.7}$$

$$\vec{x}_n(t + \Delta t) \;=\; x_n(t) + \Delta t \vec{v}_n(t + \tfrac{1}{2}\Delta t). \tag{11.8}$$

In the first time step, the velocity must be computed by means of a half-Euler step. Other higher-order schemes are often preferred or required, but the important algorithmic issues of particle methods can be addressed with this simple scheme.

In program N-Body-1, four arrays of length N represent the mass, the position, the velocity, and a quantity proportional to the acceleration of each body. Variables of type vector refer to quantities in $\mathbb{R}^\nu$. Standard vector operations are assumed to be defined.

```
program N-Body-1
declare
      m, n, k : integer ;
      w : array [0..N − 1] of real ;
      x⃗, v⃗, a⃗ : array [0..N − 1] of vector
assign
      { Half-Euler Time Step }
      ⟨ ∥ n : 0 ≤ n < N ::
          a⃗[n] := ⟨ + m : 0 ≤ m < N and m ≠ n ::
              w[m](x⃗[m] − x⃗[n])/ ∥ x⃗[m] − x⃗[n] ∥³
          ⟩
      ⟩ ;
      ⟨ ∥ n : 0 ≤ n < N :: v⃗[n] := v⃗[n] + a⃗[n]G½Δt ⟩ ;
      ⟨ ∥ n : 0 ≤ n < N :: x⃗[n] := x⃗[n] + v⃗[n]Δt ⟩ ;
      { Leap-Frog Time Stepping }
      ⟨ ; k : 1 ≤ k < K ::
          ⟨ ∥ n : 0 ≤ n < N ::
              a⃗[n] := ⟨ + m : 0 ≤ m < N and m ≠ n ::
                  w[m](x⃗[m] − x⃗[n])/ ∥ x⃗[m] − x⃗[n] ∥³
              ⟩
          ⟩ ;
          ⟨ ∥ n : 0 ≤ n < N :: v⃗[n] := v⃗[n] + a⃗[n]GΔt ⟩ ;
          ⟨ ∥ n : 0 ≤ n < N :: x⃗[n] := x⃗[n] + v⃗[n]Δt ⟩
      ⟩
end
```

Program N-Body-1 performs K time steps. Step number 0 (prior to the quantification over k) uses a half-Euler time step to set up the leap-frog time stepping. In the remainder of the discussion, we shall omit the half-Euler time step. To save floating-point multiplications, the accelerations are computed only up to the scalar factor G.

The main computational effort of program N-Body-1 occurs in the first concurrent quantification over n, where a sum of $N - 1$ terms is computed for each of the N bodies. The total number of floating-point operations is, therefore, proportional to $N(N - 1)$. The symmetry $\vec{F}_{m,n} = -\vec{F}_{n,m}$ can be used to halve the total number of floating-point operations, a small refinement left as an exercise.

The development of multicomputer program N-Body-2 is an immediate application of data distribution. All local arrays have dimension I, because one can always use a perfectly load-balanced linear data distribution by introducing bodies of zero mass such that $N = PI$.

$0..P - 1 \parallel p$ **program** N-Body-2
  declare
      i, j, k, q : integer ;
      w : array $[0..I - 1]$ of real ;
      $\vec{x}, \vec{x}_0, \vec{v}, \vec{a}$: array $[0..I - 1]$ of vector
  assign
      { Half-Euler Time Step Omitted }
      $\langle\, ; k : 1 \le k < K ::$
          $\langle\, ; j : 0 \le j < I :: \vec{a}[j] := \vec{0} \,\rangle$;
          $\langle\, ; j : 0 \le j < I :: \vec{x}_0[j] := \vec{x}[j] \,\rangle$;
          $\langle\, ; q : 0 \le q < P ::$
              send $\{\vec{x}_0[j], \vec{a}[j] : 0 \le j < I\}$ **to** $(p + 1) \bmod P$;
              receive $\{\vec{x}_0[j], \vec{a}[j] : 0 \le j < I\}$ **from** $(p - 1) \bmod P$;
              $\langle\, ; j : 0 \le j < I ::$
                  $\vec{a}[j] := \vec{a}[j] + \langle\, + i : 0 \le i < I ::$
                    $w[i](\vec{x}[i] - \vec{x}_0[j])/ \parallel \vec{x}[i] - \vec{x}_0[j] \parallel^3$
                $\rangle$
              $\rangle$
          $\rangle$;
          $\langle\, ; j : 0 \le j < I :: \vec{v}[j] := \vec{v}[j] + \vec{a}[j]G\Delta t \,\rangle$;
          $\langle\, ; j : 0 \le j < I :: \vec{x}[j] := \vec{x}[j] + \vec{v}[j]\Delta t \,\rangle$
      $\rangle$
  end

At every time step, the gravitational field is initialized to zero, and the current body locations are copied into a temporary array. In the quantification over q, the body positions and field vectors are sent from process to process to sum all contributions to the field from all bodies stored in every process. As in program N-Body-1, the last two quantifications over j update the positions and velocities.

The multicomputer efficiency of program N-Body-2 is excellent. Its load balance is almost perfect. Moreover, its communication time is a linear function of $I = N/P$ and its arithmetic time is proportional to I^2. Therefore, communication time is small compared with arithmetic time as long as $I \ll I^2$ or, equivalently, as long as $P \ll N$. We conclude that program N-Body-2 is well suited for coarse-grained computations.

11.3   Fast Particle Methods

The number of floating-point operations performed by programs N-Body-1 and N-Body-2 is a quadratic function of the number of bodies. This severely limits the maximum-feasible problem size on any computer, because doubling the number of bodies requires quadrupling the computing resources. Concurrency alone cannot solve this problem; we must reformulate the step responsible for the quadratic operation count.

The gravitational field far away from a cluster of bodies is well approximated by the gravitational field of a point mass located at the center of mass of the cluster. When trying to exploit this observation, two issues immediately arise: how to identify clusters and how to control the errors introduced by the point-mass approximation. We need a *clustering technique* and an *error-control technique*. We shall develop one clustering technique in detail. The goal of clustering is to replace a maximum number of body-body interactions by more efficient body-cluster interactions, thereby realizing a close-to-optimal operation count. For error control, we refer to the literature.

To reduce the notational burden, we shall use three new data structures, whose implementations are left unspecified. As displayed in Figure 11.2, a variable b of type *body* defines a body with mass $b.w$, location $b.\vec{x}$, velocity $b.\vec{v}$, and acceleration $b.\vec{a}$. (In fact, only a quantity proportional to the actual acceleration will be computed in $b.\vec{a}$.)

body	
w	real
$\vec{x}$	vector
$\vec{v}$	vector
$\vec{a}$	vector

FIGURE 11.2. The body data structure.

Variables of type *domain* will represent regions of space. Actual implementations are often restricted to rectangular domains, because it is difficult to represent general domains in terms of elementary data types. Although the actual representation of domains is not specified, the following operations involving domains are assumed available.

- The distance between a location, represented by a variable $\vec{x}$ of type vector, and a domain, represented by a variable $\mathcal{D}$ of type domain, is given by $d(\mathcal{D}, \vec{x})$.

- The statement

  if $\vec{x} \in \mathcal{D}$ **then** ...

  tests whether location $\vec{x}$ is inside domain $\mathcal{D}$.

- Any domain $\mathcal{D}$ can be split into a certain number of non-intersecting subdomains that cover $\mathcal{D}$. This is done by program Split-Domain with calling sequence

  program Split-Domain($\mathcal{D},S$)
  declare
      $\mathcal{D}$: domain ;
      S : set of domain

  For example, if $\mathcal{D}$ is a two-dimensional rectangular domain, program Split-Domain will split it into four quadrants. More detailed specifications of program Split-Domain depend on the actual representation of variables of type domain. The type 'set of domain' is a minor extension of previously encountered notation, since we have already frequently used sets of integers. As before, the cardinality operator # applied to set variables returns the number of elements in the set.

A variable c of type *cell* collects all bodies contained in a certain region of space into one data structure. As displayed in Figure 11.3, a cell c has mass $c.w$, location $c.\vec{x}$, domain $c.\mathcal{D}$, set of bodies $c.\mathcal{B}$, and set of cells $c.\mathcal{C}$. The domain $c.\mathcal{D}$ is a subdomain of the computational domain. The set of bodies $c.\mathcal{B}$ consists of all bodies inside the domain $c.\mathcal{D}$. The location $c.\vec{x}$ is the center of mass of the set of bodies $c.\mathcal{B}$. The mass $c.w$ is the sum of the masses of the bodies in $c.\mathcal{B}$.

cell	
w	real
$\vec{x}$	vector
$\mathcal{D}$	domain
$\mathcal{B} = \{b_0, b_1, \ldots\}$	set of body
$\mathcal{C} = \{c_0, c_1, \ldots\}$	set of cell

FIGURE 11.3. The cell data structure.

The set of cells $c.\mathcal{C}$ is the main element of the cell data structure. Self-referential data structures, which contain elements of their own type, are not immediately available in practical programming languages but can be

cell c

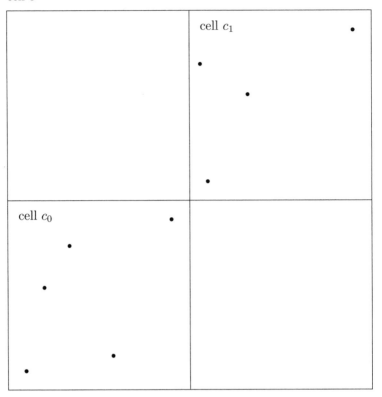

FIGURE 11.4. Clustering a cell.

implemented via pointers. We allow self-referential structures to avoid technicalities associated with pointer arithmetic. A cell defines a tree structure, and the set $c.\mathcal{C}$ contains all subcells of cell c. Cells without any subcells are called terminal cells.

With these new data structures in place, we develop a clustering technique. It is difficult to define "a best clustering," let alone compute it. All clustering algorithms compute "a reasonable clustering" based on heuristic rules. The *root cell* is defined over the computational domain $\mathbb{R}^{\nu}$ and contains all bodies of the computation. All other cells are, directly or indirectly, derived from the root cell. If a domain $c.\mathcal{D}$ contains more than N_b bodies, then it is split into a certain number of non-intersecting subdomains and a subcell is created for each subdomain that contains at least one body. In Figure 11.4, cell c is defined over a two-dimensional rectangular domain, which is split into four equal quadrants. In this case, two subcells are created, one for the lower-left and one for the upper-right quadrant.

Given a set of bodies $\mathcal{B}$ and a domain $\mathcal{D}$, program Cluster-Bodies recursively constructs the tree structure in cell c. When initially called, the domain $\mathcal{D}$ is the computational domain $\mathbb{R}^{\nu}$. In subsequent recursive calls, it is guaranteed that all bodies of $\mathcal{B}$ are located inside domain $\mathcal{D}$.

program Cluster-Bodies($\mathcal{D},\mathcal{B},c$)
declare
    $\mathcal{D}$, $\mathcal{D}_0$: domain ;
    $\mathcal{B}$, $\mathcal{B}_0$: set of body ;
    c, c_0 : cell ;
    b : body ;
    S : set of domain
assign
    $c.\mathcal{D}$, $c.\mathcal{B}$, $c.\mathcal{C} := \mathcal{D}$, $\mathcal{B}$, $\emptyset$;
    if $\#\mathcal{B} \leq N_b$ **then begin** { Terminal Cell }
        $c.w := \langle\, + b \,:\, b \in \mathcal{B} \,::\, b.w \,\rangle$;
        $c.\vec{x} := \langle\, + b \,:\, b \in \mathcal{B} \,::\, b.w\, b.\vec{x} \,\rangle / c.w$;
        return
    end ;
    { Nonterminal Cell }
    Split-Domain($\mathcal{D},S$) ;
    $\langle\, ; \mathcal{D}_0 \,:\, \mathcal{D}_0 \in S \,::$
        $\mathcal{B}_0 := \langle\, \bigcup\, b \,:\, b \in \mathcal{B} \text{ and } b.\vec{x} \in \mathcal{D}_0 \,::\, \{b\} \,\rangle$;
        if $\mathcal{B}_0 \neq \emptyset$ **then begin**
            Create-New-Cell(c_0) ;
            Cluster-Bodies($\mathcal{D}_0,\mathcal{B}_0,c_0$) ;
            $c.\mathcal{C} := c.\mathcal{C} \bigcup \{c_0\}$
        end
    $\rangle$;
    $c.w := \langle\, + c_0 \,:\, c_0 \in c.\mathcal{C} \,::\, c_0.w \,\rangle$;
    $c.\vec{x} := \langle\, + c_0 \,:\, c_0 \in c.\mathcal{C} \,::\, c_0.w\, c_0.\vec{x} \,\rangle / c.w$
end

If the number of elements of $\mathcal{B}$ is less than or equal to N_b, cell c is a terminal cell and the associated computations of mass and center of mass are trivial. Otherwise, the domain is split, and a set of bodies belonging to each subdomain is computed. For each subdomain with a nonempty set of bodies, a new cell is created with the unspecified program Create-New-Cell. A clustering tree is recursively computed for the newly created cell, which is then added to the set of subcells of c. Finally, the mass $c.w$ and the center of mass $c.\vec{x}$ is computed from the subcell quantities.

Given a clustering tree, program Compute-Field computes the gravitational field at any location $\vec{x}$ by descending recursively down the tree defined by the root cell.

program Compute-Field($c,\vec{x},\vec{a}$)
declare
    c, c_0 : cell ;
    $\vec{x}$, $\vec{a}$: vector ;
    b : body ;
    δ : real
assign
    Error-Control($\vec{x},c,\delta$) ;
    if $d(\vec{x}, c.\mathcal{D}) > \delta$ **then begin**
        $\vec{a} := \vec{a} + c.w(c.\vec{x} - \vec{x})/ \parallel c.\vec{x} - \vec{x} \parallel^3$;
        return
    end ;
    if $\#c.\mathcal{C} = 0$ **then begin**
        $\vec{a} := \vec{a}+\langle\ +\ b\ :\ b \in c.\mathcal{B}$ **and** $b.\vec{x} \neq \vec{x}$ ::
                         $b.w(b.\vec{x} - \vec{x})/ \parallel b.\vec{x} - \vec{x} \parallel^3$
                  $\rangle$;
        return
    end ;
    $\langle$; c_0 : $c_0 \in c.\mathcal{C}$ :: Compute-Field($c_0,\vec{x},\vec{a}$) $\rangle$
end

If location $\vec{x}$ is far away from cell c, the far-field approximation applies. For locations near a terminal cell, we use the N^2 algorithm. For locations near a nonterminal cell, the field is computed by summing the fields of the subcells. The error control is hidden in the unspecified program Error-Control, which computes the minimum distance δ for which the far-field approximation is sufficiently accurate.

Combining the above elements, we obtain program Fast-N-Body-1, which has an operation count of $O(N \log N)$ (under certain assumptions).

program Fast-N-Body-1($\mathcal{B}$)
declare
    $\mathcal{B}$: set of body ;
    k : integer ;
    c : cell ;
    b : body
assign
    { Half-Euler Time Step Omitted }
    $\langle$; k : $1 \leq k < K$ ::
        Cluster-Bodies($\mathbb{R}^\nu,\mathcal{B},c$) ;
        $\langle$; b : $b \in \mathcal{B}$ :: $b.\vec{a} := \vec{0}$ $\rangle$;
        $\langle$; b : $b \in \mathcal{B}$ :: Compute-Field($c,b.\vec{x},b.\vec{a}$) $\rangle$;
        $\langle$; b : $b \in \mathcal{B}$ :: $b.\vec{v} := b.\vec{v} + b.\vec{a}G\Delta t$ $\rangle$;
        $\langle$; b : $b \in \mathcal{B}$ :: $b.\vec{x} := b.\vec{x} + b.\vec{v}\Delta t$ $\rangle$
    $\rangle$
end

11.4   Concurrent Fast Particle Methods

For the fast method, the data distribution must be tied to the body locations, because a clustering algorithm must be applied to the distributed bodies. Through its impact on clustering, the data distribution may have a significant numerical effect. This is unlike the N^2 algorithm, where the data distribution leaves the numerical method intact, except for insignificant round-off errors.

Program Fast-N-Body-2 is a multicomputer implementation of program Fast-N-Body-1. The initial data distribution is arbitrary and is defined through the multicomputer parameter $\mathcal{B}$, which denotes one set of bodies per process. Occasionally, we shall stress that there are actually P sets of bodies by using an explicit process identifier on the variable $\mathcal{B}$, and we shall refer to the sets $\mathcal{B}[p]$, where $0 \leq p < P$. The computational domain $\mathbb{R}^\nu$ is covered by P non-intersecting subdomains $\mathcal{D}[p]$, and subdomain $\mathcal{D}[p]$ is mapped to process p. For now, assume that the subdomains are given.

```
0..P − 1 ∥ p program Fast-N-Body-2(B)
declare
      B  :  set of body ;
      D  :  domain ;
      k, q  :  integer ;
      c  :  cell ;
      b  :  body
assign
      { Half-Euler Time Step Omitted }
      〈 ; k : 1 ≤ k < K ::
            Redistribute-Data(D,B) ;
            Cluster-Bodies(D,B,c) ;
            〈 ; b : b ∈ B :: b.a⃗ := 0⃗ 〉 ;
            〈 ; q : 0 ≤ q < P ::
                  send B to (p + 1) mod P ;
                  receive B from (p − 1) mod P ;
                  〈 ; b : b ∈ B :: Compute-Field(c,b.x⃗,b.a⃗) 〉
            〉 ;
            〈 ; b : b ∈ B :: b.v⃗ := b.v⃗ + b.a⃗GΔt 〉 ;
            〈 ; b : b ∈ B :: b.x⃗ := b.x⃗ + b.v⃗Δt 〉
      〉
end
```

Programs Fast-N-Body-2 and N-Body-2 have a remarkably similar structure. In the quantification over q, all bodies are sent from process to process in order to sum local contributions to their gravitational field. These local contributions are computed in program Compute-Field, which uses a locally constructed clustering tree. Having computed the gravitational field at all body locations, the body velocities and locations are updated. Note that programs Cluster-Bodies and Compute-Field are not multicomputer programs; each process calls one independent instance of the sequential programs defined in Section 11.3.

The only new component is program Redistribute-Data. Here, the subdomains and associated sets of bodies are adapted to increase the load balance and to improve the clustering characteristics of the computation. Our first attempt to define a redistribution algorithm is given by program Redistribute-Bodies, which keeps all subdomains fixed. Bodies that moved out of subdomain $\mathcal{D}[p]$ are deleted from the set $\mathcal{B}[p]$, and bodies that moved into $\mathcal{D}[p]$ are added to the set $\mathcal{B}[p]$.

> $0..P-1 \parallel p$ **program** Redistribute-Bodies($\mathcal{D}$,$\mathcal{B}$)
> **declare**
> > $\mathcal{D}$: domain ;
> > $\mathcal{B}$, $\mathcal{B}_0$, $\mathcal{B}_1$: set of body ;
> > q : integer ;
> > b : body
> **assign**
> > $\mathcal{B}_0 := \langle\, \bigcup b \;:\; b \in \mathcal{B}$ **and** $b.\vec{x} \notin \mathcal{D} \;::\; \{b\} \,\rangle$;
> > $\mathcal{B} := \mathcal{B} \setminus \mathcal{B}_0$;
> > $\langle\; ;\; q \;:\; 0 \leq q < P \;::$
> > > **send** $\mathcal{B}_0$ **to** $(p+1) \bmod P$;
> > > **receive** $\mathcal{B}_0$ **from** $(p-1) \bmod P$;
> > > $\mathcal{B}_1 := \langle\, \bigcup b \;:\; b \in \mathcal{B}_0$ **and** $b.\vec{x} \in \mathcal{D} \;::\; \{b\} \,\rangle$;
> > > $\mathcal{B}_0 := \mathcal{B}_0 \setminus \mathcal{B}_1$;
> > > $\mathcal{B} := \mathcal{B} \bigcup \mathcal{B}_1$
> > $\rangle$
> **end**

The first assignment of program Redistribute-Bodies collects in the set $\mathcal{B}_0$ those bodies of $\mathcal{B}$ that moved out of domain $\mathcal{D}$. These bodies are then deleted from $\mathcal{B}$ and sent around all other processes to be added to the appropriate set of bodies.

Sometimes, the communication in program Redistribute-Bodies can be reduced. If the time step is sufficiently small, bodies move only between neighboring domains. In this case, the communication can be reduced to a boundary-exchange operation as in grid-oriented computations. We do not investigate this aspect further, because we wish to correct the more serious problem of program Redistribute-Bodies: bad load balance. With fixed subdomains, particle methods are almost always badly load-balanced, because bodies tend to cluster in a few narrow areas of the computational domain.

For the bodies to remain evenly distributed over the domains and, hence, over the processes, the domains $\mathcal{D}[p]$ must be recomputed. This is done in program Redistribute-Data, which is an implementation of the *orthogonal-recursive-bisection technique*. This method has been shown to achieve a good compromise between the competing demands of load balance, clustering properties, and time required to compute the subdomains.

$0..P-1 \parallel p$ **program** Redistribute-Data$(\mathcal{D},\mathcal{B})$
declare
$\qquad \mathcal{D}, \mathcal{D}_0, \mathcal{D}_1$: domain ;
$\qquad \mathcal{B}, \mathcal{B}_0$: set of body ;
$\qquad d$: integer
assign
$\qquad \mathcal{D} := \mathbb{R}^{\nu}$;
$\qquad \langle \ ; \ d \ : \ 0 \le d < D \ ::$
$\qquad\qquad$ Half-Space$(d,\mathcal{D},\mathcal{B},\mathcal{D}_0,\mathcal{D}_1)$;
$\qquad\qquad$ **if** $p \bar{\vee} 2^d = 0$ **then** $\mathcal{D} := \mathcal{D}_0$ **else** $\mathcal{D} := \mathcal{D}_1$;
$\qquad\qquad \mathcal{B}_0 := \langle \bigcup b \ : \ b \in \mathcal{B} \text{ and } b.\vec{x} \notin \mathcal{D} \ :: \ \{b\} \rangle$;
$\qquad\qquad \mathcal{B} := \mathcal{B} \setminus \mathcal{B}_0$;
$\qquad\qquad$ **send** $\mathcal{B}_0$ **to** $p \bar{\vee} 2^d$;
$\qquad\qquad$ **receive** $\mathcal{B}_0$ **from** $p \bar{\vee} 2^d$;
$\qquad\qquad \mathcal{B} := \mathcal{B} \bigcup \mathcal{B}_0$
$\qquad \rangle$
end

Initially, $\mathcal{D} = \mathbb{R}^{\nu}$ in all processes. Program Redistribute-Data uses recursive doubling to split domain $\mathcal{D}$ into $P = 2^D$ subdomains. In every step, domain $\mathcal{D}$ is split into two halves $\mathcal{D}_0$ and $\mathcal{D}_1$, each containing approximately the same number of bodies, until there is one domain per process. Program Half-Space, which will be developed later, computes the actual split of domain $\mathcal{D}$. Half the processes assigned to domain $\mathcal{D}$ casts off $\mathcal{D}_0$, the other half casts off $\mathcal{D}_1$. In typical recursive-doubling fashion, we decide on the basis of bit number d of p $(p \bar{\vee} 2^d)$ whether to cast off $\mathcal{D}_0$ or $\mathcal{D}_1$. All bodies located in the subdomain being cast off by process p are sent to process $p \bar{\vee} 2^d$. Because of recursive doubling, we know that process $p \bar{\vee} 2^d$ keeps the subdomain cast off by process p. The bodies cast off by process p are collected in the set $\mathcal{B}_0$, deleted from $\mathcal{B}$, and sent to process $p \bar{\vee} 2^d$. The bodies cast off by process $p \bar{\vee} 2^d$ are received by process p and added to the set $\mathcal{B}$. After D steps, each process is assigned to its own subdomain and its own set of bodies.

Program Half-Space computes a split of domain $\mathcal{D}$ into two subdomains $\mathcal{D}_0$ and $\mathcal{D}_1$ that contain an approximately equal number of bodies. A split is defined by a hyperplane. Here, we only consider hyperplanes orthogonal to one of the coordinate axes, because they are defined by simple equations of the form $x = x_s$, $y = y_s$, or $z = z_s$. In Figure 11.5, a split orthogonal to the x-axis is followed by a split orthogonal to the y-axis. The coordinate axes of successive splits are always alternated to avoid subdomains shaped like narrow strips, which would complicate the clustering.

The most interesting part of computing a split is the computation of the median of a set of coordinate values. In particular, to perform a split

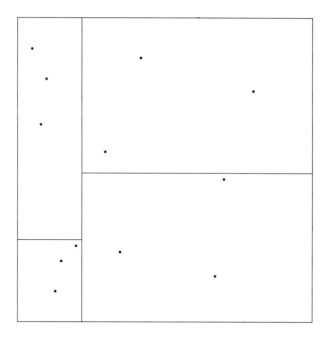

FIGURE 11.5. A domain split orthogonal to the x-axis followed by a domain split orthogonal to the y-axis.

orthogonal to the x-axis, we need the median x_s of the x-coordinates of a set of bodies. The parameters passed to program Median-Value are the step d of the recursive-doubling procedure and the set $\mathcal{X}$ of x-, y-, or z-coordinates of the bodies of $\mathcal{B}$ (for splits orthogonal to x-, y-, and z-axis, respectively). The computed median μ is a real number that is returned by program Median-Value to program Half-Space.

The recursive-doubling procedure of program Redistribute-Data implies that processes p and q are assigned to the same domain at the start of step d if bits number 0 through $d - 1$ of p and q are identical. It follows that, at the start of step d, all processes q with

$$q \in \{(p + k2^d) \bmod P \; : \; 0 \leq k < 2^{D-d}\}$$

are assigned to the same domain as process p. The bodies belonging to this domain are already distributed over the same set of processes. When computing the split of step number d in process p, we need the median of the set:

$$\bigcup_{k=0}^{2^{D-d}-1} \mathcal{X}[(p + k2^d) \bmod P].$$

The computation of the exact median is left as an exercise. Instead, program Median-Value computes a reasonable approximation sufficient for our purposes.

$0..P-1 \parallel p$ **program** Median-Value$(d,\mathcal{X},\mu)$
declare
    $\mathcal{X}$: set of real ;
    μ, μ_0 : real ;
    $d, e, q, \delta, \delta_0$: integer
assign
    $\mu :=$ random element of $\mathcal{X}$;
    $\delta := 0$;
    $\langle$; q : $0 \leq q < 2^{D-d}$ ::
        $\delta := \delta + \langle + x : x \in \mathcal{X} :: \mathrm{sign}(x - \mu) \rangle$;
        send $\{\mu, \delta\}$ **to** $(p + 2^d) \bmod P$;
        receive $\{\mu, \delta\}$ **from** $(p - 2^d) \bmod P$
    $\rangle$;
    $\delta := |\delta|$;
    $\langle$; e : $d \leq e < D$ ::
        send $\{\mu, \delta\}$ **to** $p \bar{\vee} 2^e$;
        receive $\{\mu_0, \delta_0\}$ **from** $p \bar{\vee} 2^e$;
        if $\delta_0 < \delta$ **then** $\mu, \delta := \mu_0, \delta_0$
    $\rangle$
end

Process p proposes a random element of $\mathcal{X}[p]$ as the median μ for the above set. If μ were the median, the number of elements larger than μ would equal the number of elements smaller than μ. For a given value of μ, we compute the difference δ between the number of elements larger than μ and the number of elements smaller than μ as the sum of the local differences. The local difference is computed using the sign() function. The sum of the local differences associated with each μ is computed by the round-robin communication between the processes associated with the same domain in step d. Upon completion of the sequential quantification over q, each process holds a proposed value μ and its associated global difference δ. A partial recursive-doubling procedure computes the μ-value associated with the smallest difference δ in absolute value. We accept this value of μ as an approximation for the median.

Because we did not compute exact medians, the number of bodies may not be evenly distributed. This may have an adverse impact on the load balance of the computations. However, keep in mind that, even with exact medians, load balance can be guaranteed only in a probabilistic sense, because the precise operation count in each process depends on the clustering of the bodies within the domain assigned to that process. To avoid load imbalance because of clustering effects, one can split domains on the basis of computing-time estimates obtained from the previous time step. Such strategies introduce computer dependencies, however.

11.5 Notes and References

Barnes and Hut [5] introduced the hierarchical-tree concept. Orthogonal recursive bisection, used to obtain a load-balanced computation, is due to Salmon [73]. The latter reference is also excellent as a starting point for a detailed study of concurrent fast particle methods.

The bare-bones programs Fast-N-Body-1 and Fast-N-Body-2 can be improved, refined, and made more efficient by incorporating all or some of the following techniques.

Global error control is perhaps the most significant refinement of program Compute-Field. Clearly, the important goal is to compute the global field $\vec{a}$ at location $\vec{x}$ to within a certain tolerance, say to within three or four significant digits. That does not mean that all individual contributions should be computed to the same tolerance. In particular, the contributions of distant clusters with small mass may be neglected if there are large clusters nearby. Global error control not only reduces the operation count, but is also more accurate and robust.

Accuracy can be improved by using better numerical approximations than the naive point-mass approximation for distant clusters. Better approximations require more work per interaction. However, by reducing the distance from which the cluster approximation is accurate enough, they can also reduce the total number of body-cluster interactions and, hence, the operation count. For an example of the use of high-order approximations, see Carrier, Greengard, and Rokhlin [13].

In program Compute-Field, we use clusters to compute the gravitational field at all body locations efficiently. Appel [3] proposed an algorithm that reduces the expected number of floating-point operations from $O(N \log N)$ to $O(N)$. The reduction in work is the result of cutting down the number of locations where the field is computed by using cluster-cluster interactions instead of body-cluster interactions. While cluster-cluster operations reduce the order of magnitude of the number of operations by a factor of $\log N$, the constant in front of the leading term can increase significantly because of the complexity of cluster-cluster interactions, particularly when combined with higher-order approximations.

Dynamic and adaptive time-stepping techniques may adjust the time step for each individual body. In the sun-earth-moon system, for example, it is natural to use many more time steps to compute the moon orbit, which circles the earth once a month, than for the earth orbit, which circles the sun once a year. In its extreme form, particle methods with dynamic time stepping become *discrete-event simulations*. Traditionally, however, time-stepping problems are not resolved by algorithmic remedies, but by high-order time-stepping techniques.

Exercises

Exercise 50 *Use the symmetry $\vec{F}_{m,n} = -\vec{F}_{n,m}$ to develop a program with a floating-point-operation count that is a factor of 2 less than that of program N-Body-1.*

Exercise 51 *Develop a multicomputer version of the program developed in Exercise 50.*

Exercise 52 *Adapt program Median-Value to compute exact medians.*

12
Computer Dependency

Real multicomputers differ from the prototypical model introduced in Section 1.2. As a result, program derivations that target the model environment do not address some issues that arise on real multicomputers.

Sequential computing is largely computer independent, because the basic model of a sequential computer, the *Von Neumann model*, is fixed and universally accepted; see Figure 12.1. Of course, real sequential computers deviate from this simple model; input and output would be impossible otherwise. Because high-level programming languages and operating systems succeed in hiding differences between the actual computer and the model, most users need be aware only of the model. However, users who want the highest-possible performance from one particular computer need a detailed knowledge of the underlying hardware.

The situation is fairly similar for multicomputer programming. Deviations from the multicomputer model of Figure 1.1 could remain hidden from the user by means of high-level software. However, absolute performance is considered crucial for most applications that are run on concurrent computers. Many multicomputer users want the highest-possible performance that can be reached on the available hardware. This requires special-purpose computer-dependent program transformations. Some of these transformations can be done by specialized compilers; others are the responsibility of the programmer.

Although each particular multicomputer has its own heuristic rules, it is possible to identify the main techniques to obtain highest-possible performance. Which program transformations will achieve optimal performance depends on the design chosen for each of the three basic types of com-

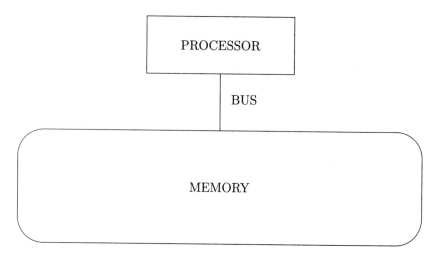

FIGURE 12.1. Prototypical Von Neumann design of a sequential computer.

puter components: *processors, memories, and connections.* For example, pipelined processors require vectorization (Section 12.1), hierarchical memories require blocking (Section 12.2), and non-idealized communication networks require process placement and custom communication algorithms (Section 12.3). Special-purpose transformations make the program less general, less portable, and less readable. Moreover, rapid changes in computer technology all but guarantee that any benefits of these special-purpose transformations be temporary. The optimizations of Sections 12.1, 12.2, and 12.3 should be used only if the increased performance is worth the considerable cost of computer-dependent programming.

Some concurrent-computer architectures differ substantially from the multicomputer model. Multiprocessors, also called shared-memory concurrent computers, do not consist of nodes. Instead, a collection of processors is connected to a collection of memory modules through a memory network. A programming model associated with multiprocessors is discussed in Section 12.4. The concept of synchronous arrays of processors leads to yet another programming model and is discussed in Section 12.5.

Benchmarking is the tool used to compare and evaluate computers and computer designs; it is the subject of Section 12.6.

12.1   Pipelined Processors

The principle of *pipelining* can be applied to all components of a computer design. Here, we only consider *pipelined floating-point processors*, which have had a major impact on scientific computing. As pictured in Figure 12.2, it is possible to execute a sequence of identical floating-point

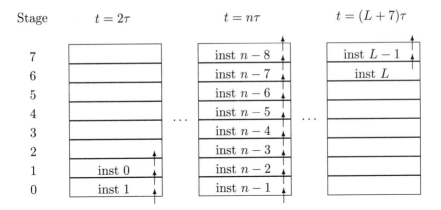

FIGURE 12.2. Prototypical floating-point pipeline with eight stages at the start $(t = 2\tau)$, in the middle $(t = n\tau)$, and at the end $(t = (L + 7)\tau)$ of a vector operation.

instructions in an assembly-line fashion by splitting floating-point instructions into several stages. Identical operations on long sequences of data occur, for example, in vector operations.

Consider a pipeline with k stages, each stage taking a time τ. Although each instruction requires k stages and a total execution time $k\tau$, one instruction is completed every time interval τ. An inefficiency occurs at the start of the vector operation, because the pipeline must be filled. It requires k cycles to complete the execution of the first instruction and to fill all the stages of the pipeline. With the last instruction, the stages of the pipeline are emptied one by one. A new vector operation can be started only after all stages are empty. The overhead of filling and emptying the pipeline is negligible if the vectors are sufficiently long.

Code vectorization is the collection of program transformations necessary to improve performance on computers with pipelined processors. On sequential computers with pipelined processors, these transformations are well understood. Many vectorization techniques have been incorporated into compilers, hence the term *vectorizing compilers*. Applying three rules of thumb during program development goes a long way to ensure efficient vectorization: write code in terms of vector operations on long vectors, take conditional statements out of loops, and reorganize data storage such that memory-access patterns are regular.

On multicomputers, pipelining and concurrency are often in conflict with one another. Long vectors, which are favored by the pipelined processors, are split up by the data distribution to achieve concurrency. Pipelined processors on multicomputers are effective only if the vectors after distribution are sufficiently long. Multicomputers with pipelined processors are, therefore, best suited for coarse-grain concurrency. In turn, this implies that each node of such a multicomputer must have a relatively large memory. As will be discussed in Section 12.2, supplying data from a large memory at a rate compatible with high processor speed requires a complex memory design, which comes with its own set of programming constraints. The requirement that memory-access patterns be regular may restrict the choice of data distribution and may lead to suboptimal load balance.

12.2 Hierarchical Memories

Computer memories must execute read and write requests from the processors at a rate compatible with processor speed. The three important characteristics of a memory module are its size, its speed, and its cost. The *size* of a memory is the amount of information it can store. For us, a convenient size measure is the number of floating-point words. Memory speed is determined by the time required to access one floating-point word; we call this the *memory-access time*. The inverse of memory-access time is the *memory band width*. The cost of a certain memory technology is reflected in the *cost per floating-point word of storage*. The following rules of thumb are generally valid:

- total cost of a memory is proportional to its size,

- high memory band width requires expensive technologies.

As a result, high band width is economical only for memories of small size. Large memories are available at a reasonable cost only from slower and less-expensive technologies.

Hierarchical-memory design is an attempt to achieve the seemingly conflicting goals of large size and high band width at a reasonable cost. This is done by combining different memory technologies into a complex system that exploits the property of *data locality*: if a read request for a certain address is received at time t, it is likely that at time $t + \Delta t$, where Δt is the cycle time of the system, a read request for a "neighboring" address will be received. For example, most implementations of vector operations access vector components in order, and memory requests of consecutive cycles will be to consecutive memory addresses.

Suppose we have available two memory modules M_1 and M_2. Module M_1 is characterized by a small access time T_1, high cost per byte C_1, and a limited size S_1. Module M_2 is slower ($T_2 > T_1$), less expensive ($C_2 < C_1$), and larger ($S_2 > S_1$). A typical hierarchical design uses M_2 as the main memory and M_1 as a *cache*; see Figure 12.3. In this mode of operation, each read request not only delivers the requested data from M_2 to processor P, but also copies from M_2 to M_1 a block of data surrounding the requested address. In subsequent cycles, there is a certain probability that read re-

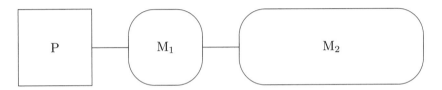

FIGURE 12.3. Prototypical cache design.

quests can be satisfied from M_1 instead of M_2 and in an access time T_1 instead of T_2. A discussion of the hardware required to detect whether or not requested data are in cache falls outside the scope of this text.

How well does it work? It all depends on the *hit ratio*, the ratio of the number of times requested data are already in cache over the total number of memory requests. If the hit ratio is h, the average access time of the hierarchical-memory system composed of M_1 and M_2 is given by:

$$T = hT_1 + (1 - h)T_2.$$

If h is almost equal to one, the access time is nearly that of the cache. The size of the hierarchical memory is equal to S_2, because the cache only contains duplicate values. The cost of the cache is amortized over the size of M_2, as can be seen from the cost per byte of the hierarchical memory:

$$C = \frac{C_1 S_1 + C_2 S_2}{S_2} = C_2 + \frac{S_1}{S_2}C_1.$$

The success of the hierarchical memory depends on the hit ratio, which is itself a function of the *cache size*, the *cache-replacement strategy*, and the *program*. Clearly, the hit ratio is some increasing function of cache size, and it tops out at $h = 1$ in the extreme case where $S_1 = S_2$. Of course, it is always the case that $S_1 \ll S_2$. When the amount of duplicate information in the cache approaches the size of the cache, old data (which may have been changed) must be transferred back to main memory M_2 to make room for new data. The criterion that decides which data to write back is called the cache-replacement strategy. In a hierarchical-memory design, cache size and cache-replacement strategy are the two key decisions that determine the hit ratio. However, these two factors are beyond the control of the user, who can only change the program. For optimal performance, the program must have the property of data locality. The user burden associated with achieving data locality can be reduced by compilers that promote data locality and by cache-replacement strategies that are sufficiently forgiving.

Usually, processors are much faster than memories of sufficient size, and the main factor that limits computing performance is memory speed, not processor speed. Because faster processors solve larger problems, their memories must be faster AND larger. This is why memory technology has a hard time keeping up with advancing processor technology and why the speed imbalance between processors and memories tends to be larger on faster computers. Caches that must bridge large speed gaps can become a heavy burden on the user. In the most extreme cases, only a user-controlled cache-replacement strategy, typically programmed in assembly language, can achieve optimal processor performance.

As an example of a program with explicit cache management, consider the implementation of the matrix–matrix-product assignment

$$C := C + AB$$

on a sequential computer with hierarchical memory. The matrices A, B, and C have compatible dimensions and are represented in program Matrix-Matrix-1 by arrays a, b, and c, respectively.

> **program** Matrix-Matrix-1
> **declare**
> k, m, n : integer ;
> a : array $[0..M - 1 \times 0..K - 1]$ of real ;
> b : array $[0..K - 1 \times 0..N - 1]$ of real ;
> c : array $[0..M - 1 \times 0..N - 1]$ of real
> **assign**
> $\langle\ \|\ (m, n)\ :\ (m, n) \in 0..M - 1 \times 0..N - 1\ ::$
> $c[m, n] := c[m, n] + \langle\ + k\ :\ k \in 0..K - 1\ ::\ a[m, k]b[k, n]\ \rangle$
> $\rangle$
> **end**

If $M = N = K$, the number of floating-point variables involved in the operation is proportional to M^2, while the number of floating-point operations is proportional to M^3. The number of memory-requests per floating-point operation is, therefore, asymptotically proportional to $1/M$. This suggests that this operation has the potential for significant data reuse from cache, for a high hit ratio, and for high performance.

Assume that two cache-management primitives **cache** and **flush** are available. The **cache** primitive copies a contiguous block of variables from main memory to cache. The **flush** primitive updates a contiguous block of variables in main memory with their current values in cache. If the cache is large enough to hold the matrix A, one column of B, and one column of C, then the matrix–matrix-product assignment can be implemented as shown by program Matrix-Matrix-2.

> **program** Matrix-Matrix-2
> **declare**
> k, m, n : integer ;
> a : array $[0..M - 1 \times 0..K - 1]$ of real ;
> b : array $[0..K - 1 \times 0..N - 1]$ of real ;
> c : array $[0..M - 1 \times 0..N - 1]$ of real

assign
    cache $\{a[m,k] : 0 \leq m < M$ **and** $0 \leq k < K\}$;
    $\langle$; n : $0 \leq n < N$ ::
        cache $\{b[k,n] : 0 \leq k < K\}$;
        cache $\{c[m,n] : 0 \leq m < M\}$;
        $\langle$; m : $0 \leq m < M$ ::
            $c[m,n] := c[m,n] + \langle + k : k \in 0..K-1 : a[m,k]b[k,n] \rangle$
        $\rangle$;
        flush $\{c[m,n] : 0 \leq m < M\}$
    $\rangle$
end

The matrix A is copied to cache by the first **cache** statement. Subsequently, the matrix C is computed column by column. In the sequential quantification over n, column n of B and C are cached. The new values of column n of C are computed and, subsequently, flushed to the corresponding main-memory locations by the **flush** statement.

In program Matrix-Matrix-2, it is assumed that matrices are stored by columns (the FORTRAN convention). If matrices are stored by rows (the C convention), a row-variant of the matrix–matrix-product program must be used. Explicit cache management is almost always complicated by such low-level, language-dependent, compiler-dependent, and/or computer-dependent technical details. Such optimizations are, therefore, reserved for a few frequently used operations, which are made available through a library of computer-dependent operations. Whereas the implementation of such libraries is not standardized, the user interfaces of these libraries are. For elementary matrix and vector operations, Basic Linear Algebra Subroutines (BLAS) is the prevailing standard. The BLAS standard divides the elementary operations into three levels: the vector–vector, the matrix–vector, and the matrix–matrix operations. Only the matrix–matrix operations can exploit cache-management optimizations, because only those operations perform $O(M^3)$ floating-point operations on $O(M^2)$ data and offer the possibility for data reuse. On systems with a hierarchical memory, a significant performance gain can be accomplished by using matrix–matrix operations in programs. The collection of program transformations that put a large fraction of floating-point operations inside matrix–matrix operations is called *blocking*.

Not unlike vectorization, blocking a multicomputer program often comes at the expense of concurrency or load balance. Moreover, blocking can be a significant global transformation of the program. Consider, for example, the LU-decomposition algorithm without pivoting as defined by program LU-2 of Section 4.1.2. To use matrix–matrix operations within each process, one must map consecutive rows and columns to the same process. For blocking, the linear distribution is thus preferred over the scatter distribution. However, we know that the linear distribution is disastrous for load balance of the LU-decomposition algorithm without pivoting. As a compromise between load balance and blocking, one can use a block-scattered distribution, which scatters blocks of a few, say five, consecutive columns over the process columns (similarly for the row distribution). Merely introducing a new data distribution is not sufficient to achieve blocking. It is also necessary to change some numerical aspects of the LU-decomposition algorithm. Instead of eliminating one row and column at a time, which leads to matrix–vector operations, blocks of rows and columns must be eliminated. In this case, a pivot is a small square matrix, say a 5×5 matrix, and all operations are matrix–matrix operations. This blocking strategy can be generalized to LU-decomposition with dynamic pivoting. In this case, the pivot is the "least-singular" submatrix of a certain set of submatrices.

While blocking is possible, albeit cumbersome, for full-matrix algorithms, it can be close to impossible for many other numerical algorithms. Consider relaxation methods, for example. The number of floating-point operations per relaxation step is proportional to the number of floating-point variables used in that step. This observation alone implies that the opportunity for data reuse and blocking is severely limited.

12.3   Communication Networks

The communication network in the multicomputer model of Figure 1.1 connects every node with every other node. It is customary to represent this network by a graph; its vertices represent nodes of the multicomputer and its edges communication channels. A network based on a complete graph, which has an edge connecting every pair of vertices, is the easiest to use but is technically infeasible if the number of multicomputer nodes P is large: each node would need $P - 1$ communication channels, and the total number of channels in the network would be $P(P - 1)/2$.

Practical communication networks between large numbers of nodes are always based on incomplete, but connected, graphs. As a result, the formula for the message-exchange time is more complicated than the one adopted thus far and must include the *network distance between nodes* as a parameter. The message-exchange time between two nodes depends on the minimum number of edges between their corresponding vertices in the graph. There is also a second effect. Messages that are simultaneously injected into a network based on an incomplete graph must compete for the limited number of channels. This leads to *network contention. Process placement* based on knowledge of the network topology can reduce or even eliminate network contention. Process placement can also reduce the average network distance that messages must traverse in the communication network.

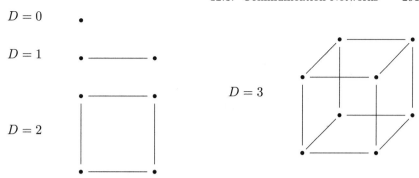

FIGURE 12.4. Hypercube graphs of dimension 0 through 3.

12.3.1 Hypercubes and Meshes

The communication networks of the first-generation multicomputers were based on hypercube graphs; see Figure 12.4 for hypercube graphs of dimension 0 through 3. A hypercube graph of dimension D is a connected graph with 2^D vertices, each vertex has D neighbors, the maximum path length between any pair of vertices is D, and the total number of edges in the graph is $D2^{D-1}$. These three basic properties make it technologically feasible to build multicomputers with many nodes connected by means of a communication network based on hypercube graphs. On a hypercube multicomputer with 2^D nodes, each node needs D communication channels, the total number of channels is $D2^{D-1}$, and the maximum network distance is D.

Each of the 2^D nodes of a D-dimensional hypercube multicomputer can be uniquely associated with an identifier p in the range $0..2^D - 1$. This identifier is called the *node number*. The binary representation of node numbers consists of D bits, and node numbers are assigned to nodes such that communication channels connect pairs of nodes with node numbers that differ in exactly one bit. In a three-dimensional hypercube multicomputer, for example, node number 5 has binary representation '101', and there are channels from node 5 to nodes $1 = $ '001', $7 = $ '111', and $4 = $ '100'. However, there is no channel between nodes $5 = $ '101' and $6 = $ '110', because they differ in two bits.

The message-exchange time between two nodes p and q not only depends on the length of the message, but also on the path length between the two nodes. On a hypercube multicomputer, the path length is the number of bits in which p and q differ. For example, a message from node $5 = $ '101' to node $6 = $ '110' is first sent from node $5 = $ '101' to node $4 = $ '100', where it is received and forwarded to node $6 = $ '110'. In this mode of operation, called *store-and-forward routing*, communication between nodes 5 and 6 is a factor of 2 slower than communication between nodes 5 and 4. Because the maximum path length in a D-dimensional hypercube is D, communication between an arbitrary pair of nodes can be up to D times slower than communication between two neighboring nodes.

Store-and-forward routing was replaced by more advanced routing mechanisms in second-generation multicomputers. *Worm-hole routing*, for example, reserves all required channels between source and destination node, such that messages exit the network only at their destination node. By eliminating unnecessary operations, worm-hole routing reduces significantly the dependence of message-exchange times on the path length and makes feasible networks that are less connected than hypercube networks. Networks based on two- or three-dimensional meshes, for example, have the advantage that the number of channels per node is a constant independent of the total number of nodes of the multicomputer. Moreover, for a given number of nodes, the total number of channels in a two- or three-dimensional mesh is much smaller than in a hypercube. This reduces cabling requirements and makes it economically feasible to use faster, more expensive, communication channels. One way to increase the performance of individual channels is to increase the *path width*, the number of bits sent simultaneously. The main objection to lower-dimensional networks is the increased risk of network contention due to increased competition for the fewer available communication channels.

The choice of network topology is a difficult hardware-design decision, which involves a compromise between connectivity and channel band width and depends on what technology is available at the moment. It is, therefore, a safe bet that network topologies of multicomputers will change. Every time the network topology is exploited in a multicomputer program, a computer dependency is created, and the expected life time of the program is reduced.

12.3.2 Process Placement

By assigning particular processes to particular nodes, communication between non-neighboring nodes and network contention can be eliminated or reduced. This program-optimization technique is called *process placement*.

A classic example of process placement is found when implementing the recursive-doubling procedure on a hypercube multicomputer. Process p of program Recursive-Doubling-2 of Section 1.7.2 only communicates with processes $p \bar{\vee} 2^d$, where $0 \le d < D$. By placing process p on node p of a D-dimensional hypercube multicomputer, all messages of the recursive-doubling procedure are exchanged via nearest-neighbor communications. The requirement that the number of processes equal the number of nodes can be relaxed somewhat. If the number of processes P is a multiple of the number of nodes 2^D, say $P = 2^D R$, place process p on node $(p \bmod 2^D)$. This places R processes on each node such that communication is between processes placed on the same node or placed on neighboring nodes.

$D=1$	Reflect	$D=2$	Reflect	$D=3$	Reflect	$D=4$
'0'=0	'0'	'00'=0	'00'	'000'=0	'000'	'0000'=0
'1'=1	'1'	'01'=1	'01'	'001'=1	'001'	'0001'=1
	'1'	'11'=3	'11'	'011'=3	'011'	'0011'=3
	'0'	'10'=2	'10'	'010'=2	'010'	'0010'=2
			'10'	'110'=6	'110'	'0110'=6
			'11'	'111'=7	'111'	'0111'=7
			'01'	'101'=5	'101'	'0101'=5
			'00'	'100'=4	'100'	'0100'=4
					'100'	'1100'=12
					'101'	'1101'=13
					'111'	'1111'=15
					'110'	'1110'=14
					'010'	'1010'=10
					'011'	'1011'=11
					'001'	'1001'=9
					'000'	'1000'=8

TABLE 12.1. Derivation of the binary-reflected Gray-code sequence.

The above trivial process placement fails for program Full-Recursive-Doubling-2 of Section 6.1, because processes p and $p+2^d$ are, in general, not hypercube neighbors. Instead, one can use process placement based on the *binary-reflected Gray-code sequence*, which is part of the lore of hypercube programming. The binary-reflected Gray-code sequence is a permutation of the index range $0..2^D - 1$. For $D = 1$, the sequence is given by the first column of Table 12.1. The sequence for $D = 1$ is reflected in the second column. The bit strings of this reflection are expanded with an extra bit to the left: the first half of the sequence receives a zero bit and the second half a one bit. These operations result in the sequence for $D = 2$ in the third column. The sequence for a general value of D is obtained by reflection of the sequence for $D-1$ and subsequent expansion with a zero and a one bit to the left of the first and second half, respectively.

Process placement for full recursive doubling with 2^D processes on a D-dimensional hypercube multicomputer places process p on node q, where q is the p-th entry of the D-dimensional binary-reflected Gray-code sequence. This places processes p and $(p+1) \bmod 2^D$ on nodes that are neighbors in the hypercube network. Processes p and $(p + 2^d) \bmod 2^D$ with $0 < d < D$ are placed on nodes that are a distance of two apart. Figure 12.5 shows this in detail for $D = 4$.

The communication structure of grid-oriented and linear-algebra computations is characterized by two- or three-dimensional process meshes, which require frequent communication between processes (p,q) and $(p\pm1, q\pm1)$ or (p,q,r) and $(p\pm1, q\pm1, r\pm1)$, respectively. On a hypercube multicomputer,

PROCESS STRUCTURE   HYPERCUBE CHANNELS

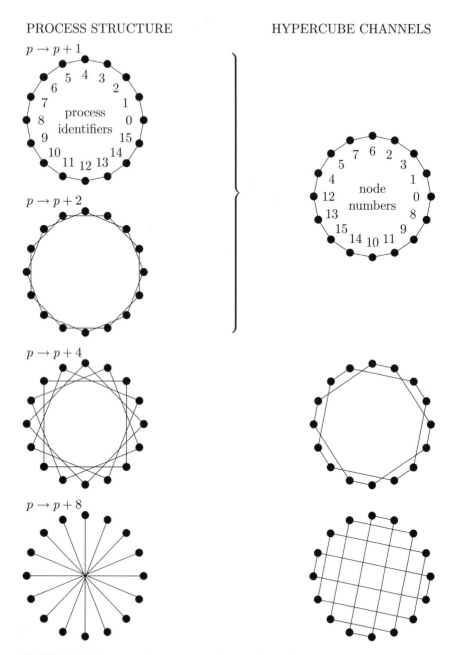

FIGURE 12.5. Process placement on a hypercube multicomputer according to the binary-reflected Gray code and network channels used to communicate between processes p and $(p + 2^d) \bmod 2^D$.

the process placement for such computations is usually a combination of Gray-code placements in all process-mesh dimensions.

In large programs, conflicts arise. Consider, for example, the multicomputer implementation of a partial-differential-equation solver that applies a relaxation method in one part of the computation and a fast Fourier transform in another part. While Gray-code process placement is optimal for the relaxation method, trivial process placement is optimal for the fast Fourier transform. This process-placement dilemma is solved by the observation that Gray-code placement applied to the fast Fourier transform leads to communication between nodes that are at most a distance of two apart in the hypercube network. Moreover, the communication paths between processes p and $p \bar{\vee} 2^d$ (see Figure 12.5) suggest that the transpose operation remains free of network contention. Although suboptimal, Gray-code placement is a good process placement for the component that computes the fast Fourier transform and, as a result, for the program as a whole.

For many programs, a near-optimal or the optimal process placement on hypercube multicomputers is known. Finding the optimal placement on low-dimensional networks like meshes is more difficult. Of course, there are easy cases: the processes of a two-dimensional-grid computation are trivially placed on the nodes of a two-dimensional-mesh multicomputer. However, the optimal placement on low-dimensional networks is often difficult to find, even for fairly common situations. For example, what is the optimal process placement for a three-dimensional-grid computation on a two-dimensional-mesh multicomputer or for recursive doubling on a two- or three-dimensional-mesh multicomputer? On low-dimensional networks, process placement is worth doing only in the simplest cases; meshes must perform well for random message traffic.

12.3.3   A Transpose Algorithm

Algorithms that require a high volume of communication lead to network contention, because messages compete with one another for the limited number of available channels. The matrix transpose is an example of such an algorithm. As seen in Sections 7.5.6, 8.4.2, 8.7.2, and 8.7.3, the matrix transpose also has applications in grid-oriented computations. Its frequent use justifies considering computer-dependent efficient implementations.

Program Transpose-1 specifies an in-place transpose of an $M \times M$ array.

> **program** Transpose-1
> **declare**
> m, n : integer ;
> a : array $[0..M - 1 \times 0..M - 1]$ of real
> **assign**
> $\langle \| \ (m, n) \ : \ (m, n) \in 0..M-1 \times 0..M-1 \ :: \ a[n, m] := a[m, n] \rangle$
> **end**

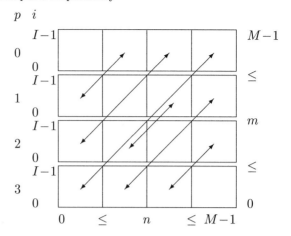

FIGURE 12.6. Matrix transpose by subblocks with $P = 4$ processes.

Program Transpose-2, is the corresponding multicomputer program. Its detailed derivation is omitted, but Figure 12.6 displays the algorithm for $P = 4$. It is assumed that M is divisible by P, say $M = PI$. The matrix rows are distributed using the perfectly load-balanced linear distribution, and local row index i corresponds to global row index $m = pI + i$. The column index n is not distributed. The matrix is considered as a $P \times P$ matrix of subblocks. The global transpose is obtained by transposing each subblock and, subsequently, transposing the matrix of subblocks. After the transpose operation, the global indices m and n switch roles. Index m is no longer distributed, but index n is distributed, and it corresponds to the local index i such that $n = pI + i$.

```
0..P − 1 ∥ p program Transpose-2
declare
       i, j, q : integer ;
       ℐ : set of integer ;
       a : array [0..I − 1 × 0..M − 1] of real
assign
       ℐ := 0..I − 1 ;
       ⟨ ; q : 0 ≤ q < P ::
              ⟨ ∥ (i, j) : (i, j) ∈ ℐ × ℐ :: a[j, qI + i] := a[i, qI + j] ⟩
       ⟩ ;
       ⟨ ; q : 0 ≤ q < P and q ≠ p ::
              send {a[i, qI + j] : (i, j) ∈ ℐ × ℐ} to q
       ⟩ ;
       ⟨ ; q : 0 ≤ q < P and q ≠ p ::
              receive {a[i, qI + j] : (i, j) ∈ ℐ × ℐ} from q
       ⟩
end
```

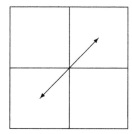 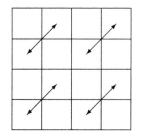 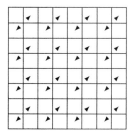

FIGURE 12.7. Three stages of a matrix transpose by recursive doubling.

In the communication step of program Transpose-2, each of the P processes injects $P - 1$ messages of length I^2 into the network. When applied to large problems, this program always leads to observable deterioration of the communication performance due to network contention.

On hypercube multicomputers, network contention is avoided by a combination of process placement and algorithm development. The local transpose of the subblocks remains as is. The subsequent transpose of the matrix of subblocks, however, is performed by recursive doubling. The procedure follows from the observation that the transpose of a matrix is obtained by transposing four submatrices, followed by a switch of the lower-left and upper-right submatrices (see Figure 12.7). This observation leads to program Transpose-3; the details of its derivation are omitted.

```
0..P − 1 ‖ p program Transpose-3
declare
      i, j, n, d, q : integer ;
      I, N : set of integer ;
      a : array [0..I − 1 × 0..M − 1] of real
assign
      I := 0..I − 1 ;
      ⟨ ; q : 0 ≤ q < P ::
            ⟨ ‖ (i, j) : (i, j) ∈ I × I :: a[j, qI + i] := a[i, qI + j] ⟩
      ⟩ ;
      ⟨ ; d : 0 ≤ d < D ::
            N := ⟨ ⋃ q : 0 ≤ q < P and (p∨̄q) ∧ 2^d = 2^d ::
                  qI..(q + 1)I − 1
            ⟩ ;
            send {a[i, n] : (i, n) ∈ I × N} to p∨̄2^d ;
            receive {a[i, n] : (i, n) ∈ I × N} from p∨̄2^d
      ⟩
end
```

The quantification that computes the index set N uses the boolean expression $(p\bar{\lor}q) \land 2^d = 2^d$, which is true only if bit number d of p and q differ.

As is the case for earlier recursive-doubling procedures, trivial process placement results in nearest-neighbor communication on a hypercube multicomputer. Moreover, no two messages ever compete for the same channel, and there is no network contention. Gray-code placement also results in contention-free communication, at most doubles the execution time of the transpose, and is preferable in the context of grid-oriented computations on a hypercube multicomputer.

12.3.4   Communication Hiding

The nodes of many multicomputers have separate processing units for computation and communication. In principle, some computation and communication tasks can be performed simultaneously. We call this type of concurrency *communication hiding*.

Communication hiding is possible upon sending and upon receiving a message. If the **send** primitive returns to the calling process before all work related to the send operation is done, it is possible for the calling process to perform arithmetic work during some part of the latency. By their very nature, arithmetic time and latency are computer-dependent quantities. Therefore, the amount of work that can be done during the latency varies considerably from one multicomputer to the next. Communication hiding at the receiving end requires a fast test to check whether or not a message has arrived. The timing problem is even more challenging here, because it is totally unpredictable how long the wait will be. If too much work is scheduled, an increasing number of waiting messages may use a significant amount of buffer space and may even lead to memory overflow. If not enough work is scheduled, the receiving process is continually testing whether new messages have arrived.

Communication hiding may introduce subtle synchronization problems. For example, a message buffer might be inadvertently changed before the **send** instruction has completed its operation. To avoid such race conditions, the programming environment must supply communication primitives that prohibit inadvertent user access to internal message buffers.

A more palatable form of communication hiding is possible when more than one process is mapped to each node. While one process is communicating, another can compute. This is an attractive option on multicomputers with a low-cost context switch. (A context switch is the operating-system call that deactivates one process and wakes up another.)

Communication hiding also occurs when message exchanges overlap with one another. Consider the broadcast of a message of length M from one process to all other processes. For long messages, one could consider splitting the message of length M into Q messages of length I, where $M = QI$. This is done in program Broadcast-1.

```
0..P − 1 ∥ p program Broadcast-1
declare
        m, q  :  integer ;
        t  :  array [0..M − 1] of real
assign
        ⟨ ; q  :  0 ≤ q < Q  ::
                if p > 0 then receive {t[m] : qI ≤ m < (q + 1)I} from p − 1 ;
                if p < P − 1 then send {t[m] : qI ≤ m < (q + 1)I} to p + 1
        ⟩
end
```

On a linear array of nodes, with process p placed on node p, the execution time of program Broadcast-1 is estimated as follows. It takes a time $(P − 1)\tau_C(I)$ for message number 0 to arrive at node $P − 1$. Subsequent messages arrive at time intervals $\tau_C(I)$. This leads to an execution-time estimate of

$$(P − 1)\tau_C(I) + (Q − 1)\tau_C(I) = (P + Q − 2)(\tau_S + \beta M/Q).$$

This expression is minimized, and we find that the optimal number of messages and the optimal message length are, respectively, given by:

$$\begin{cases} Q &= \sqrt{\frac{(P-2)\beta M}{\tau_S}} \\ I &= \sqrt{\frac{\tau_S}{\beta(P-2)}}\sqrt{M}. \end{cases}$$

Of course, it makes sense to split long messages only if $Q > 1$ and $I < M$. This requirement leads to:

$$M > \frac{\tau_S}{\beta(P − 2)}.$$

For most current multicomputers, the ratio τ_S/β is so large that splitting messages rarely makes sense.

The above analysis also holds for general multicomputers, not just linear arrays of nodes. Consider the broadcast of a message of length M on hypercube multicomputers via recursive doubling. As messages are propagated through the D-dimensional hypercube network, they traverse paths of length D (where $D = \log_2 P$). The broadcast time is thus minimized for

$$\begin{cases} Q &= \sqrt{\frac{(D-2)\beta M}{\tau_S}} \\ I &= \sqrt{\frac{\tau_S}{\beta(D-2)}}\sqrt{M}. \end{cases}$$

A comparison of the coefficients of $\sqrt{M}$ shows that the optimal message length is larger on hypercube than on linear-array multicomputers (provided both have the identical τ_S and β).

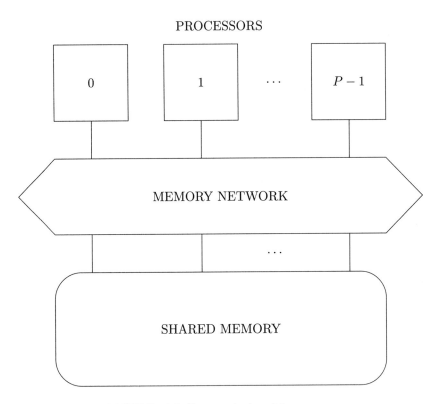

FIGURE 12.8. Prototypical multiprocessor.

12.4 Multiprocessors

Multiprocessors, also called *shared-memory concurrent computers*, present a different approach towards concurrency. Figure 12.8 pictures the prototypical multiprocessor: all processors can access one globally shared memory. On multiprocessors, processes have a shared data space, which is the software equivalent of the hardware-provided shared memory. Because multiprocessor processes can exchange information via shared variables, no explicit communication calls are required. Multiprocessor processes are not insulated from one another: one process can corrupt the data space of another. This is a significant difference with multicomputer processes, whose only information-exchange mechanism is explicit communication calls.

Different programming models inspire different program-derivation techniques. Consider the development of a multiprocessor version of the vector-sum operation, which was defined in Section 1.1 and transformed into a multicomputer program in Section 1.4. Once again, a duplication step leads to program Vector-Sum-2. The selection step is different, however. The work is distributed over the processes by introducing a partition of the index set $\mathcal{M} = 0..M - 1$. Subsequently, it is observed that the arrays x, y, and z need not be duplicated. Instead, they are shared between the processes (this is indicated in the declaration of the variables). As in the multicomputer program, the variable m remains a duplicated variable. The multiprocessor program that results is given by program Vector-Sum-5.

$0..P - 1 \parallel p$ **program** Vector-Sum-5
declare
$\qquad m \;:\; \text{integer} \;;$
$\qquad x,\, y,\, z \;:\; \text{shared array } [0..M - 1] \text{ of real}$
initially
$\qquad \langle\; ;\; m \;:\; m \in \mathcal{M}_p \;::\; x[m], y[m] = \tilde{x}_m, \tilde{y}_m \;\rangle$
assign
$\qquad \langle\; ;\; m \;:\; m \in \mathcal{M}_p \;::\; z[m] := \alpha x[m] + \beta y[m] \;\rangle$
end

The transformation to local indices is no longer required. This simplifies considerably the selection step. The absence of local indices is a standard argument favoring multiprocessors over multicomputers.

Real multiprocessors differ from the model of Figure 12.8. The main difficulty in implementing this model is memory latency. Every memory access must go through the memory network, often causing substantial delays. This stands in fundamental contrast with multicomputers, where delays are associated only with data that are actually exchanged. Because every memory access suffers from the shared-memory delay, it is crucial that the latency of the memory network be hidden. One way of achieving this is by placing some local memory in between processors and memory network (see Figure 12.9). The mode of operation of these local memories differs from

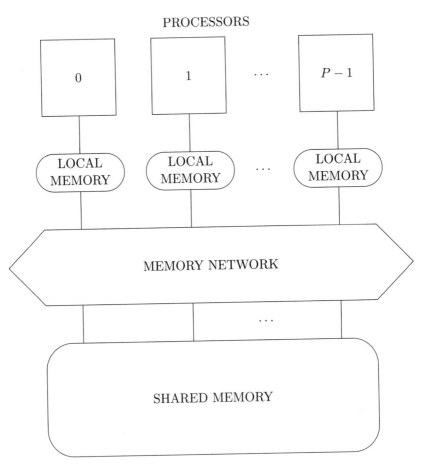

FIGURE 12.9. Prototypical multiprocessor, but local memory is associated with every processor.

one multiprocessor to the next. Some multiprocessor architectures operate the local memories as caches with the intent of making them invisible to the user. In other designs, processor and associated local memory act as a node of a multicomputer, and the memory network and shared memory act as a communication network with the added feature of data sharing.

What are the consequences of local memory on the multiprocessor programming environment? Is program Vector-Sum-5 efficient? If the local memories are operated as caches for the memory network, the index sets $\mathcal{M}_p$ must be index ranges to ensure data locality and a high hit ratio. If a linear data distribution of the array is acceptable, this simple program need not be further transformed. However, if the index sets $\mathcal{M}_p$ are not ranges, data locality can be achieved by reorganizing the data. For example, the index set $\mathcal{M}_p$ could be mapped to a local-index set $\mathcal{I}_p$. Efficiency considerations force us to use data distributions, even on concurrent computers with shared-memory capabilities.

While solving the memory-latency problem, caches introduce another problem: there are multiple copies of the same variable present in the concurrent hierarchical-memory system. In this complex memory consisting of the shared memory and the collection of all caches, which copy of the variable contains the correct value? This fundamental question arises in all programs that change and access different copies of variables concurrently. This synchronization problem is called the *cache-coherence problem*. It cannot be solved by hardware or system software in a way that is applicable on systems with many processors. The programmer must declare shared variables as cacheable or noncacheable. In program Vector-Sum-5, the arrays x and y are merely read, the array z is changed. It is thus likely that x and y are cacheable, but not z. Whether or not x and y are actually cacheable depends not only on program Vector-Sum-5, but also on all other programs acting on x and y. The global nature of this synchronization problem complicates the construction of software libraries for multiprocessors. The fact that some solutions exist hardly matters: the attempt to make the local memories transparent to the user by operating them as caches has failed!

In designs where processor and associated local memory act as a node of a multicomputer, the memory network is considered a communication network enhanced with data sharing. In program Vector-Sum-5, the arrays x, y, and z must now be distributed over the local memories. This necessitates the introduction of local indices. Hence, an important advantage of multiprocessors over multicomputers is lost. It is more appropriate to consider these architectures as multicomputers with shared-memory capabilities. The procedure to develop multicomputer programs remains intact. However, there is sometimes a convenience associated with not having to distribute explicitly some data. An example of the convenient use of such shared-memory capabilities are the pivot indices that are stored during the LU-decomposition for use in the back-solve program.

Multiprocessor computations with only one sequential process are inherently inefficient, because these sequential computations do not effectively hide the latency of the memory network. Using such computations to obtain sequential-execution times for the purpose of performance analysis always leads to an over-estimate of the efficiency and speed-up. These measures should, therefore, be used with customary caution.

12.5 Synchronous Arrays of Processors

The architecture and programming model of *synchronous arrays of processors* are inspired by the concept of *data concurrency*, which holds that the source of concurrency is the data, not the program. The control flow of data-concurrent programs is essentially sequential, but each synchronous-array instruction is applied concurrently to many data items. This philosophy is reflected in the prototypical architecture of a synchronous array, which is pictured in Figure 12.10. Each instruction is broadcast to all nodes. If an instruction specifies arithmetic on variables local to each node, it is executed immediately. However, some instructions require communication between the nodes, which is handled by the communication network.

Each synchronous-array instruction corresponds to a quantification over the concurrent or some other associative separator. For example, each of the following quantifications corresponds to one synchronous-array instruction:

$$\langle \, \| \ m \ : \ m \in \mathcal{M} \ :: \ z[m] := \alpha x[m] + \beta y[m] \, \rangle \ ;$$
$$\langle \, \| \ m \ : \ m \in \mathcal{M} \ :: \ \textbf{if } z[m] < 0 \textbf{ then } z[m] := 0 \, \rangle \ ;$$
$$\tau := \langle \, + \ m \ : \ m \in \mathcal{M} \ :: \ z[m]y[m] \, \rangle$$

The first quantification computes a vector sum, the second sets negative array entries equal to zero, and the third computes an inner product. In the execution of individual synchronous-array instructions, there is a high level of concurrency. However, the synchronous-array instructions of a program are performed sequentially. The sequential separators between the synchronous-array instructions act as barriers.

Most techniques discussed in this book are applicable to synchronous arrays. Consider the implementation of the synchronous-array instructions that correspond to the three quantifications above. The vector-sum quantification again leads us to consider data-distribution issues. Conceptually, each index m is mapped to a different node. Therefore, the array index m equals the node number, and a local index is not required. It is typical, however, that the size of the index set $\mathcal{M}$ is much larger than the available number of nodes. This is resolved by mapping the *virtual nodes* to which the vector components were assigned to *real nodes*. This map is just another notation for data distribution with local indices. Mapping several virtual nodes to each real node increases the granularity of the program and, in most cases, increases the efficiency.

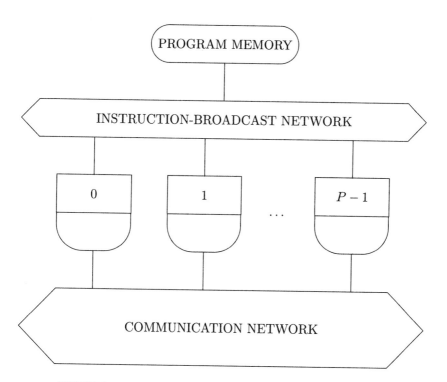

FIGURE 12.10. Prototypical synchronous array of processors.

The second quantification is implemented in two steps. First, an *ignore flag* is set in all nodes where the array entry is not less than zero. Second, in all nodes where the ignore flag is not set, the assignment $z[m] := 0$ is evaluated. The ignore flag is the basic mechanism that deals with **if**-statements that depend on local values. It is also a source for load imbalance on synchronous arrays of processors.

The inner-product quantification requires the communication network to compute the global sum. Recursive doubling can be used to implement this operation. The nodes only store array variables. Scalar variables, like τ, are stored in program memory, because they are used to determine the sequential control flow at the level of synchronous-array instructions.

The advantage of synchronous arrays of processors is that many concurrency issues remain hidden from the user. Concurrency issues are resolved in the implementation of the synchronous-array instructions. If a program can be formulated efficiently in terms of existing synchronous-array instructions, concurrency is not a user problem at all. If appropriate synchronous-array instructions do not exist, the user must implement the instruction. In this endeavor, all multicomputer techniques are applicable. The similarity of multicomputer and synchronous-array environments is also obvious from the observation that inefficiency on synchronous arrays is caused by the barriers in between every instruction, load imbalance due to idle nodes in inhomogeneous calculations, and internode communication. Except for the first one, these are the same sources of inefficiency that were encountered on multicomputers.

12.6 Benchmarking

Throughout this book, the emphasis has been on programs, not on hardware. In the practice of concurrent scientific computing, however, it is not always possible to separate one's concerns. The execution times of an implementation of a numerical method on some computer can be used for several purposes: showing how well the numerical method works, showing how able you are at developing efficient concurrent programs, or showing how well the computer works. The purpose determines the performance analysis. The first question, the performance of numerical methods, is the subject of standard numerical-analysis texts. The second question, analysis of implementations, has been our main performance focus. *Benchmarking* addresses the third question, the performance analysis of computers.

In the quantitative performance comparison of computers, the bottom line is either *absolute performance* (floating-point operations per second) or *performance-cost ratio* (floating-point operations per second per dollar). Execution times of computations remain the measures that we have at our disposal. The principles of performance analysis that were introduced in Section 1.6 remain valid. Once again, the complex space of all computations must be narrowed down to one that can be described by means of one or two parameters. This must be accomplished so that we can compare computers in an objective manner.

The computer is the study parameter and, in principle, all other parameters should be fixed. For a given problem and program, one can obtain execution times on different computers. Information about the computer hardware can then be derived from the comparison of these execution times. It is rarely the case, however, that the program can be a fixed parameter. In the supercomputing arena, virtually all programs require some computer-dependent adjustments to achieve reasonable performance. As a result, the program is usually a dependent parameter. But how dependent? In the extreme case, the program is chosen on the basis of optimality: for each computer, choose the program that solves the fixed problem in a minimal amount of time. Reality lies somewhere in between the two extremes, and the program is neither a fixed nor an optimal parameter. Typically, computer-dependent choices in some computational parameters like number of processes, process placement, etc., are allowed. Insertion of compiler directives to assist the compiler in producing better assembly-language code is another computer-dependent optimization that is often considered acceptable. How much adaptation one allows depends on the importance one attaches to the cost of software development. If willing to write computer-dependent programs for optimality, the comparison of computers should be based on near-optimal programs. Often, the program is closer to an optimal parameter when one is interested in absolute performance at whatever cost, and the program is closer to a fixed parameter when performance-cost ratio is of prime concern, because software-development costs are substantial.

Conclusions based on analyzing the execution times for just one problem are difficult to generalize. Unless one intends to use the computer for one particular application only, the comparison must be repeated for several problems. How to choose a representative set of sample problems can be the subject of considerable debate. Unfortunately, the only available guideline is to be intellectually honest.

In performance-cost analyses, there is another catch. What is the cost of a computer? Characterizing the cost of a computer by one number is a very crude approximation of reality. The *purchase cost* of a computer is not always a good indicator of the true cost. Many supercomputers demand a significant *maintenance cost*, which should be included in the cost measure. These costs are subject to the normal fluctuations of economic cycles and competitive pricing in a free-market economy. One might be tempted to use the *manufacturing cost*, but that is usually not publicly available and is itself subject to temporal fluctuations. From an engineering point of view, the *total area of silicon* in the computer is a good cost measure. Although it is difficult to obtain, it does not fluctuate and, in the long term, should be strongly correlated with the economic cost of computers.

Which cost to use depends on the goals of the performance analysis. When making a purchase decision at a particular moment in time, one should take advantage of opportunities in the market. Therefore, use the purchase cost augmented with the present cost of future maintenance. When evaluating computer designs, however, silicon area is a better cost measure. The performance per area of silicon is an engineering measure that is independent of temporal market forces. No single number can capture all aspects of reality, and performance per area of silicon is no exception to this rule. It ignores, for example, that concurrent-computer designs can take advantage of less-expensive mass-produced components. Although designs based on such components might require more silicon area, they may prove to be more economical than special-purpose designs.

Parametrizing the performance of programs and computers is a difficult endeavor. One must be aware of the pitfalls of each measure. In fact, all measures can be used to misrepresent performance data and to hide rather than clarify issues. To avoid the most obvious pitfalls, one must consider several performance measures in combination. Although quantitative measures are important, they cannot replace human judgment.

12.7   Notes and References

Computer architects must resolve conflicts between many competing goals. The design of a well-balanced computer and the hardware implementation of concepts discussed in this chapter are technically complex subjects. For introductions, see Hennessy and Patterson [48], Hwang and Briggs [55], and Stone [77]. The latter is also a good reference on multiprocessors. Hillis [51] introduces synchronous arrays of processors and data-concurrent programming.

Blocking to exploit hierarchical-memory designs is at the heart of the level 3 basic linear algebra subroutine (BLAS) package [24]. Demmel and Higham [22] discuss its numerical consequences. The transpose algorithm of Section 12.3.3 is that of Van de Velde [81]. However, concurrent-transpose algorithms have been the subject of numerous papers.

Hockney and Jesshope [53, 54] present a detailed characterization of the performance of concurrent computers and programs. Berry et al. [7] and Messina et al. [64] present practical benchmarking approaches.

Exercises

Exercise 53 *Develop a matrix–matrix-product program for a sequential computer with a hierarchical-memory system as pictured in Figure 12.3. This time, do not make the assumption that the matrix A fits into cache.*

Exercise 54 *The Jacobi and red-black Gauss-Seidel relaxation algorithms for the Poisson problem can be improved through communication hiding. Communication and computation can be overlapped as follows. The grid points on the interior boundaries are relaxed first, and their new values are sent immediately to the neighbors. Subsequently, the interior grid points are relaxed, and the new boundary values received. Write a detailed algorithm, and propose a performance analysis.*

Exercise 55 *The communication system on early hypercube multicomputers did not hide the topology of the communication network from the user. A hypercube of dimension D has $P = 2^D$ nodes, and a communication channel between node p and q exists if and only if the binary expansions of p and q differ in one bit only. The number of the differing bit was called the channel number. In early versions, only one process per node was allowed, and the communication protocol was channel rather than node oriented. The statement*

$$\text{shift}\,(c_i, b_i, s_i, c_o, b_o, s_o)$$

receives a message of size s_i from channel c_i and sends a message of size s_o to channel c_o. The received message is put in buffer b_i. The outgoing message takes its information from buffer b_o. The channel identifiers c are not the channel numbers but channel masks: when communicating through channel c in process p, one is actually communicating with process $p \bar{\vee} c$.

Develop a recursive-doubling program for hypercubes with such a channel-oriented communication library.

Exercise 56 *Develop a hypercube program for converting a row-distributed matrix into a column-distributed matrix.*

Exercise 57 *Develop a hypercube program for vector shifting using Gray-code process placement and using channel-oriented communication. Pay particular care to the segment shifts. Estimate the performance of a vector shift on a hypercube.*

Bibliography

[1] A. V. Aho, J. E. Hopcroft, and J. D. Ullman. *Data Structures and Algorithms*. Addison-Wesley, 1983.

[2] G. M. Amdahl. Validity of the single-processor approach to achieving large scale computing capabilities. In *AFIPS Conference Proceedings*, volume 30, pages 483–485, 1967.

[3] A. W. Appel. An efficient program for many-body simulation. *SIAM Journal on Scientific and Statistical Computing*, 6:85–103, 1985.

[4] O. Axelsson. A survey of preconditioned iterative methods for linear systems of algebraic equations. *BIT*, 25:166–187, 1985.

[5] J. E. Barnes and P. Hut. A hierarchical $O(N \log N)$ force-calculation algorithm. *Nature*, 324:446–449, 1986.

[6] G. M. Baudet. Asynchronous iterative methods for multiprocessors. *Journal of the ACM*, 25:226–244, 1978.

[7] M. Berry, D. Chen, P. Koss, D. Kuck, S. Lo, Y. Pang, L. Pointer, R. Roloff, A. Sameh, E. Clementi, S. Chin, D. Scheider, G. Fox, P. Messina, D. Walker, C. Hsiung, J. Schwarzmeier, K. Lue, S. Orszag, F. Seidl, O. Johnson, R. Goodrum, and J. Martin. The PERFECT club benchmarks: Effective performance evaluation of supercomputers. *The International Journal of Supercomputer Applications*, 3(3):5–40, 1989.

[8] C. H. Bischof. A parallel QR factorization algorithm with controlled local pivoting. *SIAM Journal on Scientific and Statistical Computing*, 12:36–37, 1991.

[9] P. E. Bjørstad and O. B. Widlund. Iterative methods for the solution of elliptic problems on regions partitioned into substructures. *SIAM Journal on Numerical Analysis*, 23:1097–1120, 1986.

[10] G. S. J. Bowgen and J. J. Modi. Implementation of QR factorization on the DAP using Householder transformations. *Computer Physics Communication*, 37:167–170, 1985.

[11] A. Brandt. Multi-level adaptive solutions to boundary-value problems. *Mathematics of Computation*, 31:333–390, 1977.

[12] W. L. Briggs. *A Multigrid Tutorial.* SIAM, 1987.

[13] J. Carrier, L. Greengard, and V. Rokhlin. A fast adaptive multipole algorithm for particle simulations. *SIAM Journal on Scientific and Statistical Computing*, 9:669–686, 1988.

[14] T. F. Chan, R. Glowinski, J. Periaux, and O. B. Widlund, editors. *Domain Decomposition Methods for Partial Differential Equations.* SIAM, 1990.

[15] K. M. Chandy and J. Misra. *Parallel Program Design, A Foundation.* Addison-Wesley, 1988.

[16] D. Chazan and W. Miranker. Chaotic relaxation. *Linear Algebra and Its Applications*, 2:199–222, 1969.

[17] E. Chu and J. A. George. QR factorization of a dense matrix on a shared-memory multiprocessor. *Parallel Computing*, 11:55–72, 1989.

[18] E. Chu and J. A. George. QR factorization of a dense matrix on a hypercube multiprocessor. *SIAM Journal on Scientific and Statistical Computing*, 11:990–1028, 1990.

[19] J. W. Cooley and J. W. Tukey. An algorithm for the machine calculation of complex Fourier series. *Mathematics of Computation*, 19:297–301, 1965.

[20] N. H. Decker. Note on the parallel efficiency of the Frederickson-McBryan multigrid algorithm. *SIAM Journal on Scientific and Statistical Computing*, 12:208–220, 1991.

[21] J. W. Demmel. LAPACK — a portable linear algebra library for high-performance computers. *Concurrency: Practice and Experience*, 3:655–666, 1991.

[22] J. W. Demmel and N. J. Higham. Stability of block algorithms with fast level-3 BLAS. *ACM Transactions on Mathematical Software*, 18:274–291, 1992.

[23] E. W. Dijkstra. *A Discipline of Programming*. Prentice-Hall, 1976.

[24] J. J. Dongarra, J. Du Croz, S. Hammarling, and I. S. Duff. A set of level 3 basic linear algebra subprograms. *ACM Transactions on Mathematical Software*, 16:1–17, 1990.

[25] J. J. Dongarra, J. Du Croz, S. Hammarling, and R. J. Hanson. An extended set of FORTRAN basic linear algebra subprograms. *ACM Transactions on Mathematical Software*, 14:1–17, 1988.

[26] F. W. Dorr. The direct solution of the discrete Poisson equation on a rectangle. *SIAM Review*, 12:248–263, 1970.

[27] I. S. Duff, A. M. Erisman, and J. K. Reid. *Direct Methods for Sparse Matrices*. Oxford University Press, 1986.

[28] V. Faber, O. M. Lubeck, and A. B. White, Jr. Superlinear speedup of an efficient sequential algorithm is not possible. *Parallel Computing*, 3:259–260, 1986.

[29] V. Faber, O. M. Lubeck, and A. B. White, Jr. Comments on the paper "Parallel efficiency can be greater than unity". *Parallel Computing*, 4:209–210, 1987.

[30] D. Fischer. On superlinear speedups. *Parallel Computing*, 17:695–697, 1991.

[31] G. C. Fox, S. W. Otto, and A. J. G. Hey. Matrix algorithms on a hypercube I: Matrix multiplication. *Parallel Computing*, 4:17–31, 1987.

[32] J. N. Franklin. *Matrix Theory*. Series in Applied Mathematics. Prentice-Hall, 1968.

[33] P. O. Frederickson and O. A. McBryan. Normalized convergence rates for the PSMG method. *SIAM Journal on Scientific and Statistical Computing*, 12:221–229, 1991.

[34] R. W. Freund. A transpose-free quasi-minimal residual algorithm for non-Hermitian linear systems. *SIAM Journal on Scientific Computing*, 14:470–482, 1993.

[35] R. W. Freund, M. H. Gutknecht, and Nachtigal N. M. An implementation of the look-ahead Lanczos-algorithm for non-Hermitian matrices. *SIAM Journal on Scientific Computing*, 14:137–158, 1993.

[36] R. W. Freund and N. M. Nachtigal. QMR: A quasi-minimal residual method for non-Hermitian linear systems. *Numerische Mathematik*, 60:315–340, 1991.

[37] K. A. Gallivan, R. J. Plemmons, and A. H. Sameh. Parallel algorithms for dense linear algebra computations. *SIAM Review*, 32:54–135, 1990.

[38] J. A. George and J. W.-H. Liu. *Computer Solution of Large Sparse Positive Definite Systems*. Prentice-Hall, 1981.

[39] G. H. Golub and D. P. O'Leary. Some history of the conjugate gradient and Lanczos methods. *SIAM Review*, 31:50–102, 1989.

[40] G. H. Golub and C. F. Van Loan. *Matrix Computations*. Johns Hopkins, second edition, 1989.

[41] J. L. Gustafson. Reevaluating Amdahl's law. *Communications of the ACM*, 31:532–533, 1988.

[42] W. Hackbusch. *Multi-Grid Methods and Applications*, volume 4 of *Springer Series in Computational Mathematics*. Springer-Verlag, 1985.

[43] W. Hackbusch and U. Trottenberg. *Multigrid Methods*, volume 960 of *Lecture Notes in Mathematics*. Springer-Verlag, 1982.

[44] D. L. Harrar. Orderings, multicoloring, and consistently ordered matrices. *SIAM Journal on Matrix Analysis and Applications*, 14:259–278, 1993.

[45] L. Hart and S. McCormick. Asynchronous multilevel adaptive methods for solving partial differential equations on multiprocessors: Basic ideas. *Parallel Computing*, 12:131–144, 1989.

[46] M. T. Heath, E. Ng, and B. W. Peyton. Parallel algorithms for sparse linear systems. *SIAM Review*, 33:420–460, 1991.

[47] D. E. Heller. A survey of parallel algorithms in numerical linear algebra. *SIAM Review*, 20:740–777, 1978.

[48] J. L. Hennessy and D. A. Patterson. *Computer Architecture: A Quantitative Approach*. Morgan Kaufmann Publishers, Inc., 1990.

[49] M. R. Hestenes and E. Stiefel. Methods of conjugate gradients for solving linear systems. *Journal of Research of the National Bureau of Standards*, 49:409–436, 1952.

[50] N. J. Higham. Exploiting fast matrix multiplication within the level 3 BLAS. *ACM Transactions on Mathematical Software*, 16:352–368, 1990.

[51] W. D. Hillis. *The Connection Machine*. The MIT Press, 1985.

[52] C. A. R. Hoare. *Communicating Sequential Processes*. Prentice-Hall, 1984.

[53] R. W. Hockney and C. R. Jesshope. *Parallel Computers — Architecture, Programming and Algorithms*. Adam Hilger, 1981.

[54] R. W. Hockney and C. R. Jesshope. *Parallel Computers 2*. Adam Hilger, 1988.

[55] K. Hwang and F. A. Briggs. *Computer Architecture and Parallel Processing*. McGraw-Hill, 1984.

[56] R. Janßen. A note on superlinear speedup. *Parallel Computing*, 4:211–213, 1987.

[57] S. K. Kim and A. T. Chronopoulos. An efficient parallel algorithm for extreme eigenvalues of sparse nonsymmetric matrices. *The International Journal of Supercomputer Applications*, 6(1):98–111, 1992.

[58] D. E. Knuth. *The Art of Computer Programming*. Addison-Wesley, second edition, 1973. Volume 1. Fundamental algorithms. Volume 2. Semi-numerical algorithms. Volume 3. Sorting and searching.

[59] C. Lanczos. An iteration method for the solution of the eigenvalue problem of linear differential and integral operators. *Journal of Research of the National Bureau of Standards*, 45:255–282, 1950.

[60] C. L. Lawson, R. J. Hanson, D. R. Kincaid, and F. T. Krogh. Basic linear algebra subprograms for Fortran usage. *ACM Transactions on Mathematical Software*, 5:308–323, 1979.

[61] N. K. Madsen, G. H. Rodrigue, and J. I. Karush. Matrix multiplication by diagonals on a vector/parallel processor. *Information Processing Letters*, 5:41–45, 1976.

[62] O. A. McBryan and E. F. Van de Velde. Hypercube algorithms and implementations. *SIAM Journal on Scientific and Statistical Computing*, 8:s227–s287, 1987.

[63] O. A. McBryan and E. F. Van de Velde. Matrix and vector operations on hypercube parallel processors. *Parallel Computing*, 5:117–126, 1987.

[64] P. Messina, C. Baillie, E. Felten, P. Hipes, R. Williams, A. Alagar, A. Kamrath, R. Leary, W. Pfeiffer, J. Rogers, and D. Walker. Benchmarking advanced architecture computers. *Concurrency: Practice and Experience*, 2:195–255, 1990.

[65] W. L. Miranker. A survey of parallelism in numerical analysis. *SIAM Review*, 13:524–547, 1971.

[66] J. J. Modi. *Parallel Algorithms and Matrix Computation*. Oxford University Press, 1988.

[67] D. P. O'Leary and R. E. White. Multi-splittings of matrices and parallel solution of linear systems. *SIAM Journal on Algebraic and Discrete Methods*, 6:630–640, 1985.

[68] J. M. Ortega and R. G. Voigt. Solution of partial differential equations on vector and parallel computers. *SIAM Review*, 27:149–240, 1985.

[69] J. M. Ortega, R. G. Voigt, and C. H. Romine. A bibliography on parallel and vector numerical algorithms. Technical Report 6, ICASE, 1988.

[70] T. S. Papatheodorou and Y. G. Saridakis. Parallel algorithms and architectures for multisplitting iterative methods. *Parallel Computing*, 12:171–182, 1989.

[71] D. Parkinson. Parallel efficiency can be greater than unity. *Parallel Computing*, 3:261–262, 1986.

[72] M. C. Pease. An adaptation of the fast Fourier transform for parallel processing. *Journal of the ACM*, 15:252–264, 1968.

[73] J. K. Salmon. *Parallel Hierarchical N-Body Methods*. PhD thesis, California Institute of Technology, 1990. Available as Report CRPC-90-14.

[74] C. L. Seitz. The Cosmic Cube. *Communications of the ACM*, 28:22–33, 1985.

[75] G. M. Shroff. A parallel algorithm for the eigenvalues and eigenvectors of a general complex matrix. *Numerische Mathematik*, 58:779–805, 1991.

[76] H. S. Stone. Parallel tridiagonal equation solvers. *ACM Transactions on Mathematical Software*, 1:289–307, 1975.

[77] H. S. Stone. *High Performance Computer Architecture*. Addison-Wesley, 1987.

[78] G. Szegö. *Orthogonal Polynomials*, volume 23 of *Colloquium Publication*. American Mathematical Society, 1939.

[79] C. Temperton. Direct methods for the solution of the discrete Poisson equation: Some comparisons. *Journal of Computational Physics*, 31:1–20, 1979.

[80] C. Temperton. Self-sorting in-place fast Fourier transforms. *SIAM Journal on Scientific and Statistical Computing*, 12:808–823, 1991.

[81] E. F. Van de Velde. Data redistribution and concurrency. *Parallel Computing*, 16:125–138, 1990.

[82] E. F. Van de Velde. Experiments with multicomputer LU-decomposition. *Concurrency: Practice and Experience*, 2:1–26, 1990.

[83] E. F. Van de Velde and J. Lorenz. Adaptive data distribution for concurrent continuation. *Numerische Mathematik*, 62:269–294, 1992.

[84] C. F. Van Loan. *Computational Frameworks for the Fast Fourier Transform*. SIAM, 1992.

[85] J. H. Wilkinson. *The Algebraic Eigenvalue Problem*. Oxford University Press, 1965.

[86] N. Wirth. *Algorithms + Data Structures = Programs*. Prentice-Hall, 1976.

[87] P. H. Worley. The effect of time constraints on scaled speedup. *SIAM Journal on Scientific and Statistical Computing*, 11:838–858, 1990.

[88] D. M. Young. *Iterative Solution of Large Linear Systems*. Academic Press, 1971.

Index

319